19th SIGBioMed Workshop on Biomedical Language Processing (BioNLP 2020)

Online
9 July 2020

ISBN: 978-1-7138-1383-5

BioNLP 2020

The 19th SIGBioMed Workshop on Biomedical Language Processing

Proceedings of the Workshop

July 9, 2020

BioNLP 2020: Research unscathed by COVID-19

Kevin Bretonnel Cohen, Dina Demner-Fushman, Sophia Ananiadou, Junichi Tsujii

The past year has been more than exciting for natural language processing in general, and for biomedical natural language processing in particular. A gradual accretion of studies of reproducibility and replicability in natural language processing, biomedical and otherwise had been making it clear that the reproducibility crisis that has hit most of the rest of science is not going to spare text mining or its related fields. Then, in March of 2020, much of the world ground to a sudden halt.

The outbreak of the COVID-19 disease caused by the novel coronavirus SARS-CoV-2 made computational work more obviously relevant than it had perhaps ever been before. Suddenly, newscasters were arguing about viral clades, the daily news was full of stories about modelling, and your neighbor had heard of PCR. But, some of us did not really see a role for natural language processing in the brave new world of computational instant reactions to an international pandemic.

That was wrong.

In mid-late March of 2020, a joint project between the Allen Artificial Intelligence Institute (Ai2), the National Library of Medicine (NLM), and the White House Office of Science and Technology Policy (OSTP) released CORD-19, a corpus of work on the SARS-CoV-2 virus, on COVID-19 disease, and on related coronavirus research. It was immediately notable for its inclusion of "gray literature" from preprint servers, which mostly have been neglected in text mining research, as well as for its flexibility with regards to licensing of content types. Perhaps most importantly, it was released in conjunction with a number of task types, including one related to ethics–although the value of medical ethics has been widely obvious since the Nazi "medical" experimentation horrors of the Second World War, the worldwide pandemic has made the value of medical **ethicists** more apparent to the general public than at any time since. Those task type definitions enabled the broader natural language processing community to jump into the fray quite quickly, and initial results have been quick to arrive.

Meanwhile, the pandemic did nothing to slow research in biomedical natural language processing on any other topic, either. That can be seen in the fact that this year the Association for Computational Linguistics SIGBIOMED workshop on biomedical natural language processing received 73 submissions. The unfortunate effect of the pandemic was the cancellation of the physical workshop, which would have allowed acceptance of all high-quality submissions as posters, if not for podium presentations. Indeed, the poster sessions at BioNLP have been continuously growing in size, due to the large number of high-quality submissions that the workshop receives annually. Unfortunately, because this year the Association for Computational Linguistics annual meeting will take place online only, there will be no poster session for the workshop. Consequently, only a handful of submissions could be accepted for presentation.

Transitioning of the traditional conferences to online presentations at the beginning of the COVID-19 pandemic showed that the traditional presentation formats are not as engaging remotely as they are in the context of in-person sessions. We are therefore exploring a new form of presentation, hoping it will be more engaging, interactive, and informative: 22 papers (about 30% of the submissions) will be presented in panel-like sessions. Papers will be grouped by similarity of topic, meaning that participants with related interests will be able to interact regarding their papers with a hopefully optimal number of people on line at the same time. As we write this introduction, the conference plans and platform are still evolving, as are the daily lives of much of the planet, so we hope that you will join us in planning for the worst, while hoping for the best.

Panel Discussions

papers referenced in this section are included in this volume, unless otherwise indicated

Session 1: High accuracy information retrieval, spin and bias
Invited talk and discussion lead: Kirk Roberts

Presentations: The exploration of Information Retrieval approaches enhanced with linguistic knowledge continues in the work that allows life-science researchers to search PubMed and the CORD-19 collection using patterns over dependency graphs (Taub-Tabib et al.) Representing biomedical relationships in the literature by encoding dependency structure with word embeddings promises to improve retrieval of relationships and literature-based discovery (Paullada et al.) Word embeddings trained on biomedical research articles and the tests based on their associations and coherence, among others, allow detecting and quantifying gender bias over time (Rios et al.) A BioBERT model fine-tuned for relation extraction might assist in detecting spin in reporting the results of randomized clinical trials (Koroleva et al.) Finally, a novel sequence-to-set approach to generating terms for pseudo-relevance feedback is evaluated (Das et al.)

Session 2: Clinical Language Processing
Invited talk and discussion lead: Tim Miller

Presentations: Not surprisingly, much of the potentially reproducible work in the clinical domain is based on the Medical Information Mart for Intensive Care (MIMIC) data (Johnson et al., 2016). Kovaleva et al. used the MIMIC-CXR data to explore Visual Dialog for radiology and prepare the first publicly available silver- and gold-standard datasets for this task. Searle et al. present a MIMIC-based silver standard for automatic clinical coding and warn that frequently assigned codes in MIMIC-III might be undercoded. Mascio et al. used MIMIC and the Shared Annotated Resources (ShARe)/CLEF dataset in four classification tasks: disease status, temporality, negation, and uncertainty. Temporality is explored in-depth by Lin et al., and Wang et al. explore approaches to a clinical Semantic Textual Similarity (STS) task. Xu et al. apply reinforcement learning to deal with noise in clinical text for readmission prediction after kidney transplant.

Session 3: Language Understanding
Invited talk and discussion lead: Anna Rumshisky

Presentations: Bringing clinical problems and poetry together, this creative work seeks to better understand dyslexia through a self-attention transformer and Shakespearean sonnets (Bleiweiss). Detection of early stages of Alzheimer's disease using unsupervised clustering is explored with 10 years of President Ronald Reagan's speeches (Wang et al.). Stavropoulos et al. introduce BIOMRC, a cloze-style dataset for biomedical machine reading comprehension, along with new publicly available models, and provide a leaderboard for the task. Another type of question answering – answering questions that can be answered by electronic medical records–is explored by Rawat et al. Hur et al. study veterinary records to identify reasons for administration of antibiotics. DeYoung et al. expand the Evidence Inference dataset and evaluate BERT-based models for the evidence inference task.

Session 4: Named Entity Recognition and Knowledge Representation
Invited talk and discussion lead: Hoifung Poon

Invited talk: Machine Reading for Precision Medicine

The advent of big data promises to revolutionize medicine by making it more personalized and effective, but big data also presents a grand challenge of information overload. For example, tumor sequencing has become routine in cancer treatment, yet interpreting the genomic data requires painstakingly curating knowledge from a vast biomedical literature, which grows by thousands of papers every day. Electronic medical records contain valuable information to speed up clinical trial recruitment and drug development, but curating such real-world evidence from clinical notes can take hours for a single patient. Natural language processing (NLP) can play a key role in interpreting big data for precision medicine. In particular, machine reading can help unlock knowledge from text by substantially improving curation efficiency. However, standard supervised methods require labeled examples, which are expensive and time-consuming to produce at scale. In this talk, Dr. Poon presents Project Hanover, where the team overcomes the annotation bottleneck by combining deep learning with probabilistic logic, and by exploiting self-supervision from readily available resources such as ontologies and databases. This enables the researchers to extract knowledge from millions of publications, reason efficiently with the resulting knowledge graph by learning neural embeddings of biomedical entities and relations, and apply the extracted knowledge and learned embeddings to supporting precision oncology.

Hoifung Poon is the Senior Director of Precision Health NLP at Microsoft Research and an affiliated professor at the University of Washington Medical School. He leads Project Hanover, with the overarching goal of structuring medical data for precision medicine. He has given tutorials on this topic at top conferences such as the Association for Computational Linguistics (ACL) and the Association for the Advancement of Artificial Intelligence (AAAI). His research spans a wide range of problems in machine learning and natural language processing (NLP), and his prior work has been recognized with Best Paper Awards from premier venues such as the North American Chapter of the Association for Computational Linguistics (NAACL), Empirical Methods in Natural Language Processing (EMNLP), and Uncertainty in AI (UAI). He received his PhD in Computer Science and Engineering from University of Washington, specializing in machine learning and NLP.

Presentations: Nejadgholi et al. analyze errors in NER and introduce an F-score that models a forgiving user experience. Peng et al. study NER, relation extraction, and other tasks with a multi-tasking learning approach. Amin et al. explore multi-instance learning for relation extraction. ShafieiBavani et al. also explore relation and event extraction, but in the context of simultaneously predicting relationships between all mention pairs in a text. Chang et al. provide a benchmark for knowledge graph embedding models on the SNOMED-CT knowledge graph and emphasize the importance of knowledge graphs for learning biomedical knowledge representation.

Acknowledging the community

As always, we are profoundly grateful to the authors who chose BioNLP for presenting their innovative research. The authors' willingness to continue sharing their work through BioNLP consistently makes the workshop noteworthy and stimulating. We are equally indebted to the program committee members (listed elsewhere in this volume) who produced thorough reviews on a tight review schedule and with an admirable level of insight, despite the timeline being even shorter than usual and the workload higher, while at the same time handling the unprecedented changes in their work and life caused by the COVID-19 pandemic.

References

Johnson, A., Pollard, T., Shen, L. et al. MIMIC-III, a freely accessible critical care database. Sci Data 3, 160035 (2016). https://doi.org/10.1038/sdata.2016.35

Wang, L.L., Lo,K., Chandrasekhar, Y. et al. Cord-19: The covid-19 open research dataset. ArXiv, abs/2004.10706, 2020.

Roland Roller, DFKI GmbH, Berlin, Germany
Diana Sousa, University of Lisbon, Portugal
Karin Verspoor, The University of Melbourne, Australia
Davy Weissenbacher, University of Pennsylvania, USA
W John Wilbur, US National Library of Medicine
Shankai Yan, US National Library of Medicine
Chrysoula Zerva, National Centre for Text Mining and University of Manchester, UK
Ayah Zirikly, Clinical Center, National Institutes of Health, USA
Pierre Zweigenbaum, LIMSI - CNRS, France

Additional Reviewers:

Jingcheng Du, School of Biomedical Informatics, UTHealth

Invited Speakers:

Hoifung Poon, Microsoft Research
Tim Miller, Boston Childrens Hospital and Harvard Medical School
Kirk Roberts, School of Biomedical Informatics, UTHealth
Anna Rumshisky, University of Massachusetts Lowell

Table of Contents

Conference Program

Thursday July 9, 2020

08:30–08:40   **Opening remarks**

08:40–10:30   **Session 1: High accuracy information retrieval, spin and bias**

08:40–09:10   *Invited Talk – Kirk Roberts*

09:10–09:20   *Quantifying 60 Years of Gender Bias in Biomedical Research with Word Embeddings*
Anthony Rios, Reenam Joshi and Hejin Shin

09:20–09:30   *Sequence-to-Set Semantic Tagging for Complex Query Reformulation and Automated Text Categorization in Biomedical IR using Self-Attention*
Manirupa Das, Juanxi Li, Eric Fosler-Lussier, Simon Lin, Steve Rust, Yungui Huang and Rajiv Ramnath

09:30–09:40   *Interactive Extractive Search over Biomedical Corpora*
Hillel Taub Tabib, Micah Shlain, Shoval Sadde, Dan Lahav, Matan Eyal, Yaara Cohen and Yoav Goldberg

09:40–09:50   *Improving Biomedical Analogical Retrieval with Embedding of Structural Dependencies*
Amandalynne Paullada, Bethany Percha and Trevor Cohen

09:50–10:00   *DeSpin: a prototype system for detecting spin in biomedical publications*
Anna Koroleva, Sanjay Kamath, Patrick Bossuyt and Patrick Paroubek

10:00–10:30   *Discussion*

10:30–10:45   *Coffee Break*

10:45–13:00   Session 2: Clinical Language Processing

10:45–11:15   *Invited Talk – Tim Miller*

11:15–11:25   *Towards Visual Dialog for Radiology*
Olga Kovaleva, Chaitanya Shivade, Satyananda Kashyap, Karina Kanjaria, Joy Wu,
Deddeh Ballah, Adam Coy, Alexandros Karargyris, Yufan Guo, David Beymer
Beymer, Anna Rumshisky and Vandana Mukherjee Mukherjee

11:25–11:35   *A BERT-based One-Pass Multi-Task Model for Clinical Temporal Relation Extraction*
Chen Lin, Timothy Miller, Dmitriy Dligach, Farig Sadeque, Steven Bethard and
Guergana Savova

11:35–11:45   *Experimental Evaluation and Development of a Silver-Standard for the MIMIC-III
Clinical Coding Dataset*
Thomas Searle, Zina Ibrahim and Richard Dobson

11:45–11:55   *Comparative Analysis of Text Classification Approaches in Electronic Health
Records*
Aurelie Mascio, Zeljko Kraljevic, Daniel Bean, Richard Dobson, Robert Stewart,
Rebecca Bendayan and Angus Roberts

11:55–12:05   *Noise Pollution in Hospital Readmission Prediction: Long Document Classification
with Reinforcement Learning*
Liyan Xu, Julien Hogan, Rachel E. Patzer and Jinho D. Choi

12:05–12:15   *Evaluating the Utility of Model Configurations and Data Augmentation on Clinical
Semantic Textual Similarity*
Yuxia Wang, Fei Liu, Karin Verspoor and Timothy Baldwin

12:15–12:45   *Discussion*

12:45–13:30   *Lunch*

13:30–15:30 **Session 3: Language Understanding**

13:30–14:00 ***Invited Talk – Anna Rumshisky***

14:00–14:10 *Entity-Enriched Neural Models for Clinical Question Answering*
Bhanu Pratap Singh Rawat, Wei-Hung Weng, So Yeon Min, Preethi Raghavan and Peter Szolovits

14:10–14:20 *Evidence Inference 2.0: More Data, Better Models*
Jay DeYoung, Eric Lehman, Benjamin Nye, Iain Marshall and Byron C. Wallace

14:20–14:30 *Personalized Early Stage Alzheimer's Disease Detection: A Case Study of President Reagan's Speeches*
Ning Wang, Fan Luo, Vishal Peddagangireddy, Koduvayur Subbalakshmi and Rajarathnam Chandramouli

14:30–14:40 *BioMRC: A Dataset for Biomedical Machine Reading Comprehension*
Dimitris Pappas, Petros Stavropoulos, Ion Androutsopoulos and Ryan McDonald

14:40–14:50 *Neural Transduction of Letter Position Dyslexia using an Anagram Matrix Representation*
Avi Bleiweiss

14:50–15:00 *Domain Adaptation and Instance Selection for Disease Syndrome Classification over Veterinary Clinical Notes*
Brian Hur, Timothy Baldwin, Karin Verspoor, Laura Hardefeldt and James Gilkerson

15:00–15:30 ***Discussion***

15:30–15:45 ***Coffee Break***

Quantifying 60 Years of Gender Bias in Biomedical Research with Word Embeddings

Anthony Rios[1], Reenam Joshi[2], and Hejin Shin[3]
[1]Department of Information Systems and Cyber Security
[2]Department of Computer Science
[3]Library Systems
University of Texas at San Antonio
{Anthony.Rios, Reenam.Joshi, Hejin.Shin}@utsa.edu

Abstract

Gender bias in biomedical research can have an adverse impact on the health of real people. For example, there is evidence that heart disease-related funded research generally focuses on men. Health disparities can form between men and at-risk groups of women (i.e., elderly and low-income) if there is not an equal number of heart disease-related studies for both genders. In this paper, we study temporal bias in biomedical research articles by measuring gender differences in word embeddings. Specifically, we address multiple questions, including, How has gender bias changed over time in biomedical research, and what health-related concepts are the most biased? Overall, we find that traditional gender stereotypes have reduced over time. However, we also find that the embeddings of many medical conditions are as biased today as they were 60 years ago (e.g., concepts related to drug addiction and body dysmorphia).

1 Introduction

It is important to develop gender-specific best-practice guidelines for biomedical research (Holdcroft, 2007). If research is heavily biased towards one gender, then the biased guidance may contribute towards health disparities because the evidence drawn-on may be questionable (i.e., not well studied). For example, there is more research funding for the study of heart disease in men (Weisz et al., 2004). Therefore, the at-risk populations of older women in low economic classes are not as well-investigated. Therefore, this opens up the possibility for an increase in the health disparities between genders.

Among informatics researchers, there has been increased interest in understanding, measuring, and overcoming bias associated with machine learning methods. Researchers have studied many applica-

tion areas to understand the effect of bias. For example, Kay et al. (2015) found that the Google image search application is biased (Kay et al., 2015). Specifically, they found an unequal representation of gender stereotypes in image search results for different occupations (e.g., all police images are of men). Likewise, ad-targeting algorithms may include characteristics of sexism and racism (Datta et al., 2015; Sweeney, 2013). Sweeney (2013) found that the names of black men and women are likely to generate ads related to arrest records. In healthcare, much of the prior work has studied the bias in the diagnosis process made by doctors (Young et al., 1996; Hartung and Widiger, 1998). There have also been studies about ethical considerations about the use of machine learning in healthcare (Cohen et al., 2014).

It is possible to analyze and measure the presence of gender bias in text. Garg et al. (2018) analyzed the presence of well-known gender stereotypes over the last 100 years. Hamberg (2008) shown that gender blindness and stereotyped preconceptions are the key cause for gender bias in medicine. Heath et al. (2019) studied the gender-based linguistic differences in physician trainee evaluations of medical faculty. Salles et al. (2019) measured the implicit and explicit gender bias among health care professionals and surgeons. Feldman et al. (2019) quantified the exclusion of females in clinical studies at scale with automated data extraction. Recently, researchers have studied methods to quantify gender bias using word embeddings trained on biomedical research articles (Kurita et al., 2019). Kurita et al. (2019) shown that the resulting embeddings capture some well-known gender stereotypes. Moreover, the embeddings exhibit the stereotypes at a lower rate than embeddings trained on other corpora (e.g., Wikipedia). However, to the best of our knowledge, there has not been an automated temporal study in the change

1

Proceedings of the BioNLP 2020 workshop, pages 1–13
Online, July 9, 2020 ©2020 Association for Computational Linguistics

of gender bias.

In this paper, we look at the temporal change of gender bias in biomedical research. To study social biases, we make use of word embeddings trained on different decades of biomedical research articles. The two main question driving this work are, In what ways has bias changed over time, and Are there certain illnesses associated with a specific gender? We leverage three computational techniques to answer these questions, the Word Embedding Association Test (WEAT) (Caliskan et al., 2017), the Embedding Coherence Test (ECT) (Dev and Phillips, 2019), and Relational Inner Product Association (RIPA) (Ethayarajh et al., 2019). To the best of our knowledge, this will be the first temporal analysis of bias of word embeddings trained on biomedical research articles. Moreover, to the best of our knowledge, this is the first analysis that measures the gender bias associated with individual biomedical words.

Our work is most similar to Garg et al. (2018). Garg et al. (2018) study the temporal change of both gender and racial biases using word embeddings. Our work substantially differs in three ways. First, this paper is focused on biomedical literature, not general text corpora. Second, we analyze gender stereotypes using three distinct methods to see if the bias is robust to various measurement techniques. Third, we extend the study beyond gender stereotypes. Specifically, we look at bias in sets of occupation words, as well as bias in mental health-related word sets. Moreover, we quantify the bias of individual occupational and mental health-related words.

In summary, the paper makes the following contributions:

- We answer the question; How has the usage of gender stereotypes changed in the last 60 years of biomedical research? Specifically, we look at the change in well-known gender stereotypes (e.g., *Math vs Arts*, *Career vs Family*, *Intelligence vs Appearance*, and occupations) in biomedical literature from 1960 to 2020.

- The second contribution answers the question; What are the most gender-stereotyped words for each decade during the last 60 years, and have they changed over time? This contribution is more focused than simply looking at traditional gender stereotypes. Specifically,

we analyze two groups of words: occupations and mental health disorders. For each group, we measure the overall change in bias over time. Moreover, we measure the individual bias associated with each occupation and mental health disorder.

2 Related Work

In this section, we discuss research related to the three major themes of this paper: gender disparities in healthcare, biomedical word embeddings, and bias in natural language processing (NLP).

2.1 Gender Disparities in Healthcare.

There is evidence of gender disparities in the healthcare system, from the diagnosis of mental health disorders to differences in substance abuse. An important question is, Do similar biases appear in biomedical research? In this work, while we explore traditional gender stereotypes (e.g., *Intelligence vs Appearance*), we also measure potential bias in the occupations and mental health-related disorders associated with each gender.

With regard to mental health, as an example, affecting more than 17 million adults in the United States (US) alone, major depression is one of the most common mental health illnesses (Pratt and Brody, 2014). Depression can cause people to lose pleasure in daily life, complicate other medical conditions, and possibly lead to suicide (Pratt and Brody, 2014). Moreover, depression can occur to anyone, at any age, and to people of any race or ethnic group. While treatment can help individuals suffering from major depression, or mental illness in general, only about 35% of individuals suffering from severe depression seek treatment from mental health professionals. It is common for people to resist treatment because of the belief that depression is not serious, that they can treat themselves, or that it would be seen as a personal weakness rather than a serious medical illness (Gulliver et al., 2010). Unfortunately, while depression can affect anyone, women are almost twice as likely as men to have had depression (Albert, 2015). Moreover, depression is generally higher among certain demographic groups, including, but not limited to, Hispanic, non-Hispanic black, low income, and low education groups (Bailey et al., 2019). The focus of this paper is to understand the impact of these mental health disparities in word embeddings trained on biomedical corpora.

2.2 Biomedical Word Embeddings.

Word embeddings capture the distributional nature between words (i.e., words that appear in similar contexts will have a similar vector encoding). Over the years, there have been multiple methods of producing word embeddings, including, but not limited to, latent semantic analysis (Deerwester et al., 1990), Word2Vec (Mikolov et al., 2013a,b), and GLOVE (Pennington et al., 2014). Moreover, pre-trained word embeddings have been shown to be useful for a wide variety of downstream biomedical NLP tasks (Wang et al., 2018), such as text classification (Rios and Kavuluru, 2015), named entity recognition (Habibi et al., 2017), and relation extraction (He et al., 2019). In Chiu et al. (2016), the authors study a standard methodology to train good biomedical word embeddings. Essentially, they study the impact of the various Word2Vec-specific hyperparameters. In this paper, we use the strategies proposed in Chiu et al. (2016) to train optimal decade-specific biomedical word embeddings.

2.3 Bias and Natural Language Processing.

Unfortunately, because word embeddings are learned using naturally occurring data, implicit biases expressed in text will be transferred to the vectors. Bias (and fairness) is an important topic among natural language processing researchers. Bias has been found in word embeddings (Bolukbasi et al., 2016; Zhao et al., 2018, 2019), text classification models (Dixon et al., 2018; Park et al., 2018; Badjatiya et al., 2019; Rios, 2020), and in machine translation systems (Font and Costa-jussà, 2019; Escudé Font, 2019). In general, each paper generally focuses on either testing whether bias exists in various models, or on removing bias from classification models for specific applications.

Much of the work on measuring (gender) bias using word embeddings neither studies the temporal aspect (i.e., how bias changes over time) nor focuses on biomedical research (Chaloner and Maldonado, 2019). For example, Caliskan et al. (2017) studied the bias in groups of words—focusing on traditional gender stereotypes. Kurita et al. (2019) expanded on Caliskan et al. (2017) to generalize to contextual word embeddings. Garg et al. (2018) developed a technique to study 100 years of gender and racial bias using word embeddings. They evaluated the bias over time using the US Census as a baseline to compare embedding bias to demographic and occupation shifts. There has

Year	# Articles
1960-1969	1,479,370
1970-1979	2,305,257
1980-1989	3,322,556
1990-1999	4,109,739
2000-2010	6,134,431
2010-2020	8,686,620
Total	26,037,973

Table 1: The total number of articles in each decade.

also been work on measuring bias in sentence embeddings (May et al., 2019). Furthermore, there has been a significant amount of research that explores different ways to measure bias in word embeddings (Caliskan et al., 2017; Dev and Phillips, 2019; Ethayarajh et al., 2019). In this work, we make use of many of the bias measurement techniques (Caliskan et al., 2017; Dev and Phillips, 2019; Ethayarajh et al., 2019) to apply them to the biomedical domain.

3 Dataset

We analyze PubMed-indexed titles and abstracts published anytime between 1960 and 2020. The total number of articles per decade are shown in Table 1. The text is lower-cased and tokenized using the SimpleTokenizer available in GenSim (Khosrovian et al., 2008). We find that the total number of papers have grown substantially each decade, from 1.4 million indexed articles in the 1960s to 8.6 million in the 2010s. Yet, the rate of growth stayed relatively stable each decade.

4 Method

We train the Skip-Gram model on PubMed-indexed titles and abstracts from 1960 to 2020. The hyperparameters of the Skip-Gram model are optimized independently for each decade. Next, given the best set of embeddings for each decade, we explore three different techniques to measure bias: the Word Embedding Association Test (WEAT), the Embedding Coherence Test (ECT), and the Relational Inner Product Association (RIPA). Each method allows us to quantify bias in different ways, such as comparing multiple sets of words (e.g., comparing the bias with respect to *Career vs Family*), comparing a single set of words (e.g., occupations), and measuring the bias of individual words (e.g., nurse). In this section, we briefly discuss the procedure we used to train the word embeddings,

3

Attribute Words	Male vs Female	X	male, man, boy, brother, he, him, his, son, father, uncle, grandfather
		Y	female, woman, girl, sister, she, her, hers, daughter, mother, aunt, grandmother
Target Words	Career vs Family	A	executive, management, professional, corporation, salary, office, business, career
		B	home, parents, children, family, cousins, marriage, wedding, relatives
	Math vs Art	A	math, algebra, geometry, calculus, equations, computation, numbers, addition
		B	poetry, art, Shakespeare, dance, literature, novel, symphony, drama
	Science vs Art	A	science, technology, physics, chemistry, Einstein, NASA, experiment, astronomy
		B	poetry, art, Shakespeare, dance, literature, novel, symphony, drama
	Intelligence vs Appearance	A	precocious, resourceful, inquisitive, genius, inventive, astute, adaptable, reflective, discerning, intuitive, inquiring, judicious, analytical, apt, venerable, imaginative, shrewd, thoughtful, wise, smart, ingenious, clever, brilliant, logical, intelligent
		B	alluring, voluptuous, blushing, homely, plump, sensual, gorgeous, slim, bald, athletic, fashionable, stout, ugly, muscular, slender, feeble, handsome, healthy, attractive, fat, weak, thin, pretty, beautiful, strong
	Weak vs Strong	A	power, strong, confident, dominant, potent, command, assert, loud, bold, succeed, triumph, leader, shout, dynamic, winner
		B	weak, surrender, timid, vulnerable, weakness, wispy, withdraw, yield, failure, shy, follow, lose, fragile, afraid, loser

Table 2: Attribute and Target words words used by WEAT to measure the presence of traditional gender stereotypes in biomedical literature.

as well as provide descriptions of each of the bias measurement techniques.

4.1 Word2Vec Model Training.

We train a Skip-Gram model using GenSim (Khosrovian et al., 2008). Following Chiu et al. (2016), we search over the following key hyper-parameters: Negative sample size, sub-sampling, minimum-count, learning rate, vector dimension, and context window size. See Chiu et al. (2016, Table 2) for more details.

To find the best model, as we search over the various hyper-parameters, we make use of the UMLS-Sim dataset (McInnes et al., 2009). UMLS-Sim consists of 566 medical concept pairs for measuring similarity. The degree of association between terms in UMLS-Sim was rated by four medical residents from the University of Minnesota medical school. All these clinical terms correspond to Unified Medical Language System (UMLS) concepts included in the Metathesaurus (Bodenreider, 2004). Evaulation is performed using Spearman's rho rank correlation between a vector of cosine similarities between each of the 566 pairs of words and their respective medical-resident ratings. Intuitively, the ranking of the pairs using cosine similarity, from most similar pairs to the least, should be similar to the human (medical expert) annotations.

4.2 Word Embedding Association Test

The implicit bias test measures unconscious prejudice (Greenwald et al., 1998). WEAT is a generalization of the implicit bias test for word embeddings, measuring the association between two sets of target concepts and two sets of attributes. We use the same target and attribute sets from Kurita et al. (2019). We list the targets and attributes in Table 2. The attribute sets of words are related to the groups in which the embeddings are biases towards or against, e.g., *Male vs Female*. The words in the target categories—*Career vs Family*, *Math vs Arts*, *Science vs Arts*, *Intelligence vs Appearance*, and *Strength vs Weakness*—represent the specific types of biases. For example, using the attributes and targets, we want to know whether the learned embeddings that represent *men* are more related to *career* than the *female*-related words (i.e., test if female words are more related to family, than male words).

Formally, let X and Y be equal-sized sets of *target* concept embeddings and let A and B be sets of *attribute* embeddings. To measure the bias, we follow Caliskan et al. (2017), which defines the following *test statistic* that is the difference between the sums over the respective target concepts,

$$s(X, Y, A, B) = \left[\sum_{x \in X} s(x, A, B) \right] - \left[\sum_{y \in Y} s(y, A, B) \right]$$

where $s(w, A, B)$ measures the association between a single target word w (e.g., career) with each of the attribute (gendered) words as

$$s(w, A, B) = \left[\sum_{a \in A} \cos(\vec{\mathbf{w}}, \vec{\mathbf{a}})\right] - \left[\sum_{b \in B} \cos(\vec{\mathbf{w}}, \vec{\mathbf{b}})\right],$$

such that $\cos()$ represents the cosine similarity between two vectors. $\vec{\mathbf{w}} \in \mathbb{R}^d$, $\vec{\mathbf{a}} \in \mathbb{R}^d$, and $\vec{\mathbf{b}} \in \mathbb{R}^d$ represents the word embedding for x, y, and w, respectively. Similarly, d is the dimension of each word embedding. Instead of using the test statistic directly, to measure bias, we use the *effect size*. Effect size is a normalized measure of the separation of the two distributions, defined as

$$\frac{\mu_{x \in X}\big[s(x, A, B)\big] - \mu_{y \in Y}\big[s(y, A, B)\big]}{\sigma_{w \in X \cup Y}s(w, A, B)}$$

where $\mu_{x \in X}$ and $\mu_{y \in Y}$ represent the mean score over target words for a specific attribute word. Likewise, $\sigma_{w \in X \cup Y}$ is the standard deviation of the scores for the word w in the union of X and Y. Intuitively, a positive score means that the attribute words in X (e.g., male, man, boy) are more similar to the target words A (e.g., strong, power, dominant) than Y (e.g., female, woman, girl). Moreover, larger effects represent more biased embeddings.

As previously stated, the Attribute and Target words are from Kurita et al. (2019). It is important to note that the list is manually curated. Moreover, the bias measurement can change depending on the exact list of words. RIPA is more robust to slight changes to the attribute words than WEAT (Ethayarajh et al., 2019).

4.3 Embedding Coherence Test.

We also explore a second method of measuring bias, the Embedding Coherence Test (ECT) (Dev and Phillips, 2019). Unlike WEAT, it compares the attribute Words (e.g., *Male vs Female*) with a single target set (e.g., *Career*). Thus, we do not need two contrasting target sets (e.g., *Career vs Family*) to measure bias. We take advantage of this to measure bias associated with occupations and mental health-related disorders. Specifically, we use a total of 290 occupation words and 222 mental health-related words. The occupation words come from prior work measuring per-word bias (Dev and Phillips, 2019). To form a list of mental health words, we use the Diagnostic and Statistical Manual of Mental Disorders (DSM-5), a taxonomic and

Year	Sim	Pair Cnt
1960-1969	.6586	101
1970-1979	.6715	207
1980-1989	.7033	277
1990-1999	.7282	265
2000-2010	.7078	272
2010-2020	.6867	306

Table 3: Quality of the embeddings trained for each decade, measured using the UMLS-Sim dataset. Sim represents Spearman's rho ranking correlation. Pair count is the number of UMLS-Sim's word-pairs that were present in that decades embeddings.

diagnostic tool published by the American Psychiatric Association (Association et al., 2013). For each mental health disorder in DSM-5, which are generally multi-word expressions, we split it into individual words. Next, we manually remove uninformative adjective and function words. For example, the disorder "Specific learning disorder, with impairment in mathematics" is tokenized into the following words: "learning", "disorder", "impairment", and "mathematics". A complete listing of the occupational and mental health words can be found in the appendix.

Formally, ECT first computes the mean vectors for the attribute word sets X and Y, defined as

$$\vec{\mathbf{v}}_X = \frac{1}{|X|} \sum_{x \in X} \vec{\mathbf{x}}$$

where $\vec{\mathbf{v}}_X \in \mathbb{R}^d$ and $|X|$ represents the number of words in category X. $\vec{\mathbf{v}}_Y$ is calculated similarly.

For both $\vec{\mathbf{v}}_X$ and $\vec{\mathbf{v}}_Y$, ECT computes the (cosine) similarities with all vectors $a \in A$, i.e., the cosine similarity is calculated between each target word a and $\vec{\mathbf{v}}_X$ and stored in $s_X \in \mathbb{R}^{|A|}$. The two resultant vectors of similarity scores, s_X (for X) and s_Y (for Y) are used to obtain the final ECT score. It is the Spearman's rank correlation between the rank orders of s_X and s_Y—the higher the correlation, the lower the bias. Intuitively, if the correlation is high, then the rank of target words based on similarity is correlated when calculated for the both X and Y (i.e., male and female).

4.4 Relational Inner Product Association.

While ECT only requires a single target set, both WEAT and ECT [1] calculate the bias between sets

---

[1] The cosine similarities from ECT *can* be used to measure scores for individual words, but it is not as robust as RIPA (Ethayarajh et al., 2019).

5

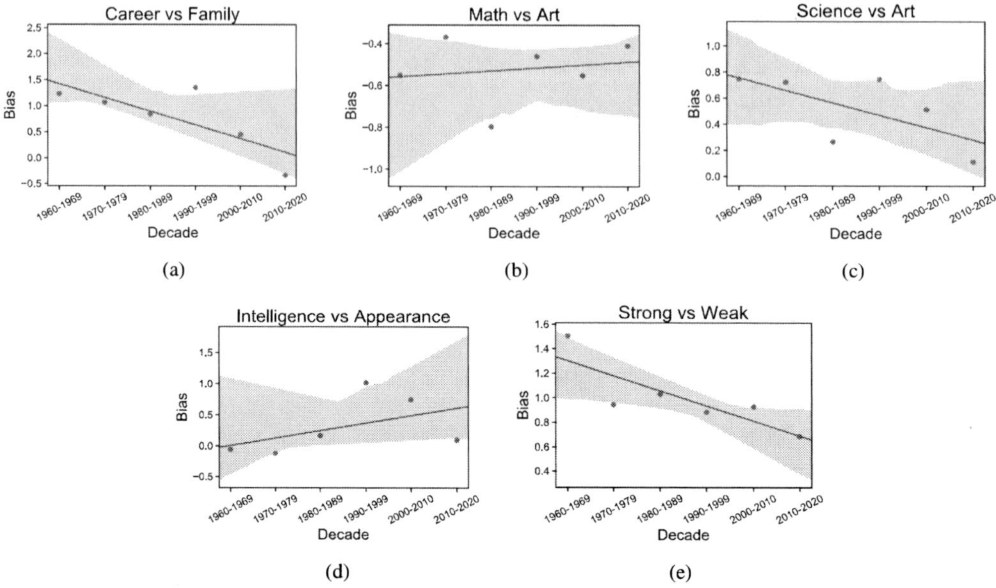

Figure 1: Each subfigure plots the bias measures using WEAT for one of five gender stereotypes: (a) Career vs Family, (b) Math vs Art, (c) Science vs Art, (d) Intelligence vs Appearance, and (e) Strong vs Weak. A bias score of zero represents no bias, i.e., no measurable difference between the two target categories for each gender. The shaded area of each subplot represents the bootstrap estimated 95% confidence interval.

of words. However, neither approach calculates a robust bias score for individual words. To study the most gender biased words over time, we make use of RIPA (Ethayarajh et al., 2019). Intuitively, RIPA uses a single vector to represent gender, then each word is scored by taking the dot product between the gender embedding and its respective embedding. The sign of the score will determine whether the embedding is more male or female-related.

The major aspect of RIPA is creating the gender embedding. Formally, given S, a non-empty set of ordered word pairs (x, y) (e.g., ('man', 'woman'), ('male', 'female')) that defines the gender association, we take the first principal component of all the difference vectors $\{\vec{\mathbf{x}} - \vec{\mathbf{y}} | (x, y) \in S\}$, which we call the relation vector $\vec{\mathbf{g}} \in \mathbb{R}^d$—that would be a one-dimensional bias subspace. Then, for some word vector $\vec{\mathbf{w}} \in \mathbb{R}^d$ the dot product is taken with $\vec{\mathbf{g}}$ to measure bias.

5 Results

In this section, we present the results of our study in four parts. First, we report the embedding quality using UMLS-sim. Second, we study the temporal bias of traditional gender stereotypes, such as *Career vs Family* and *Strong vs Weak*. Ideally, we want to understand how, and which, stereotypes

have changed over time. To understand the biased stereotypes, we make use of the WEAT method. Third, we look at whether occupational and mental health-related words are biased, and how the bias has changed over time. For this result, we only use a single set of target words. Thus, we make use of ECT. Fourth, we use RIPA to find the most biased words for each gender in each decade.

5.1 Embedding Quality.

In Table 3, we report the quality of each decade's embeddings based on the UMLS-sim dataset. Overall, we find that the quality consistently improves until the 1990s, however, we see drops in the 2000s and 2010s. We hypothesize that the reason for the decrease in embedding quality is because of the growth of research articles indexed on PubMed. Intuitively, word embeddings are only able to capture a single sense of a word. However, given the breadth of articles PubMed indexes—from machine learning (e.g., BioNLP) to biomaterials—multiple word meanings are being stored in a single vector. Thus, the overall quality begins to drop.

5.2 Traditional Gender Stereotypes.

In Figure 1, we plot the bias scores reported using WEAT. Remember, a large positive score means

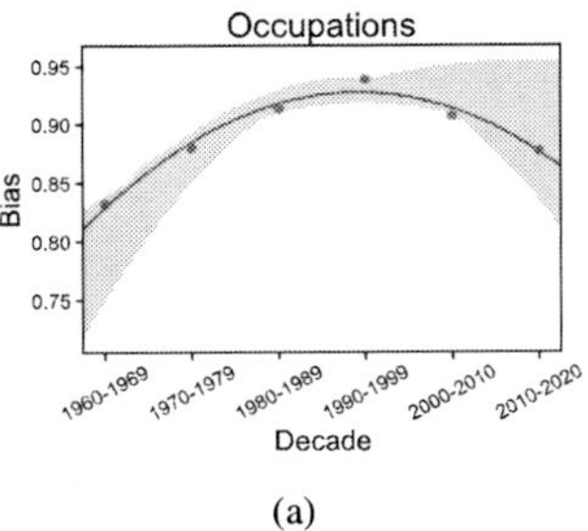
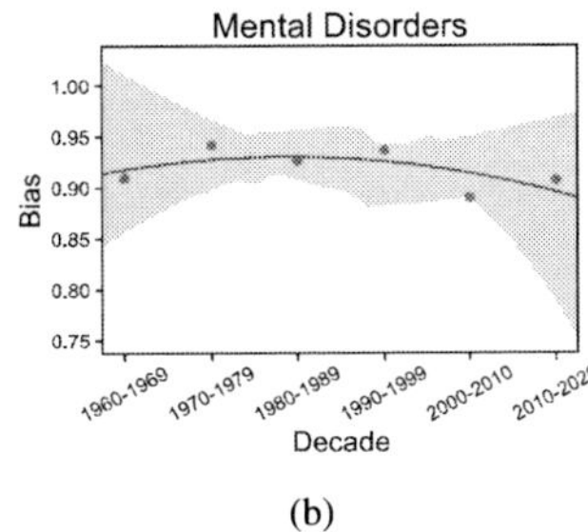

Figure 2: ECT bias estimates for both the set of occupation and mental disorder words. The shaded area of each subplot represents the bootstrap estimated 95% confidence interval.

that the *male* words are more similar to the targets *A* (e.g., career) than the *female* words. There is no measurable bias with a value of zero. Overall, we find that the results from the WEAT test vary depending on the stereotype. For *Career vs Family*, in Figure 1a, we find a steady linear decrease in bias each decade—with the exception of the 1990s. We also find similar linear decreases in bias for both *Science vs Art* and *Strong vs Weak* (Figures 1c and 1e). In Figure 1b, for *Math vs Art*, however, the bias stays relatively static, i.e., it does not dramatically change over time. Moreover, the WEAT score for *Math vs Art* is negative, meaning that the female words are more similar to math than the male words. Likewise, for *Intelligence vs Appearance* (Figure 1d), we see relatively little bias from 1960 to 1989, however, in the 1990s and 2000s, we had a substantial jump in the bias score.

Our evaluation supports prior work evaluating bias in biomedical word embeddings (e.g., *Strong vs Weak* is the most biased stereotype in biomedical literature) (Chaloner and Maldonado, 2019, Table 2). However, we also find differences when measuring bias over time. For example, we find that from 2010 to 2019 there is not a lot evidence for the *Career vs Family* stereotype in biomedical corpora, matching the results from Chaloner and Maldonado (2019, Table 2). Yet, this is only a recent phenomenon. The embeddings trained on articles published from 1990 to 1999 exhibit a *Career vs Family* bias score greater than 1.5. Overall, comparing to Chaloner and Maldonado (2019, Table 2), this means that the bias in recently published biomedical literature may not be as strong as what is found in general text corpora. But, if we exclude the most recent decade's embeddings, the bias in biomedical literature becomes much stronger. Future work should explore comparing the temporal bias in general text corpora to what is found in biomedical literature.

5.3 Occupational and Mental Health Bias.

In Figure 2, we report the gender bias results from ECT on two categories: occupations (e.g., doctor, nurse, teacher) and mental health disorders (e.g., depression, alcoholism, PTSD). Again, unlike WEAT, ECT calculates bias scores on a single target set of words. Therefore, we do not need two contrasting target word sets (e.g., *Math vs Art*), instead we can focus on bias for a single set (e.g., *Math*). Also, the larger the score, the lower the bias—a score of one would represent no difference between male and female words for that specific target set. Interestingly, we find that the ECT scores follow a similar pattern as found in Table 3, the better the embedding quality, the lower the bias.

Comparing Figures 2a and 2b, we find that the word embeddings for both occupations and mental disorders have relatively little bias in the 1990s. Furthermore, while there was small variation, mental disorders experienced little change in bias decade-by-decade. Yet, occupation-related words had a substantial amount of bias in the 1960s and 1970s. Moreover, we find that the bias related to occupations experienced more change, than mental disorders, starting 0.83 in the 1960s and increase by more than ten points to 0.94 in the 1990s. Whereas, mental disorder-related bias scores only ranged from 0.90 to 0.94.

5.4 Biased Words.

In Figure 2, we analyze the bias of individual occupational and mental health-related words. We found a substantial change in the bias of occupational-related words.

We found little change in the bias of mental health-related words since the 1960s. Yet, while

	Male					Female				
	1970-1979	1980-1989	1990-1999	2000-2010	2010-2020	1970-1979	1980-1989	1990-1999	2000-2010	2010-2020
Occupations										
1	promoter	conductor	chef	dentist	mediator	teacher	housewife	neurosurgeon	swimmer	priest
2	collector	chef	baker	counselor	promoter	professor	teenager	pediatrician	baker	fisherman
3	investigator	biologist	astronaut	librarian	dentist	counselor	bishop	educator	butcher	teenager
4	principal	collector	swimmer	pharmacist	principal	physician	lawyer	teenager	medic	chef
5	baker	dad	prisoner	teenager	collector	pediatrician	pediatrician	counselor	barber	writer
6	researcher	singer	mechanic	bishop	cop	consultant	athlete	neurologist	physicist	nanny
7	character	chemist	character	acquaintance	conductor	doctor	physician	consultant	soldier	historian
8	mechanic	butler	worker	cardiologist	substitute	student	pathologist	dentist	baron	president
9	analyst	mechanic	soldier	promoter	coach	lawyer	educator	athlete	director	inventor
10	conductor	promoter	analyst	attorney	employee	pathologist	carpenter	doctor	singer	housewife
Mental Disorders										
1	caffeine	cannabis	separation	lacunar	lacunar	dysmorphic	factitious	binge	dissociative	munchausen
2	restrictive	hypnotic	restrictive	bulimia	circadian	psychogenic	dysmorphic	nervosa	coordination	mutism
3	attachment	caffeine	coordination	erectile	nicotine	anorexia	nervosa	bulimia	separation	factitious
4	separation	coordination	dyskinesia	gambling	gambling	adolescent	mutism	opioid	parasitosis	dysmorphic
5	circadian	hallucinogen	conversion	bereavement	phencyclidine	nervosa	bulimia	hypersomnia	terror	hysteria
6	coordination	dependence	mathematics	binge	ocpd	mutism	tourette	narcolepsy	hysteria	cotard
7	benzodiazepine	attachment	attachment	nervosa	cocaine	infancy	infancy	anorexia	conversion	claustrophobia
8	dependence	mathematics	residual	mood	insomnia	munchausen	episode	panic	malingering	ekbom
9	selective	restrictive	parasitosis	depressive	sleep	factitious	anorexia	korsakoff	tic	diogenes
10	conversion	pdd	developmental	polysubstance	caffeine	disorder	munchausen	factitious	munchausen	encopresis

Table 4: The top ten words with the largest RIPA scores (i.e., the most biased) across each decade. The RIPA scores are reported for both occupations and mental health disorders. While all the listed words are biased, they are ranked starting with the most biased word to the least.

we found little change in mental health bias overall, are there at least a few disorders that changed over time? Moreover, we found a slight bias in mental health terms, therefore, What are the biased terms in each group? We look at the most gender biased occupational and mental health-related terms for each decade in Table 4. Because of space limitations, we only display the gendered words from the 1970s to the 2010s. The words from the 1960s can be found in the appendix. The word-level scores were generated using RIPA. First, for occupations, the words vary between male and female. For example, in the 1970s, male-related words include "mechanic", "principal", and "investigator". The female-related words include "teacher", "counselor", and "pediatrician". Interestingly, the jobs associated with men such as "principal" and "researcher" are positions with power over the jobs associated with woman. For example "principals" (male) have power over "teachers" (female) and "researchers" (male) have power over "students" (female). We also find other well-known occupations appear to be gender-related. For instance, "butler" in the 1980s is associated to male while "nanny" is related to female in the 2010s.

With regard to mental health, we find that disorders associated with well-known gender disparities appear to be biased using RIPA (Organization, 2013). For example, through the last 60 years, words associated with addictions are male-related, e.g., "caffeine", "cannabis", "nicotine", and "gambling". Similarly, disorders related to appearance are more female-related, e.g., "dysmorphic" [2] and "anorexia". We also find that disorders related to emotions are more female-related, such as "munchausen" [3], "hysteria" [4], and "terror". Interestingly, we find that the word "hysteria" is heavily biased in the 2010s. Even though the diagnosis of female hysteria substantially fell in the 1900s (Micale, 1993), it still seems to be a biased term. We want to note that this could simply be caused by research studying mental health diagnosis bias in women, however, the underlying cause of why the term is biased in the 2010s is left for future work.

6 Discussion

In this section, we discuss the impact of the results on two stakeholders of this research: BioNLP researchers and general biomedical researchers. Furthermore, we discuss the limitations of focusing on binary gender (*Male vs Female*).

---

[2] Dysmorphia is a mental health disorder in which you can't stop thinking about one or more perceived defects or flaws

[3] Munchausen is a mental disorder in which a person repeatedly and deliberately acts as if he or she has a physical or mental illness

[4] Hysteria is a (biased) catch-all for symptoms including, but not limited to, nervousness, hallucinations, and emotional outbursts

6.1 Impact on BioNLP researchers.

The results in this paper are important for BioNLP research in two ways. First, we have produced decade-specific word embeddings. [5] Therefore, BioNLP research can use the embeddings to study other historical phenomenon in biomedical research articles. Second, the analysis of historical bias in biomedical research in this paper can be applied to other domains, beyond occupations and mental disorders.

6.2 Impact on Biomedical Researchers.

With regard to general biomedical researchers (e.g., medical researchers and biologist), this work can provide a way to measure which demographics current research is leaning towards in an automated fashion. As discussed in Holdcroft (2007), if research is heavily focused on a single gender, then health disparities can increase. Treatments should be explored equally for all at-risk patients. Furthermore,with the use of contextual word embeddings (Scheuerman et al., 2019), implicit bias measurement techniques can be used as part of the writing process to avoid gendered language when it is not necessary (e.g., using singular they vs he/she).

6.3 A Note About Gender.

Similar to prior work measuring gender bias (Chaloner and Maldonado, 2019), we focus on binary gender. However, it is important to note that the results for binary gender do not necessarily generalize to other genders, including, but not limited to, binary trans people, non-binary people, gender non-conforming people (Scheuerman et al., 2019). Therefore, we want to explicitly note that **our research does not necessarily generalize beyond binary gender**. In future work, we recommend that researcher's studies should be performed for other genders, beyond simply studying *Male vs Female*.

How can this study be expanded beyond binary gender? The three bias measurement techniques studied in this paper (i.e., WEAT, ECT, and RIPA) require sets of words representing a single gender (e.g., boy, men, male). Unfortunately, there is not a large number of words to represent every gender of interest. A promising area of research is to explore bias in contextual word embeddings. With the use of contextual word embeddings (Kurita et al.,

2019), we can measure the bias of individual words across many contexts. Thus, we can possibly overcome the problem of a limited number of words per gender.

7 Conclusion

In this paper, we studied the historical bias present in word embeddings from 1960 to 2020. In summary, we found that while some biases have shown a consistently decrease over time (e.g., *Strong vs Weak*), others have stayed relatively static, or worse, increased (e.g., *Intelligence vs Appearance*). Moreover, we found that the gender bias towards occupations has substantially changed over time, showing that in the past, there was more gender bias associated with certain jobs.

There are two major avenues for future work. First, this work quantified various aspects of gender bias over time. However, we do not know why the bias is present in the word embeddings. For example, is the word "hysteria" biased in 2010 because researchers are associating it with women implicitly, or is it that researchers are studying the historical usage of the diagnosis to ensure the diagnosis is not made because of implicit bias in the future? Thus, our future work will focus on causal studies of bias in biomedical literature. Second, we simply independently trained Skip-Gram word embeddings for each decade. However, recent work has shown that dynamic embeddings, rather than static (decade-specific), perform better with regard to analyzing public perception over time (Gillani and Levy, 2019). Future work will focus on developing new techniques to study bias temporally. Moreover, many techniques may depend on the magnitude of the bias, therefore, we plan to analyze the circumstances in which one embedding approach may measure bias (e.g., Skip-Gram) better than another (e.g., dynamic embeddings).

Acknowledgements

We would like to thank the anonymous reviewers for their invaluable help improving this manuscript. This material is based upon work supported by the National Science Foundation under Grant No. 1947697.

References

Paul R Albert. 2015. Why is depression more prevalent in women? *Journal of psychiatry & neuroscience: JPN*, 40(4):219.

---

[5] `https://github.com/AnthonyMRios/Gender-Bias-PubMed`

American Psychiatric Association et al. 2013. *Diagnostic and statistical manual of mental disorders (DSM-5®)*. American Psychiatric Pub.

Pinkesh Badjatiya, Manish Gupta, and Vasudeva Varma. 2019. Stereotypical bias removal for hate speech detection task using knowledge-based generalizations. In *The World Wide Web Conference*, pages 49–59.

Rahn Kennedy Bailey, Josephine Mokonogho, and Alok Kumar. 2019. Racial and ethnic differences in depression: current perspectives. *Neuropsychiatric disease and treatment*, 15:603.

Olivier Bodenreider. 2004. The unified medical language system (umls): integrating biomedical terminology. *Nucleic acids research*, 32(suppl_1):D267–D270.

Tolga Bolukbasi, Kai-Wei Chang, James Y Zou, Venkatesh Saligrama, and Adam T Kalai. 2016. Man is to computer programmer as woman is to homemaker? debiasing word embeddings. In *Advances in neural information processing systems*, pages 4349–4357.

Aylin Caliskan, Joanna J Bryson, and Arvind Narayanan. 2017. Semantics derived automatically from language corpora contain human-like biases. *Science*, 356(6334):183–186.

Kaytlin Chaloner and Alfredo Maldonado. 2019. Measuring gender bias in word embeddings across domains and discovering new gender bias word categories. In *Proceedings of the First Workshop on Gender Bias in Natural Language Processing*, pages 25–32.

Billy Chiu, Gamal Crichton, Anna Korhonen, and Sampo Pyysalo. 2016. How to train good word embeddings for biomedical nlp. In *Proceedings of the 15th workshop on biomedical natural language processing*, pages 166–174.

I Glenn Cohen, Ruben Amarasingham, Anand Shah, Bin Xie, and Bernard Lo. 2014. The legal and ethical concerns that arise from using complex predictive analytics in health care. *Health affairs*, 33(7):1139–1147.

Amit Datta, Michael Carl Tschantz, and Anupam Datta. 2015. Automated experiments on ad privacy settings. *Proceedings on Privacy Enhancing Technologies*, 2015(1):92–112.

Scott Deerwester, Susan T Dumais, George W Furnas, Thomas K Landauer, and Richard Harshman. 1990. Indexing by latent semantic analysis. *Journal of the American society for information science*, 41(6):391–407.

Sunipa Dev and Jeff Phillips. 2019. Attenuating bias in word vectors. In *The 22nd International Conference on Artificial Intelligence and Statistics*, pages 879–887.

Lucas Dixon, John Li, Jeffrey Sorensen, Nithum Thain, and Lucy Vasserman. 2018. Measuring and mitigating unintended bias in text classification. In *Conference on AI, Ethics, and Society*.

Joel Escudé Font. 2019. Determining bias in machine translation with deep learning techniques. Master's thesis, Universitat Politècnica de Catalunya.

Kawin Ethayarajh, David Duvenaud, and Graeme Hirst. 2019. Understanding undesirable word embedding associations. In *Proceedings of the 57th Annual Meeting of the Association for Computational Linguistics*, pages 1696–1705.

Sergey Feldman, Waleed Ammar, Kyle Lo, Elly Trepman, Madeleine van Zuylen, and Oren Etzioni. 2019. Quantifying sex bias in clinical studies at scale with automated data extraction. *JAMA network open*, 2(7):e196700–e196700.

Joel Escudé Font and Marta R Costa-jussà. 2019. Equalizing gender biases in neural machine translation with word embeddings techniques. *arXiv preprint arXiv:1901.03116*.

Nikhil Garg, Londa Schiebinger, Dan Jurafsky, and James Zou. 2018. Word embeddings quantify 100 years of gender and ethnic stereotypes. *Proceedings of the National Academy of Sciences*, 115(16):E3635–E3644.

Nabeel Gillani and Roger Levy. 2019. Simple dynamic word embeddings for mapping perceptions in the public sphere. In *Proceedings of the Third Workshop on Natural Language Processing and Computational Social Science*, pages 94–99.

Anthony G Greenwald, Debbie E McGhee, and Jordan LK Schwartz. 1998. Measuring individual differences in implicit cognition: the implicit association test. *Journal of personality and social psychology*, 74(6):1464.

Amelia Gulliver, Kathleen M Griffiths, and Helen Christensen. 2010. Perceived barriers and facilitators to mental health help-seeking in young people: a systematic review. *BMC psychiatry*, 10(1):113.

Maryam Habibi, Leon Weber, Mariana Neves, David Luis Wiegandt, and Ulf Leser. 2017. Deep learning with word embeddings improves biomedical named entity recognition. *Bioinformatics*, 33(14):i37–i48.

Katarina Hamberg. 2008. Gender bias in medicine. *Women's Health*, 4(3):237–243.

Cynthia M Hartung and Thomas A Widiger. 1998. Gender differences in the diagnosis of mental disorders: Conclusions and controversies of the dsm–iv. *Psychological bulletin*, 123(3):260.

Bin He, Yi Guan, and Rui Dai. 2019. Classifying medical relations in clinical text via convolutional neural networks. *Artificial intelligence in medicine*, 93:43–49.

Janae K Heath, Gary E Weissman, Caitlin B Clancy, Haochang Shou, John T Farrar, and C Jessica Dine. 2019. Assessment of gender-based linguistic differences in physician trainee evaluations of medical faculty using automated text mining. *JAMA network open*, 2(5):e193520–e193520.

Anita Holdcroft. 2007. Gender bias in research: how does it affect evidence based medicine? *Journal of the Royal Society of Medicine*, 100(1):2.

Matthew Kay, Cynthia Matuszek, and Sean A Munson. 2015. Unequal representation and gender stereotypes in image search results for occupations. In *Proceedings of the 33rd Annual ACM Conference on Human Factors in Computing Systems*, pages 3819–3828. ACM.

Keyvan Khosrovian, Dietmar Pfahl, and Vahid Garousi. 2008. Gensim 2.0: a customizable process simulation model for software process evaluation. In *International conference on software process*, pages 294–306. Springer.

Keita Kurita, Nidhi Vyas, Ayush Pareek, Alan W Black, and Yulia Tsvetkov. 2019. Quantifying social biases in contextual word representations. In *1st ACL Workshop on Gender Bias for Natural Language Processing*.

Chandler May, Alex Wang, Shikha Bordia, Samuel Bowman, and Rachel Rudinger. 2019. On measuring social biases in sentence encoders. In *Proceedings of the 2019 Conference of the North American Chapter of the Association for Computational Linguistics: Human Language Technologies, Volume 1 (Long and Short Papers)*, pages 622–628.

Bridget T McInnes, Ted Pedersen, and Serguei VS Pakhomov. 2009. Umls-interface and umls-similarity: open source software for measuring paths and semantic similarity. In *AMIA annual symposium proceedings*, volume 2009, page 431. American Medical Informatics Association.

Mark S Micale. 1993. On the" disappearance" of hysteria: A study in the clinical deconstruction of a diagnosis. *Isis*, 84(3):496–526.

Tomas Mikolov, Kai Chen, Greg Corrado, and Jeffrey Dean. 2013a. Efficient estimation of word representations in vector space. *ICLR Workshop*.

Tomas Mikolov, Ilya Sutskever, Kai Chen, Greg S Corrado, and Jeff Dean. 2013b. Distributed representations of words and phrases and their compositionality. In *Advances in neural information processing systems*, pages 3111–3119.

World Health Organization. 2013. *Gender Disparities in Mental Health*. World Health Organization.

Ji Ho Park, Jamin Shin, and Pascale Fung. 2018. Reducing gender bias in abusive language detection. In *Proceedings of the 2018 Conference on Empirical Methods in Natural Language Processing*, pages 2799–2804.

Jeffrey Pennington, Richard Socher, and Christopher D Manning. 2014. Glove: Global vectors for word representation. In *Proceedings of the 2014 conference on empirical methods in natural language processing (EMNLP)*, pages 1532–1543.

LA Pratt and DJ Brody. 2014. Depression and obesity in the us adult household population, 2005-2010. *NCHS data brief*, (167):1–8.

Anthony Rios. 2020. FuzzE: Fuzzy fairness evaluation of offensive language classifiers on african-american english. In *Proceedings of the AAAI Conference on Artificial Intelligence*, volume 34.

Anthony Rios and Ramakanth Kavuluru. 2015. Convolutional neural networks for biomedical text classification: application in indexing biomedical articles. In *Proceedings of the 6th ACM Conference on Bioinformatics, Computational Biology and Health Informatics*, pages 258–267.

Arghavan Salles, Michael Awad, Laurel Goldin, Kelsey Krus, Jin Vivian Lee, Maria T Schwabe, and Calvin K Lai. 2019. Estimating implicit and explicit gender bias among health care professionals and surgeons. *JAMA network open*, 2(7):e196545–e196545.

Morgan Klaus Scheuerman, Jacob M. Paul, and Jed R. Brubaker. 2019. How computers see gender: An evaluation of gender classification in commercial facial analysis services. *Proc. ACM Hum.-Comput. Interact.*, 3(CSCW).

Latanya Sweeney. 2013. Discrimination in online ad delivery. *Queue*, 11(3):10.

Yanshan Wang, Sijia Liu, Naveed Afzal, Majid Rastegar-Mojarad, Liwei Wang, Feichen Shen, Paul Kingsbury, and Hongfang Liu. 2018. A comparison of word embeddings for the biomedical natural language processing. *Journal of biomedical informatics*, 87:12–20.

Daniel Weisz, Michael K Gusmano, and Victor G Rodwin. 2004. Gender and the treatment of heart disease in older persons in the united states, france, and england: a comparative, population-based view of a clinical phenomenon. *Gender medicine*, 1(1):29–40.

Terry Young, Rebecca Hutton, Laurel Finn, Safwan Badr, and Mari Palta. 1996. The gender bias in sleep apnea diagnosis: are women missed because they have different symptoms? *Archives of internal medicine*, 156(21):2445–2451.

Jieyu Zhao, Tianlu Wang, Mark Yatskar, Ryan Cotterell, Vicente Ordonez, and Kai-Wei Chang. 2019. Gender bias in contextualized word embeddings. *arXiv preprint arXiv:1904.03310*.

Jieyu Zhao, Yichao Zhou, Zeyu Li, Wei Wang, and Kai-Wei Chang. 2018. Learning gender-neutral word embeddings. In *Proc. of EMNLP*, pages 4847–4853.

A 1960s Most Biased Words

Male:

- physician
- doctor
- president
- dentist
- psychiatrist
- surgeon
- student
- nurse
- worker
- professor

Female:

- substitute
- principal
- editor
- baker
- character
- author
- pharmacist
- scientist
- therapist
- teacher

B Mental Health-Related Terms

[abuse, acute, adaptation, adjustment, adolescent, adult, affective, agoraphobia, alcohol, alcoholic, alzheimer, amnesia, amnestic, amphetamine, anorexia, anosognosia, anterograde, antisocial, anxiety, anxiolytic, asperger, atelophobia, attachment, attention, atypical, autism, autophagia, avoidant, avoidant, restrictive, barbiturate, behavior, benzodiazepine, bereavement, bibliomania, binge, bipolar, body, borderline, brief, bulimia, caffeine, cannabis, capgras, catalepsy, catatonia, catatonic, childhood, circadian, claustrophobia, cocaine, cognitive, communication, compulsive, condition, conduct, conversion, coordination, cotard, cyclothymia, daydreaming, defiant, deficit, delirium, delusion, delusional, delusions, dependence, depersonalization, depression, depressive, derealization, dermatillomania, desynchronosis, deux, developmental, diogenes, disease, disorder, dissociative, dyscalculia, dyskinesia, dyslexia, dysmorphic, eating, ejaculation, ekbom, encephalitis, encopresis, enuresis, epilepsy, episode, erectile, erotomania, exhibitionism, factitious, fantastica, fetishism, fregoli, fugue, functioning, gambling, ganser, grandiose, hallucinogen, hallucinosis, histrionic, huntington, hyperactivity, hypersomnia, hypnotic, hypochondriasis, hypomanic, hysteria, ideation, identity, impostor, induced, infancy, insomnia, intellectual, intermittent, intoxication, kleptomania, korsakoff, lacunar, lethargica, love, major, maladaptive, malingering, mania, mathematics, megalomania, melancholia, misophonia, mood, munchausen, mutism, narcissistic, narcolepsy, nervosa, neurocysticercosis, neurodevelopmental, nicotine, nightmare, nos, obsessive, obsessive–compulsive, ocd, ocpd, oneirophrenia, opioid, oppositional, orthorexia, pain, panic, paralysis, paranoid, parasitosis, parasomnia, parkinson, partialism, pathological, pdd, perception, persecutory, personality, pervasive, phencyclidine, phobia, phobic, phonological, physical, pica, polysubstance, posttraumatic, pseudologia, psychogenic, psychosis, psychotic, ptsd, pyromania, reactive, residual, retrograde, rumination, schizoaffective, schizoid, schizophrenia, schizophreniform, schizotypal, seasonal, sedative, selective, separation, sexual, sleep, sleepwalking, social, sociopath, somatic, somatization, somatoform, stereotypic, stockholm, stress, stuttering, substance, suicidal, suicide, tardive, terror, tic, tourette, transient, transvestic, tremens, trichotillomania, truman, withdrawal, wonderland]

C Occupations

[detective, ambassador, coach, officer, epidemiologist, rabbi, ballplayer, secretary, actress, manager, scientist, cardiologist, actor, industrialist, welder, biologist, undersecretary, captain, economist, politician, baron, pollster, environmentalist, photographer, mediator, character, housewife, jeweler, physicist, hitman, geologist, painter, employee, stockbroker, footballer, tycoon, dad, patrolman, chancellor, advocate, bureaucrat, strategist, pathologist, psychologist, campaigner, magistrate, judge, illustrator, surgeon, nurse, missionary, stylist, solicitor, scholar, naturalist, artist, mathematician, businesswoman, investigator, curator, soloist, servant, broadcaster, fisherman, landlord, housekeeper, crooner, archaeologist, teenager, councilman, attorney, choreographer, principal, parishioner, therapist, administrator, skipper, aide, chef, gangster, astronomer, educator, lawyer, midfielder, evangelist, novelist, senator, collector, goalkeeper, singer, acquaintance, preacher, trumpeter, colonel, trooper, understudy, paralegal, philosopher, councilor, violinist, priest, cellist, hooker, jurist, commentator, gardener, journalist, warrior, cameraman, wrestler, hairdresser, lawmaker, psychiatrist, clerk, writer, handyman, broker, boss, lieutenant, neurosurgeon, protagonist, sculptor, nanny, teacher, homemaker, cop, planner, laborer, programmer, philanthropist, waiter, barrister, trader, swimmer, adventurer, monk, bookkeeper, radiologist, columnist, banker, neurologist, barber, policeman, assassin, marshal, waitress, artiste, playwright, electrician, student, deputy, researcher, caretaker, ranger, lyricist, entrepreneur, sailor, dancer, composer, president, dean, comic, medic, legislator, salesman, observer, pundit, maid, archbishop, firefighter, vocalist, tutor, proprietor, restaurateur, editor, saint, butler, prosecutor, sergeant, realtor, commissioner, narrator, conductor, historian, citizen, worker, pastor, serviceman, filmmaker, sportswriter, poet, dentist, statesman, minister, dermatologist, technician, nun, instructor, alderman, analyst, chaplain, inventor, lifeguard, bodyguard, bartender, surveyor, consultant, athlete, cartoonist, negotiator, promoter, socialite, architect, mechanic, entertainer, counselor, janitor, firebrand, sportsman, anthropologist, performer, crusader, envoy, trucker, publicist, commander, professor, critic, comedian, receptionist, financier, valedictorian, inspector, steward, confesses, bishop, shopkeeper, ballerina, diplomat, parliamentarian, author, sociologist, photojournalist, guitarist, butcher, mobster, drummer, astronaut, protester, custodian, maestro, pianist, pharmacist, chemist, pediatrician, lecturer, foreman, cleric, musician, cabbie, fireman, farmer, headmaster, soldier, carpenter, substitute, director, cinematographer, warden, marksman, congressman, prisoner, librarian, magician, screenwriter, provost, saxophonist, plumber, correspondent, organist, baker, doctor, constable, treasurer, superintendent, boxer, physician, infielder, businessman, protege]

Sequence-to-Set Semantic Tagging for Complex Query Reformulation and Automated Text Categorization in Biomedical IR using Self-Attention

Manirupa Das[†], Juanxi Li[†], Eric Fosler-Lussier[†],
Simon Lin[‡], Steve Rust[‡], Yungui Huang[‡] & Rajiv Ramnath[†]
The Ohio State University[†] & Nationwide Children's Hospital[‡]
{das.65, li.8767, fosler.1, ramnath.6}@osu.edu
Simon.Lin, Steve.Rust, Yungui.Huang}@nationwidechildrens.org

Abstract

Novel contexts, comprising a set of terms referring to one or more concepts, may often arise in complex querying scenarios such as in evidence-based medicine (EBM) involving biomedical literature. These may not explicitly refer to entities or canonical concept forms occurring in a fact-based knowledge source, e.g. the UMLS ontology. Moreover, hidden associations between related concepts meaningful in the current context, may not exist within a single document, but across documents in the collection. Predicting semantic concept tags of documents can therefore serve to associate documents related in unseen contexts, or categorize them, in information filtering or retrieval scenarios. Thus, inspired by the success of sequence-to-sequence neural models, we develop a novel *sequence-to-set* framework with attention, for learning document representations in a unique unsupervised setting, using no human-annotated document labels or external knowledge resources and only corpus-derived term statistics to drive the training. This can effect term transfer within a corpus for semantically tagging a large collection of documents. Our sequence-to-set modeling approach to predict semantic tags , gives to the best of our knowledge, the state-of-the-art for both, an **unsupervised** query expansion (QE) task for the **TREC CDS 2016** challenge dataset when evaluated on an Okapi BM25–based document retrieval system; and also over the MLTM system baseline (Soleimani and Miller, 2016), for both **supervised** and **semi-supervised** multi-label prediction tasks with **del.icio.us** and **Ohsumed** datasets. We make our code and data publicly available [1].

1 Introduction

Recent times have seen an upsurge in efforts towards personalized medicine where clinicians tai-

lor their medical decisions to the individual patient, based on the patient's genetic information, other molecular analysis, and the patient's preference. This often requires them to combine clinical experience with evidence from scientific research, such as that available from biomedical literature, in a process known as evidence-based medicine (EBM). Finding the most relevant recent research however, is challenging not only due to the volume and the pace at which new research is being published, but also due to the complex nature of the information need, arising for example, out of a clinical note which may be used as a query. This calls for better automated methods for natural language understanding (NLU), e.g. to derive a set of key terms or *related concepts* helpful in appropriately transforming a complex query, by reformulation so as to be able to handle and possibly resolve medical jargon, lesser-used acronyms, misspelling, multiple subject areas and often multiple references to the same entity or concept, and retrieve the most related, yet most comprehensive set of useful results.

At the same time, tremendous strides have been made by recent neural machine learning models in reasoning with texts on a wide variety of NLP tasks. In particular, sequence-to-sequence (seq2seq) neural models often employing *attention* mechanisms, have been largely successful in delivering the state-of-the-art for tasks such as machine translation (Bahdanau et al., 2014), (Vaswani et al., 2017), handwriting synthesis (Graves, 2013), image captioning (Xu et al., 2015), speech recognition (Chorowski et al., 2015) and document summarization (Cheng and Lapata, 2016). Inspired by these successes, we aimed to harness the power of sequential *encoder-decoder* architectures with attention, to train end-to-end differentiable models that are able to learn the best possible representation of input documents in a collection while being predictive of a set of *key terms* that best describe the docu-

[1] https://github.com/mcoqzeug/seq2set-semantic-tagging

Proceedings of the BioNLP 2020 workshop, pages 14–27
Online, July 9, 2020 ©2020 Association for Computational Linguistics

ment. These will be later used to *transfer* a relevant but diverse set of key terms from the most related documents, for "semantic tagging" of the original input documents so as to aid in downstream query refinement for IR by pseudo-relevance feedback (Xu and Croft, 2000).

To this end and to the best of our knowledge, we are the first to employ a novel, completely unsupervised end-to-end neural attention-based document representation learning approach, using no external labels, in order to achieve the most meaningful term transfer between related documents, i.e. semantic tagging of documents, in a "pseudo-relevance feedback"–based (Xu and Croft, 2000) setting for unsupervised query expansion. This may also be seen as a method of document expansion as a means for obtaining query refinement terms for downstream IR. The following sections give an account of our specific architectural considerations in achieving an end-to-end neural framework for semantic tagging of documents using their repre sentations, and a discussion of the results obtained from this approach.

2   Related Work

Pseudo-relevance feedback (PRF), a *local context analysis* method for automatic query expansion (QE), is extensively studied in information retrieval (IR) research as a means of addressing the word mismatch between queries and documents. It adjusts a query relative to the documents that initially appear to match it, with the main assumption that the top-ranked documents in the first retrieval result contain many useful terms that can help discriminate relevant documents from irrelevant ones (Xu and Croft, 2000), (Cao et al., 2008). It is motivated by *relevance feedback* (RF), a well-known IR technique that modifies a query based on the relevance judgments of the retrieved documents (Salton et al., 1990). It typically adds common terms from the relevant documents to a query and re-weights the expanded query based on term frequencies in the relevant documents relative to the non-relevant ones. Thus in PRF we find an initial set of most relevant documents, then assuming that the top k ranked documents are relevant, RF is done as before, without manual interaction by the user. The added terms are, therefore, common terms from the top-ranked documents.

To this end, (Cao et al., 2008) employ term classification for retrieval effectiveness, in a "supervised"

setting, to select most relevant terms. (Palangi et al., 2016) employ a deep sentence embedding approach using LSTMs and show improvement over standard sentence embedding methods, but as a means for directly deriving encodings of queries and documents for use in IR, and not as a method for QE by PRF. In another approach, (Xu et al., 2017) train autoencoder representations of queries and documents to enrich the feature space for learning-to-rank, and show gains in retrieval performance over pre-trained rankers. But this is a fully supervised setup where the queries are *seen* at train time. (Pfeiffer et al., 2018) also use an autoencoder-based approach for actual query refinement in pharmacogenomic document retrieval. However here too, their document ranking model uses the encoding of the query and the document for training the ranker, hence the queries are *not unseen* with respect to the document during training. They mention that their work can be improved upon by the use of seq2seq-based approaches. In this sense, i.e. with respect to QE by PRF and learning a sequential document representation for document ranking, our work is most similar to (Pfeiffer et al., 2018). However the queries are completely unseen in our case and we use only the documents in the corpus, to train our neural document language models from scratch in a completely unsupervised way.

Classic sequence-to-sequence models like (Sutskever et al., 2014) demonstrate the strength of recurrent models such as the LSTM in capturing short and long range dependencies in learning effective encodings for the end task. Works such as (Graves, 2013), (Bahdanau et al., 2014), (Rocktäschel et al., 2015), further stress the key role that attention, and multi-headed attention (Vaswani et al., 2017) can play in solving the end task. We use these insights in our work.

According to the detailed report provided for this dataset and task in (Roberts et al., 2016) all of the systems described perform **direct query reweighting** aside from **supervised term expansion** and are highly tuned to the clinical queris in this dataset. In a related medical IR challenge (Roberts et al., 2017) the authors specifically mention that with only six partially annotated queries for system development, it is likely that systems were either under- or over-tuned on these queries. Since the setup of the seq2set framework is an attempt to model the PRF based query expansion method of its closest related work (Das et al., 2018) where the effort is also to train a neural generalized language model for unsupervised semantic tagging, we choose this system as the benchmark to compare against to our end-to-end approach for the same task.

3 Methodology

Drawing on sequence-to-sequence modeling approaches for text classification, e.g. textual entailment (Rocktäschel et al., 2015) and machine translation (Sutskever et al., 2014), (Bahdanau et al., 2014) we adapt from these settings into a *sequence-to-set* framework, for learning representations of input documents, in order to derive a meaningful set of terms, or *semantic tags* drawn from a closely related set of documents, that expand the original documents. These document expansion terms are then used downstream for query reformulation via PRF, for unseen queries. We employ an end-to-end framework for unsupervised representation learning of documents using TFIDF-based *pseudo-labels* (Figure 1(a))and a separate cosine similarity-based ranking module for semantic tag inference (Figure 1(b)).

We employ various methods such as doc2vec, Deep Averaging, sequential models such as LSTM, GRU, BiGRU, BiLSTM, BiLSTM with Attention and Self-attention, detailed in Figure 1(c)-(f), see Appendix A, for learning fixed-length input document representations in our framework. We apply methods like DAN (Iyyer et al., 2015), LSTM, and BiLSTM as our baselines and formulate attentional models including a self-attentional Transformer-based one (Vaswani et al., 2017) as our proposed augmented document encoders.

Further, we hypothesize that a sequential, bi-directional or attentional encoder coupled with a decoder, i.e. a sigmoid or softmax prediction layer, that conditions on the encoder output v (similar to an approach by (Kiros et al., 2015) for learning a neural probabilistic language model), would enable learning of the optimal semantic tags in our unsupervised query expansion setting, while modeling directly for this task in an end-to-end neural framework. In our setup the decoder predicts a meaningful set of concept tags that best describe a document according to the training objective. The following sections describe our setup.

3.1 The *Sequence-to-Set* Semantic Tagging Framework

Task Definition: For each query document d_q in a given a collection of documents $D = \{d_1, d_2, ..., d_N\}$, represented by a set of k keywords or labels, e.g. k terms in d_q derived from $top-|V|$ TFIDF-scored terms, find an ***alternate*** set of k most relevant terms coming from documents *"most related"* to d_q from elsewhere in the collection. These serve as *semantic tags* for expanding d_q.

In the **unsupervised** task setting described later, a document to be tagged is regarded as a query document d_q; its semantic tags are generated via PRF, and these terms will in turn be used for PRF–based expansion of unseen queries in downstream IR. Thus d_q could represent an original complex query text or a document in the collection.

In the following sections we describe the building blocks used in the setup for the baseline and proposed models for sequence-to-set semantic tagging as described in the task definition.

3.2 Training and Inference Setup

The overall architecture for sequence-to-set semantic tagging consists of two phases, as depicted in the block diagrams in Figures 1(a) and 1(b): the first, for training of input representations of documents; and the second for inference to achieve *term transfer* for semantic tagging. As shown in Figure 1(a), the proposed model architecture would first learn the appropriate feature representations of documents in a first pass of training, by taking in the tokens of an input document sequentially, using a document's pre-determined $top-k$ TFIDF-scored terms as the ***pseudo***-class labels for an input instance, i.e. prediction targets for a *sigmoid* layer for multi-label classification. The training objective is to maximize probability for these k terms, or $y_p = (t_1, t_2, ..t_k) \in V$, i.e.

$$\arg\max_{\theta} P(y_p = (t_1, t_2, ..t_k) \in V | v; \theta) \quad (1)$$

given the document's encoding v. For computational efficiency, we take V to be the list of top-10K TFIDF-scored terms from our corpus, thus $|V| = 10,000$. k is taken as 3, so each document is initially labeled with 3 terms. The sequential model is then trained with the k-hot 10K-dimensional label vector as targets for the sigmoid classification layer, employing a couple of alternative training

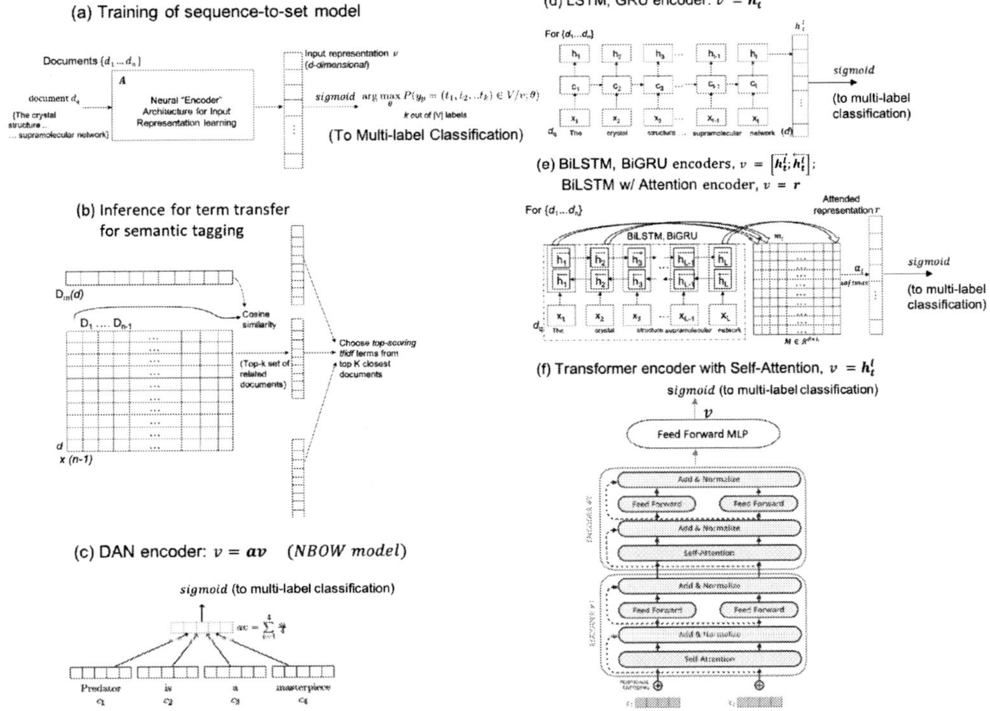

Figure 1: Overview of *Sequence-to-Set* Framework. (a) Method for training document or query representations, (b) Method for Inference via **term transfer** for semantic tagging; Document Sequence Encoders: (c) Deep Averaging encoder; (d) LSTM last hidden state, GRU encoders; (e) BiLSTM last hidden state, BiGRU (shown in dotted box), BiLSTM attended hidden states encoders; and (f) Transformer self-attentional encoder [source: (Alammar, 2018)].

objectives. The first, typical for multi-label classification, minimizes a categorical cross-entropy loss, which for a single training instance with ground-truth label *set*, y_p, is:

$$L_{CE}(\hat{y_p}) = \sum_{i=1}^{|V|} y_i \log(\hat{y_i}) \tag{2}$$

Since our goal is to obtain the most meaningful document representations most predictive of their assigned terms, and that can also be predictive of semantic tags not present in the document, we also consider a language model–based loss objective converting our decoder to a neural language model. Thus, we employ a training objective that maximizes the conditional log likelihood of the label terms L_d of a document d_q, given the document's representation v, i.e. $P(L_d|d_q)$ (where $y_p = L_d \in V$). This amounts to minimizing the negative log likelihood of the label representations

conditioned on the document encoding. Thus,

$$P(L_d|d_q) = \prod_{l \in L_d} P(l|d_q) = -\sum_{l \in L_d} \log(P(l|d_q)) \tag{3}$$

Since $P(l|d_q) \propto \exp(v_l \cdot v)$, where v_l and v are the label and document encodings, it is equivalent to minimizing:

$$L_{LM}(\hat{y_p}) = -\sum_{l \in L_d} \log(\exp(v_l \cdot v)) \tag{4}$$

Equation (4) represents our language model–style loss objective. We run experiments training with both losses (Equations (2) & (4)) as well as a variant that is a summation of both, with a hyperparameter α used to tune the language model component of the total loss objective.

4 Task Settings

4.1 Unsupervised Task – Semantic Tagging for Query Expansion

We now describe the setup and results for experiments run on our unsupervised task setting of semantic tagging of documents for PRF–based query expansion.

4.1.1 Dataset – TREC CDS 2016

The 2016 TREC CDS challenge dataset, makes available actual electronic health records (EHR) of patients (de-identified), in the form of case reports, typically describing a challenging medical case. Such a case report represents a **query** in our system, having a complex information need. There are 30 queries in this dataset, corresponding to such case reports, at 3 levels of granularity **Note**, **Description** and **Summary** text as described in (Roberts et al., 2016). The target document collection is the Open Access Subset of PubMed Central (PMC), containing 1.25 million articles consisting of *title*, *keywords*, *abstract* and *body* sections. We use a subset of 100K of these articles for which human relevance judgments are made available by TREC, for training. Final evaluation however is done on an ElasticSearch index built on top of the entire collection of 1.25 million PMC articles.

4.1.2 Unsupervised Task Experiments

We ran several sets of experiments with various document encoders, employing pre-trained and fine-tuned embedding schemes like skip-gram (Mikolov et al., 2013a) and Probabilistic Fast-Text (Athiwaratkun et al., 2018), see Appendix B. The experimental setup used is the same as the Phrase2VecGLM (Das et al., 2018), the only other known system for this dataset, that performs "unsupervised semantic tagging of documents by PRF", for downstream query expansion. Thus we take this system as the current state-of-the-art system baseline, while our *non-attention-based* document encoding models constitute our standard baselines. Our document-TFIDF representations–based query expansion forms yet another baseline. Summary text UMLS (Lindberg et al., 1993; Bodenreider, 2004) terms for use in our augmented models is available to us via the UMLS Java Metamap API (Demner-Fushman et al., 2017). The first was a set of experiments with our different models using the `Summary Text` as the base query. Following this we ran experiments with our models using the `Summary Text + Sum. UMLS terms` as the "augmented" query. We use the Adam optimizer (Kingma and Ba, 2014) for training our models. After several rounds of hyper-paramater tuning, $batch\_size$ was set to 128, $dropout$ to 0.3, the prediction layer was fixed to $sigmoid$, the loss function switched between cross-entropy and summation of cross entropy and LM losses, and models trained with early stopping.

Results from various *Seq2Set* encoder models on **base** (`Summary Text`) and **augmented** (`Summary Text + Summary-based UMLS terms`) query, are outlined in Table 1. Evaluating on base query, a *Seq2Set*-Transformer model beats all other *Seq2Set* encoders, and also the TFIDF, MeSH QE terms and Expert QE terms baselines. On the augmented query, the *Seq2Set*-BiGRU and *Seq2Set*-Transformer models outperform all other *Seq2Set* encoders, and *Seq2Set*-Transformer outperforms all non-ensemble baselines and the Phrase2VecGLM unsupervised QE ensemble system baseline significantly, with P@10 of **0.4333**. Best performing *supervised QE* systems for this dataset, tuned on all 30 queries, range between *0.35–0.4033* P@10 (Roberts et al., 2016), better than unsupervised QE systems on base query, but surpassed by the best *Seq2Set*-based models such as *Seq2Set*-Transformer on augmented query, even without ensemble. Semantic tags from a best-performing model, do appear to pick terms relating to certain conditions, e.g.: *<query_doc original pseudo-label terms:['obesity', 'diabetes', 'pulmonary-hypertension', 'children'], semantic tags: ['dyslipidaemia', 'hyperglycemia', 'bmi', 'subjects'] >*.

4.2 Supervised Task – Automated Text Categorization

The Seq2set framework's unsupervised semantic tagging setup is primarily applicable in those settings where no pre-existing document labels are available. In such a scenario, of unsupervised semantic tagging of a large document collection, the Seq2set framework therefore consists of separate training and inference steps to infer tags from other documents after encodings have been learnt. We therefore conduct a series of extensive evaluations in the manner described in the previous section, using a downstream QE task in order to validate our method. However, when a tagged document

Unsupervised QE Systems (Base Query)	P@10
BM25+*Seq2Set*-doc2vec (**baseline**)	0.0794
BM25+*Seq2Set*-TFIDF Terms (**baseline**)	0.2000
BM25+**MeSH** QE Terms (**baseline**)	0.2294
BM25+**Human Expert** QE Terms (**baseline**)	0.2511
BM25+unigramGLM+Phrase2VecGLM *ensemble* (**system baseline**)	**0.2756**
BM25+*Seq2Set*-Transformer (L_{CE}) (**model**)	**0.2861***
Supervised QE Systems (Base Query)	
BM25+ETH Zurich-*ETHSummRR*	0.3067
BM25+Fudan Univ.DMIIP-*AutoSummary1*	**0.4033**
Unsupervised QE Systems (Augmented Query)	
BM25+*Seq2Set*-doc2vec (**baseline**)	0.1345
BM25+*Seq2Set*-TFIDF Terms (**baseline**)	0.3000
BM25+unigramGLM+Phrase2VecGLM *ensemble* (**system baseline**)	**0.3091**
BM25+*Seq2Set*-BiGRU (LM only loss) (**model**)	**0.3333***
BM25+*Seq2Set*-Transformer ($L_{CE} + L_{LM}$) (**model**)	**0.4333***

Table 1: Results on IR for best *Seq2set* models, in an ***unsupervised PRF*–based** QE setting. Boldface indicates statistical significance @p<<0.01 over previous.

collection is available where the set of document labels are already known, we can learn to predict tags from this set of known labels on a new set of similar documents. Thus, in order to generalize our Seq2set approach to such other tasks and setups, we therefore aim to validate the performance of our framework on such a labeled dataset of tagged documents, which is equivalent to adapting the Seq2set framework for a supervised setup. In this setup we therefore only need to use the training module of the Seq2set framework shown in Figure 1(a), and measure tag prediction performance on a held out set of documents. For this evaluation, we therefore choose to work with the popular Delicious (del.icio.us) folksonomy dataset, same as that used by (Soleimani and Miller, 2016) in order to do an appropriate comparison with their MLTM framework that is also evaluated on a similar document multi-label prediction task.

4.2.1 Dataset – del.icio.us

The Delicious dataset contains tagged web pages retrieved from the social bookmarking site, del.icio.us. There are 20 common tags used as class labels: *reference, design, programming, internet, computer, web, java, writing, English, grammar, style, language, books, education, philosophy, politics, religion, science, history* and *culture*. The training set consists of 8250 documents and the test set consists of 4000 documents.

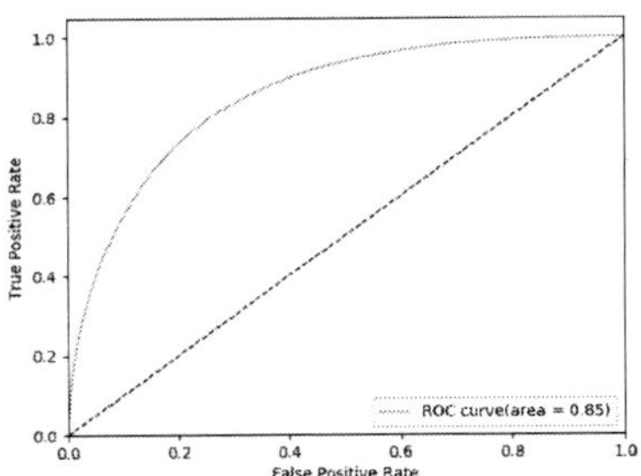

Figure 2: Seq2Set–**supervised** on **del.icio.us**, best **Transformer** model–based encoder

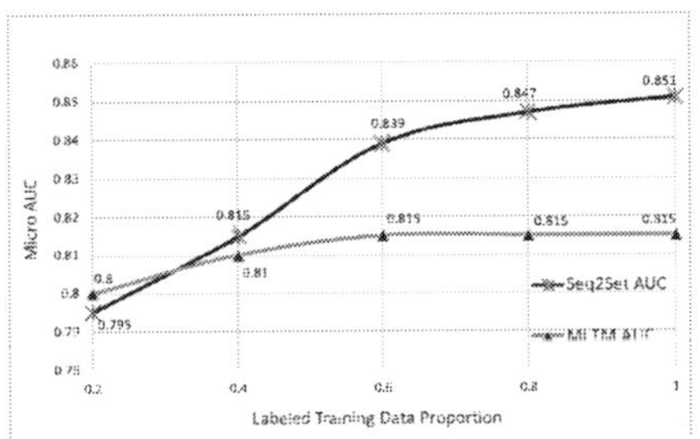

Figure 3: A comparison of document labeling performance of Seq2set versus MLTM

4.2.2 Supervised Task Experiments

We then run Seq2set-based training for our 8 different encoder models on the training set for the 20 labels, and perform evaluation on the test set measuring sentence-level ROC AUC on the labeled documents in the test set.

Figure 2 shows the ROC AUC for the best performing Transformer model from the Seq2set framework on the del.icio.us dataset, which was trained with a sigmoid–based prediction layer on cross entropy loss with a batch size of 64 and dropout set to 0.3. This best model got an ROC AUC of **0.85**, statistically significantly surpassing MLTM (AUC 0.81 @ $p << 0.001$) for this task and dataset.

Figure 3 also shows a comparison of the ROC AUC scores obtained with training Seq2set and MLTM based models for this task with various labeled data proportions. Here again we see that Seq2set has clear advantage over the current MLTM state-of-the-art, statistically significantly surpassing it ($p << 0.01$) when trained with greater than 25% of the labeled dataset.

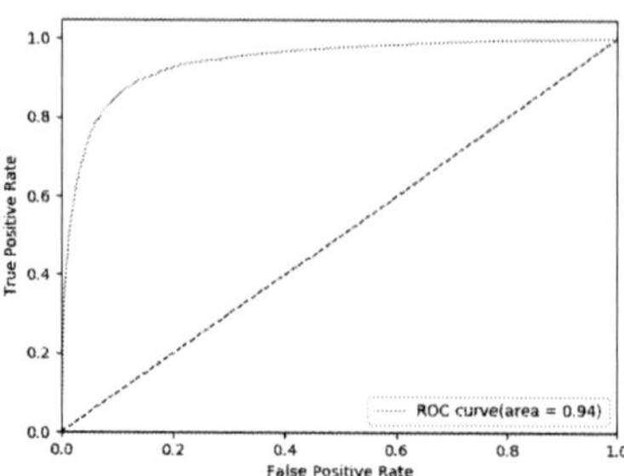

Figure 4: Seq2Set–**semi-supervised** on **Ohsumed**, best **Transformer** model–based encoder w/ Cross Entropy–based Softmax prediction; 4 layers, 10 attention heads, dropout=0

4.3 Semi-Supervised Text Categorization

We then seek to further validate how well the Seq2set framework can leverage large scale pre-training on unlabeled data given only a small amount of labeled data for training, to be able to improve prediction performance on a held out set of these known labels. This amounts to a semi-supervised setup–based evaluation of the Seq2set framework. In this setup, we perform the training and evaluation of Seq2set similar to the supervised setup, except we have an added step of pre-training the multi-label prediction on large amounts of un-labeled document data in exactly the same way as the unsupervised setup.

4.3.1 Dataset – Ohsumed

We employ the Ohsumed dataset available from the TREC Information Filtering tracks of years 87-91 and the version of the labeled **Ohsumed** dataset used by (Soleimani and Miller, 2016) for evaluation, to have an appropriate comparison with their MLTM system also evaluated for this dataset. The version of the Ohsumed dataset due to (Soleimani and Miller, 2016) consists of 11122 training and 5388 test documents, each assigned to one or multiple labels of 23 MeSH diseases categories. Almost half of the documents have more than one label.

4.3.2 Semi-Supervised Task Experiments

We first train and test our framework on the labeled subset of the **Ohsumed** data from (Soleimani and Miller, 2016) similar to the supervised setup described in the previous section. This evaluation gives a statistically significant ROC AUC of 0.93 over the 0.90 AUC for the MLTM system of (Soleimani and Miller, 2016) for a Transformer–

based Seq2set model performing best. Next we experiment with the semi-supervised setting where we first train the Seq2set framework models on a large number of documents that do not have pre-existing labels. This pre-training is performed in exactly a similar fashion as the unsupervised setup. Thus we first preprocess the Ohsumed data from years 87-90 to obtain a top-1000 TFIDF score–based vocabulary of tags, pseudo-labeling all the documents in the training set with these. Our training and evaluation for the semi-supervised setup consists of 3 phases: **Phase 1**: We employ our seq2set framework (using each one of our encoder models) for multi-label prediction on this pseudo-labeled data, having an output prediction layer of 1000 having a penultimate fully-connected layer of dimension 23, same as the number of labels in the Ohsumed dataset; **Phase 2**: After pre-training with pseudolabels we discard the final layer and continue to train labeled Ohsumed dataset from 91 by 5-fold cross-validation with early stopping. **Phase 3**: This is the final evaluation step of our semi-supervised trained Seq2set model on the labeled Ohsumed test dataset used by (Soleimani and Miller, 2016). This constitutes simply inferring predicted tags using the trained model on the test data. As shown in Figure 4, our evaluation of the Seq2set framework for the Ohsumed dataset, comparing supervised and semi-supervised training setups, yields an ROC AUC of **0.94** for our best performing **semi-supervised**–trained model of Fig. 4, compared to the various supervised trained models for the same dataset that got a best ROC AUC of 0.93. The top performing semi-supervised model again involves a Transformer–based encoder using a softmax layer for prediction, with 4 layers, 10 attention heads, and no dropout. Thus, the best results on the semi-supervised training experiments (ROC AUC 0.94) **statistically significantly out-performs** ($p \ll 0.01$) the **MLTM** system base-line (ROC AUC 0.90) on the Ohsumed dataset, while also clearly surpassing the top-performing supervised Seq2set models on the same dataset. This demonstrates that our *Seq2set* framework is able to leverage the benefits of data augmentation in the semi-supervised setup by training with large amounts of unlabeled data on top of limited labeled data.

5 Conclusion

We develop a novel *sequence-to-set* end-to-end encoder-decoder–based neural framework for multi-label prediction, by training document representations using no external supervision labels, for pseudo-relevance feedback–based unsupervised semantic tagging of a large collection of documents. We find that in this unsupervised task setting of PRF-based semantic tagging for query expansion, a multi-term prediction training objective that jointly optimizes both prediction of the TFIDF–based document pseudo-labels and the log likelihood of the labels given the document encoding, surpasses previous methods such as Phrase2VecGLM (Das et al., 2018) that used neural generalized language models for the same. Our initial hypothesis that bi-directional or self-attentional models could learn the most efficient semantic representations of documents when coupled with a loss more effective than cross-entropy at reducing language model perplexity of document encodings, is corroborated in all experimental setups. We demonstrate the effectiveness of our novel framework in every task setting, viz. for **unsupervised** QE via PRF-based semantic tagging for a downstream medical IR challenge task; as well as for both, **supervised** and **semi-supervised** task settings, where Seq2set statistically significantly outperforms the state-of-art MLTM baseline (Soleimani and Miller, 2016) on the same held out set of documents as MLTM, for multi-label prediction on a set of known labels, for automated text categorization; achieving to the best of our knowledge, the current state-of-the-art for multi-label prediction on documents, with or without known labels. We therefore demonstrate the effectiveness of our Sequence-to-Set framework for multi-label prediction, on any set of documents, applicable especially towards the automated categorization, filtering and semantic tagging for QE-based retrieval, of biomedical literature for EBM. Future directions would involve experiments replacing TDIDF labels with more meaningful terms (using unsupervised term extraction) for query expansion, initialization with pre-trained embeddings for the biomedical domain such as BlueBERT (Peng et al., 2019) and BioBERT (Lee et al., 2020), and multi-task learning with closely related tasks such as biomedical Named Entity Recognition and Relation Extraction to learn better document representations and thus more meaningful semantic tags of documents useful for downstream EBM tasks.

References

Martín Abadi, Paul Barham, Jianmin Chen, Zhifeng Chen, Andy Davis, Jeffrey Dean, Matthieu Devin, Sanjay Ghemawat, Geoffrey Irving, Michael Isard, et al. 2016. Tensorflow: A system for large-scale machine learning. In *12th {USENIX} Symposium on Operating Systems Design and Implementation ({OSDI} 16)*, pages 265–283.

Jay Alammar. 2018. The illustrated transformer. Online; posted June 27, 2018.

Ben Athiwaratkun, Andrew Gordon Wilson, and Anima Anandkumar. 2018. Probabilistic fasttext for multi-sense word embeddings. *arXiv preprint arXiv:1806.02901*.

Dzmitry Bahdanau, Kyunghyun Cho, and Yoshua Bengio. 2014. Neural machine translation by jointly learning to align and translate. *arXiv preprint arXiv:1409.0473*.

Olivier Bodenreider. 2004. The unified medical language system (umls): integrating biomedical terminology. *Nucleic acids research*, 32(suppl_1):D267–D270.

Guihong Cao, Jian-Yun Nie, Jianfeng Gao, and Stephen Robertson. 2008. Selecting good expansion terms for pseudo-relevance feedback. In *Proceedings of the 31st annual international ACM SIGIR conference on Research and development in information retrieval*, pages 243–250. ACM.

Jianpeng Cheng and Mirella Lapata. 2016. Neural summarization by extracting sentences and words. *arXiv preprint arXiv:1603.07252*.

Jan K Chorowski, Dzmitry Bahdanau, Dmitriy Serdyuk, Kyunghyun Cho, and Yoshua Bengio. 2015. Attention-based models for speech recognition. In *Advances in neural information processing systems*, pages 577–585.

Junyoung Chung, Caglar Gulcehre, KyungHyun Cho, and Yoshua Bengio. 2014. Empirical evaluation of gated recurrent neural networks on sequence modeling. *arXiv preprint arXiv:1412.3555*.

Junyoung Chung, Caglar Gulcehre, Kyunghyun Cho, and Yoshua Bengio. 2015. Gated feedback recurrent neural networks. In *International Conference on Machine Learning*, pages 2067–2075.

Manirupa Das, Eric Fosler-Lussier, Simon Lin, Soheil Moosavinasab, David Chen, Steve Rust, Yungui Huang, and Rajiv Ramnath. 2018. Phrase2vecglm: Neural generalized language model–based semantic tagging for complex query reformulation in medical ir. In *Proceedings of the BioNLP 2018 workshop*, pages 118–128.

Dina Demner-Fushman, Willie J Rogers, and Alan R Aronson. 2017. Metamap lite: an evaluation of a new java implementation of metamap. *Journal of the American Medical Informatics Association*, 24(4):841–844.

Alex Graves. 2013. Generating sequences with recurrent neural networks. *arXiv preprint arXiv:1308.0850*.

Sepp Hochreiter and Jürgen Schmidhuber. 1997. Long short-term memory. *Neural computation*, 9(8):1735–1780.

Mohit Iyyer, Varun Manjunatha, Jordan Boyd-Graber, and Hal Daumé III. 2015. Deep unordered composition rivals syntactic methods for text classification. In *Proceedings of the 53rd Annual Meeting of the ACL*, volume 1, pages 1681–1691.

Yoon Kim. 2014. Convolutional neural networks for sentence classification. *arXiv preprint arXiv:1408.5882*.

Diederik P Kingma and Jimmy Ba. 2014. Adam: A method for stochastic optimization. *arXiv preprint arXiv:1412.6980*.

Ryan Kiros, Yukun Zhu, Ruslan R Salakhutdinov, Richard Zemel, Raquel Urtasun, Antonio Torralba, and Sanja Fidler. 2015. Skip-thought vectors. In *Advances in neural information processing systems*, pages 3294–3302.

Quoc V Le and Tomas Mikolov. 2014. Distributed representations of sentences and documents. In *ICML*, volume 14, pages 1188–1196.

Jinhyuk Lee, Wonjin Yoon, Sungdong Kim, Donghyeon Kim, Sunkyu Kim, Chan Ho So, and Jaewoo Kang. 2020. Biobert: a pre-trained biomedical language representation model for biomedical text mining. *Bioinformatics*, 36(4):1234–1240.

Donald AB Lindberg, Betsy L Humphreys, and Alexa T McCray. 1993. The unified medical language system. *Yearbook of Medical Informatics*, 2(01):41–51.

Tomas Mikolov, Kai Chen, Greg Corrado, and Jeffrey Dean. 2013a. Efficient estimation of word representations in vector space. *arXiv preprint arXiv:1301.3781*.

Tomas Mikolov, Ilya Sutskever, Kai Chen, Greg S Corrado, and Jeff Dean. 2013b. Distributed representations of words and phrases and their compositionality. In *Advances in neural information processing systems*, pages 3111–3119.

Hamid Palangi, Li Deng, Yelong Shen, Jianfeng Gao, Xiaodong He, Jianshu Chen, Xinying Song, and Rabab Ward. 2016. Deep sentence embedding using long short-term memory networks: Analysis and application to information retrieval. *IEEE/ACM Transactions on Audio, Speech and Language Processing (TASLP)*, 24(4):694–707.

Yifan Peng, Shankai Yan, and Zhiyong Lu. 2019. Transfer learning in biomedical natural language processing: An evaluation of bert and elmo on ten benchmarking datasets. In *Proceedings of the 2019 Workshop on Biomedical Natural Language Processing (BioNLP 2019)*, pages 58–65.

Matthew E Peters, Mark Neumann, Mohit Iyyer, Matt Gardner, Christopher Clark, Kenton Lee, and Luke Zettlemoyer. 2018. Deep contextualized word representations. *arXiv preprint arXiv:1802.05365*.

Jonas Pfeiffer, Samuel Broscheit, Rainer Gemulla, and Mathias Göschl. 2018. A neural autoencoder approach for document ranking and query refinement in pharmacogenomic information retrieval. In *Proceedings of the BioNLP 2018 workshop*, pages 87–97.

Radim Řehůřek and Petr Sojka. 2010. Software Framework for Topic Modelling with Large Corpora. In *Proceedings of the LREC 2010 Workshop on New Challenges for NLP Frameworks*, pages 45–50, Valletta, Malta. ELRA. http://is.muni.cz/publication/884893/en.

Kirk Roberts, Anupama E Gururaj, Xiaoling Chen, Saeid Pournejati, William R Hersh, Dina Demner-Fushman, Lucila Ohno-Machado, Trevor Cohen, and Hua Xu. 2017. Information retrieval for biomedical datasets: the 2016 biocaddie dataset retrieval challenge. *Database*, 2017.

Kirk Roberts, Ellen Voorhees, Dina Demner-Fushman, and William R. Hersh. 2016. Overview of the trec 2016 clinical decision support track. Online; posted August-2016.

Tim Rocktäschel, Edward Grefenstette, Karl Moritz Hermann, Tomáš Kočiský, and Phil Blunsom. 2015. Reasoning about entailment with neural attention. *arXiv preprint arXiv:1509.06664*.

Gerard Salton, Chris Buckley, and Maria Smith. 1990. On the application of syntactic methodologies in automatic text analysis. *Information Processing & Management*, 26(1):73–92.

Hossein Soleimani and David J Miller. 2016. Semi-supervised multi-label topic models for document classification and sentence labeling. In *Proceedings of the 25th ACM international on conference on information and knowledge management*, pages 105–114. ACM.

Ilya Sutskever, Oriol Vinyals, and Quoc V Le. 2014. Sequence to sequence learning with neural networks. In *Advances in neural information processing systems*, pages 3104–3112.

Ashish Vaswani, Noam Shazeer, Niki Parmar, Jakob Uszkoreit, Llion Jones, Aidan N Gomez, Łukasz Kaiser, and Illia Polosukhin. 2017. Attention is all you need. In *Advances in Neural Information Processing Systems*, pages 5998–6008.

Bo Xu, Hongfei Lin, Yuan Lin, and Kan Xu. 2017. Learning to rank with query-level semi-supervised autoencoders. In *Proceedings of the 2017 ACM on Conference on Information and Knowledge Management*, pages 2395–2398. ACM.

Jinxi Xu and W Bruce Croft. 2000. Improving the effectiveness of information retrieval with local context analysis. *ACM Transactions on Information Systems (TOIS)*, 18(1):79–112.

Kelvin Xu, Jimmy Ba, Ryan Kiros, Kyunghyun Cho, Aaron Courville, Ruslan Salakhudinov, Rich Zemel, and Yoshua Bengio. 2015. Show, attend and tell: Neural image caption generation with visual attention. In *International conference on machine learning*, pages 2048–2057.

A Sequence-based Document Encoders

We describe below the different neural models that we use for the sequence encoder, as part of our encoder-decoder architecture for deriving semantic tags for documents.

A.1 doc2vec encoder

doc2vec is the unsupervised algorithm due to (Le and Mikolov, 2014), that learns fixed-length representations of variable length documents, representing each document by a dense vector trained to predict surrounding words in contexts sampled from each document. We derive these doc2vec encodings by pre-training on our corpus. We then use them directly as features for inferring semantic tags per Figure 1(b) without training them within our framework against the loss objectives. We expect this to be a strong document encoding baseline in capturing the semantics of documents. TFIDF Terms is our other baseline where we don't train within the framework but rather use the top-k neighbor documents' TFIDF pseudo-labels as the semantic tags for the query document.

A.2 Deep Averaging Network encoder

The Deep Averaging Network (DAN) for text classification due to (Iyyer et al., 2015) Figure 1 (c), is formulated as a neural bag of words encoder model for mapping an input sequence of tokens X to *one* of k labels. v is the output of a composition function g, in this case *averaging*, applied to the sequence of word embeddings v_w for $w \in X$. For our multi-label classification problem, v is fed to a *sigmoid* layer to obtain scores for each independent classification. We expect this to be another strong document encoder given results in the literature and it proves in practice to be.

A.2.1 LSTM and BiLSTM encoders

LSTMs (Hochreiter and Schmidhuber, 1997), by design, encompass memory cells that can store information for a long period of time and are therefore capable of learning and remembering over long and variable sequences of inputs. In addition to three types of gates, i.e. *input, forget*, and *output* gates, that control the flow of information into and out of these cells, LSTMs have a hidden state vector h_t^l, and a memory vector c_t^l. At each time step, corresponding to a token of the input document, the LSTM can choose to read from, write to, or reset the cell using explicit gating mechanisms. Thus

the LSTM is able to learn a language model for the entire document, encoded in the hidden state of the final timestep, which we use as the document encoding to give to the prediction layer. By the same token, owing to the bi-directional processing of its input, a BiLSTM-based document representation is expected to be even more robust at capturing document semantics than the LSTM, with respect to its prediction targets. Here, the document representation used for final classification is the concatenated hidden state outputs from the final step, $[\overrightarrow{h}_t^l; \overleftarrow{h}_t^l]$, depicted by the dotted box in Fig. 1(e).

A.3 BiLSTM with Attention encoder

In addition, we also propose a BiLSTM with attention-based document encoder, where the output representation is the *weighted combination* of the concatenated hidden states at each time step. Thus we learn an *attention-weighted* representation at the final output as follows. Let $X \in^{(d \times L)}$ be a matrix consisting of output vectors $[h_1, \ldots, h_L]$ that the Bi-LSTM produces when reading L tokens of the input document. Each word representation h_i is obtained by concatenating the forward and backward hidden states, i.e. $h_i = [\overrightarrow{h_i}; \overleftarrow{h_i}]$. d is the size of embeddings and hidden layers. The attention mechanism produces a vector α of attention weights and a weighted representation r of the input, via:

$$M = \tanh(WX), \quad M \in^{(d \times L)} \quad (5)$$

$$\alpha = \mathrm{softmax}(w^T M), \quad \alpha \in^L \quad (6)$$

$$r = X\alpha^T, \quad r \in^d \quad (7)$$

Here, the intermediate attention representation m_i (i.e. the ith column vector in M) of the ith word in the input document is obtained by applying a non-linearity on the matrix of output vectors X, and the attention weight for the ith word in the input is the result of a weighted combination (parameterized by w) of values in m_i. Thus $r \in^d$ is the *attention*$-$weighted representation of the word and phrase tokens in an input document used in optimizing the training objective in downstream multi-label classification, as shown by the final attended representation r in Figure 1(e).

A.4 GRU and BiGRU encoders

A Gated Recurrent Unit (GRU) is a type of recurrent unit in recurrent neural networks (RNNs) that aims at tracking long-term dependencies while keeping the gradients in a reasonable range. In contrast to the LSTM, a GRU has only 2 gates: a reset gate and an update gate. First proposed by (Chung et al., 2014), (Chung et al., 2015) to make each recurrent unit to adaptively capture dependencies of different time scales, the GRU, however, does not have any mechanism to control the degree to which its state is exposed, exposing the whole state each time. In the LSTM unit, the amount of the memory content that is seen, or used by other units in the network is controlled by the output gate, while the GRU exposes its full content without any control. Since the GRU has simpler structure, models using GRUs generally converge faster than LSTMs, hence they are faster to train and may give better performance in some cases for sequence modeling tasks. The BiGRU has the same structure as GRU except constructed for bi-directional processing of the input, depicted by the dotted box in Fig. 1(e).

A.5 Transformer self-attentional encoder

Recently, the Transformer encoder-decoder architecture due to (Vaswani et al., 2017), based on a *self-attention* mechanism in the encoder and decoder, has achieved the state-of-the-art in machine translation tasks at a fraction of the computation cost. Based entirely on attention, and replacing the recurrent layers commonly used in encoder-decoder architectures with *multi-headed* self-attention, it has outperformed most previously reported ensembles on the task. Thus we hypothesize that this self-attention-based model could learn the most efficient semantic representations of documents for our unsupervised task. Since our models use $tensorflow$ (Abadi et al., 2016), a natural choice was document representation learning using the Transformer model's available $tensor2tensor$ API. We hoped to leverage apart from the computational advantages of this model, the capability of capturing semantics over varying lengths of context in the input document, afforded by multi-headed self-attention, Figure 1(f). Self-attention is realized in this architecture, by training 3 matrices, made up of vectors, corresponding to a Query vector, a Key Vector and a Value vector for each token in the input sequence. The output of each self-attention layer is a summation of weighted Value vectors that passes on to a feed-forward neural network. Position-based encoding to replace recurrences help to lend more parallelism to computations and make things faster.

Multi-headed self-attention further lends the model the ability to focus on different positions in the input, with multiple sets of Query/Key/Value weight matrices, which we hypothesize should result in the most effective document representation, among all the models, for our downstream task.

A.6 CNN encoder

Inspired by the success of (Kim, 2014) in employing CNN architectures successfully for achieving gains in NLP tasks we also employ a CNN-based encoder in the seq2set framework. (Kim, 2014) train a simple CNN with a layer of convolution on top of pre-trained word vectors, as a sequence of length n embeddings concatenated to form a matrix input. Filters of different sizes, representing various context windows over neighboring words, are then applied to this input, over each possible window of words in the sequence to obtain feature maps. This is followed by a max-over-time pooling operation to take maximum value of the feature map as the feature corresponding to a particular filter. The model then combines these features to form a penultimate layer which is passed to a fully connected softmax layer whose output is the probability distribution over labels. In case of seq2set these features are passed to sigmoid layer for final multi-label prediction used cross entropy loss or a combination of cross-entropy and LM losses. We use filters of sizes 2, 3, 4 and 5. Like our other encoders, we fine-tune the document representations learnt.

B Embedding Algorithms Experimented with

We describe here the various algorithms used to train word embeddings for use in our models.

Skip-Gram word2vec: We generate word embeddings trained with the skip-gram model with negative sampling (Mikolov et al., 2013b) with dimension settings of 50 with a context window of 4, and also 300, with a context window of 5, using the $gensim$ package [2] (Řehůřek and Sojka, 2010).

Probabilistic FastText: The Probabilistic FastText (PFT) word embedding model of (Athiwaratkun et al., 2018) represents each word with a Gaussian mixture density, where the mean of a mixture component given by the sum of n-grams, can capture multiple word senses, sub-word structure,

---

[2]https://radimrehurek.com/gensim/

and uncertainty information. This model outperforms the n-gram averaging of FastText getting state-of-the-art performance on several word similarity and disambiguation benchmarks. The probabilistic word representations with flexible sub-word structures, can achieve multi-sense representations that also give rich semantics for rare words. This makes them very suitable to generalize for rare and out-of-vocabulary words motivating us to opt for PFT-based word vector pre-training [3] over regular FastText.

ELMo: Another consideration was to use embeddings that can explicitly capture the language model underlying sentences within a document. ELMo (Embeddings from Language Models) word vectors (Peters et al., 2018) presented such a choice where the vectors are derived from a bidirectional LSTM trained with a coupled language model (LM) objective on a large text corpus. The representations are a function of all of the internal layers of the biLM. Using linear combinations of the vectors derived from each internal state has shown marked improvements over various downstream NLP tasks, because the higher-level LSTM states capture context-dependent aspects of word meaning (e.g., they can be used without modification to perform well on supervised word sense disambiguation tasks) while lower- level states model aspects of syntax. Using the API [4] we generate ELMo embeddings fine-tuned for our corpus with dimension settings of 50 and 100 using only the top layer final representations. A discussion of the results from each set of experiments is outlined in the following section and summarized in Table 1.

C   Experimental Considerations and Hyperparameter Settings

Of the metrics available, P@10 gives the number of relevant items returned in the top-10 results and NDCG looks at precision of the returned items at the correct rankings. For our particular dataset domain, the number of relevant results returned in the top-10 is more important, hence Table 1 reports results ranked in ascending order of P@10.

The PRF setting shown in the results table means that, we take the top 10-15 documents returned by an ElasticSearch (ES) index for each of the 30 Summary Text queries in our dataset, and subsequently use the semantic tags assigned to each of these top documents as the terms for query expansion for the original query. We then re-run these expanded queries through the ES index to record the retrieval performance. Thus the queries our system is evaluated on, are *not seen* at the time of training our models, but only during evaluation, hence it is *unsupervised QE*.

Similar to Das et al. (2018), for the feedback loop based query expansion method, we had two separate human judgment–based baselines, one using the MeSH terms available from PMC for the top 15 documents returned in a first round of querying with Summary text, and the other based on human expert annotations of the 30 query topics, made available by the authors.

Since we had mixed results initially with our models, we explored various options to increase the training signal. First was by the use of *neighbor-document's* labels in our label vector for training. In this scheme, we used the 3-hot TFIDF label representation for each document to pick a list of $n-$ nearest neighbors to it. We then included the labels of those nearest documents into the label vector for the original document for use during training. We experimented with choices $3, 5, 10, 15, 20, 25$ for n. We observed improvements with incorporating neighbor labels.

Next we experimented with incorporating word neighbors into the input sequence for our models. We did this in two different ways, the first was to average all the neighbors and concatenate with the original token embedding, the other was to average all of the embeddings together. The word itself was always weighted more than the neighbors. This scheme also gave improvement.

Finally we experimented with incorporating embeddings pre-trained by latest state-of-the-art methods (Appendix B) as the input tokens for our models. After several rounds of hyper-parameter tuning, $batch\_size$ was set to 128, and $dropout$ to 0.3. We also performed a small grid search into the space of hyperparameters like number of hidden layers varied as $2, 3, 4$, and α varied as $[1.0, 10.0, 100.0, 1000.0, 10000.0]$, determining the best settings for each encoder.

A glossary of acronyms and parameters used in training of our models is as follows: sg=skip-gram; pft=Probablistic FastText; $elmo$=ELMo; d=embedding dimension; kln=number of "neighbor documents'" labels; nl=number of hidden layers in the model; h= Number of multi-attention

---

[3] https://github.com/benathi/multisense-prob-fasttext
[4] https://allennlp.org/elmo

heads; bs=batch size; dp=dropout; ep=no. of epochs; α=weight parameter for language model loss component.

Best-performing model settings: Our best performing models on the base query was a Transformer encoder with 10 attention heads: $nh = 10$, loss: cross-entropy + LM loss with $\alpha = 1000.0$, input embedding: 50-d pft, $bs = 64$ and $dp = 0.3$; and a GRU encoder for the ensemble with parameters, loss: LM only loss with $\alpha = 1000.0$, input embedding: 50-d pft, $nl = 4$, $kln = 10$, $bs = 64$ and $dp = 0.2$.

For augmented query, our best performing models were: (1) a BiGRU trained with parameters, loss function: LM only loss with $\alpha = 1.0$, input embedding: 50-d $skip - gram$, $nl = 3$, $kln = 5$, $bs = 128$ and $dp = 0.3$, and (2) a Transformer trained with parameters, loss function: cross-entropy + LM loss with $\alpha = 1000.0$, input embedding: 50-d $skip - gram$, $nl = 4$, $kln = 5$, $bs = 128$ and $dp = 0.3$.

While we obtain significant improvement over the compared baselines with our best-performing models, we believe further gains are possible by a more targeted search through the parameter space.

Interactive Extractive Search over Biomedical Corpora

Hillel Taub-Tabib[1] **Micah Shlain**[1,2] **Shoval Sadde**[1] **Dan Lahav**[3]
Matan Eyal[1] **Yaara Cohen**[1] **Yoav Goldberg**[1,2]

[1] Allen Institute for AI, Tel Aviv, Israel
[2] Bar Ilan University, Ramat-Gan, Israel
[3] Tel Aviv University, Tel-Aviv, Israel
`{hillelt,micahs}@allenai.org`

Abstract

We present a system that allows life-science researchers to search a linguistically annotated corpus of scientific texts using patterns over dependency graphs, as well as using patterns over token sequences and a powerful variant of boolean keyword queries. In contrast to previous attempts to dependency-based search, we introduce a light-weight query language that does not require the user to know the details of the underlying linguistic representations, and instead to query the corpus by providing an example sentence coupled with simple markup. Search is performed at an interactive speed due to efficient linguistic graph-indexing and retrieval engine. This allows for rapid exploration, development and refinement of user queries. We demonstrate the system using example workflows over two corpora: the PubMed corpus including 14,446,243 PubMed abstracts and the CORD-19 dataset[1], a collection of over 45,000 research papers focused on COVID-19 research. The system is publicly available at `https://allenai.github.io/spike`

1 Introduction

Recent years have seen a surge in the amount of accessible Life Sciences data. Search engines like Google Scholar, Microsoft Academic Search or Semantic Scholar allow researchers to search for published papers based on keywords or concepts, but search results often include thousands of papers and extracting the relevant information from the papers is a problem not addressed by the search engines. This paradigm works well when the information need can be answered by reviewing a number of papers from the top of the search results. However, when the information need requires extraction of information nuggets from many papers

(e.g. *all chemical-protein interactions* or *all risk factors for a disease*) the task becomes challenging and researchers will typically resort to curated knowledge bases or designated survey papers in case ones are available.

We present a search system that works in a paradigm which we call Extractive Search, and which allows rapid information seeking queries that are aimed at extracting facts, rather than documents. Our system combines three query modes: boolean, sequential and syntactic, targeting different stages of the analysis process, and different extraction scenarios. Boolean queries (§4.1) are the most standard, and look for the existence of search terms, or groups of search terms, in a sentence, regardless of their order. These are very powerful for finding relevant sentences, and for co-occurrence searches. Sequential queries (§4.2) focus on the order and distance between terms. They are intuitive to specify and are very effective where the text includes "anchor-words" near the entity of interest. Lastly, syntactic queries (§4.4) focus on the linguistic constructions that connect the query words to each other. Syntactic queries are very powerful, and can work also where the concept to be extracted does not have clear linear anchors. However, they are also traditionally hard to specify and require strong linguistic background to use. Our systems lowers their barrier of entry with a specification-by-example interface.

Our proposed system is based on the following components.

Minimal but powerful query languages. There is an inherent trade-off between simplicity and control. On the one extreme, web search engines like Google Search offer great simplicity, but very little control, over the exact information need. On the other extreme, information extraction pattern-specification languages like UIMA Ruta offer great precision and control, but also expose a low-level

---

[1] https://pages.semanticscholar.org/coronavirus-research

Proceedings of the BioNLP 2020 workshop, pages 28–37
Online, July 9, 2020 ©2020 Association for Computational Linguistics

view of the text and come with over hundred-page manual.[2]

Our system is designed to offer *high degree of expressivity*, while remaining simple to grasp: the syntax and functionality can be described in a few paragraphs. The three query languages are designed to share the same syntax to the extent possible, to facilitate knowledge transfer between them and to ease the learning curve.

Linguistic Information, Captures, and Expansions. Each of the three query types are linguistically informed, and the user can condition not only on the word forms, but also on their lemmas, parts-of-speech tags, and identified entity types. The user can also request to *capture* some of the search terms, and to *expand* them to a linguistic context. For example, in a boolean search query looking for a sentence that contains the lemmas "treat" and "treatment" ('lemma=treat|treatment'), a chemical name ('entity=SIMPLE_CHEMICAL') and the word "infection" ('infection'), a user can mark the chemical name and the word "infection" as *captures*. This will yield a list of chemical/infection pairs, together with the sentence from which they originated, all of which contain the words relating to treatments. Capturing the word "infection" is not very useful on its own: all matches result in the exact same word. But, by *expanding* the captured word to its surrounding linguistic environment, the captures list will contain terms such as "PEDV infection", "acyclovir-resistant HSV infection" and "secondary bacterial infection". Running this query over PubMed allows us to create a large and relatively focused list in just a few seconds. The list can then be downloaded as a CSV file for further processing. The search becomes *extractive*: we are not only looking for documents, but also, by use of captures, *extract information* from them.

Sentence Focus, Contextual Restrictions. As our system is intended for extraction of information, it works at the sentence level. However, each sentence is situated in a context, and we allow secondary queries to condition on that context, for example by looking for sentences that appear in paragraphs that contain certain words, or which appear in papers with certain words in their titles, in papers with specific MeSH terms, in papers whose abstracts include specific terms, etc. This combines the focus and information density of a sentence,

which is the main target of the extraction, with the rich signals available in its surrounding context.

Interactive Speed. Central to the approach is an indexed solution, based on (Valenzuela-Escárcega et al., 2020), that allows to perform all types of queries efficiently over very large corpora, while getting results almost immediately. This allows the users to interactively refine their queries and improve them based on the feedback from the results. This contrasts with machine learning based solutions that, even neglecting the development time, require substantially longer turnaround times between query and results from a large corpus.

2 Existing Information Discovery Approaches

The primary paradigm for navigating large scientific collections such as MEDLINE/PubMed[3] is document-level search.

The most immediate document-level searching technique is boolean search ("keyword search"). However, these methods suffer from an inability to capture the concepts aimed for by the user, as biomedical terms may have different names in different sub-fields and as the user may not always know exactly what they are looking for. To overcome this issue several databases offer semantic searching by exploiting MeSH terms that indicate related concepts. While in some cases MeSH terms can be assigned automatically, e.g (Mork et al., 2013), in others obtaining related concepts require a manual assignment which is laborious to obtain.

Beyond the methods incorporated in the literature databases themselves, there are numerous external tools for biomedical document searching. Thalia (Soto et al., 2018) is a system for semantic searching over PubMed. It can recognize different types of concepts occurring in Biomedical abstracts, and additionally enables search based on abstract metadata; LIVIVO (Müller et al., 2017) takes the task of vertically integrating information from divergent research areas in the life sciences; SWIFT-Review[4] offers iterative screening by re-ranking the results based on the user's inputs.

All of these solutions are focused on the document level, which can be limiting: they often surface hundreds of papers or more, requiring careful reading, assessing and filtering by the user, in order to locate the relevant facts they are looking for.

---

[2]https://uima.apache.org/d/ruta-current/tools.ruta.book.pdf

[3]https://www.ncbi.nlm.nih.gov/pubmed/
[4]https://www.sciome.com/swift-review/

To complement document searching, some systems facilitate automatic extraction of biomedical concepts, or patterns, from documents. Such systems are often equipped with analysis capabilities of the extracted information. For example, NaCTem has created systems that extract biomedical entities, relations and events.[5]; ExaCT and RobotReviewer (Kiritchenko et al., 2010; Marshall et al., 2015) take a RCT report and retrieve sentences that match certain study characteristics.

To improve the development of automatic document selection and information extraction the BioNLP community organized a series of shared tasks (Kim et al., 2009, 2011; Nédellec et al., 2013; Segura Bedmar et al., 2013; Deléger et al., 2016; Chaix et al., 2016; Jin-Dong et al., 2019). The tasks address a diverse set of biomed topics addressed by a range of NLP-based techniques. While effective, such systems require annotated training data and substantial expertise to produce. As such, they are restricted to several "head" information extraction needs, those that enjoy a wide community interest and support. The long tail of information needs of "casual" researchers remain mostly un-addressed.

3 Interactive IE Approach

Existing approaches to information extraction from bio-medical data suffer from significant practical limitations. Techniques based on supervised training require extensive data collection and annotation (Kim et al., 2009, 2011; Nédellec et al., 2013; Segura Bedmar et al., 2013; Deléger et al., 2016; Chaix et al., 2016), or a high degree of technical savviness in producing high quality data sets from distant supervision (Peng et al., 2017; Verga et al., 2017; Wang et al., 2019). On the other hand, rule based engines are generally too complex to be used directly by domain experts and require a linguist or an NLP specialist to operate. Furthermore, both rule based and supervised systems typically operate in a pipeline approach where an NER engine identifies the relevant entities and subsequent extraction models identify the relations between them. This approach is often problematic in real world biomedical IE scenarios, where relevant entities often cannot be extracted by stock NER models.

To address these limitations we present a system allowing domain experts to interactively query linguistically annotated datasets of scientific re-

search papers, using a novel multifaceted query language which we designed, and which supports boolean search, sequential patterns search, and by-example syntactic search (Shlain et al., 2020), as well as specification of search terms whose matches should be captured or expanded. The queries can be further restricted by contextual information.

We demonstrate the system on two datasets: a comprehensive dataset of PubMed abstracts and a dataset of full text papers focused on COVID-19 research.

Comparison to existing systems. In contrast to document level search solutions, the results returned by our system are sentences which include highlighted spans that directly answer the user's information need. In contrast to supervised IE solutions, our solution does not require a lengthy process of data collection and labeling or a precise definition of the problem settings.

Compared to rule based systems our system differentiates itself in a number of ways: (i) our query engine automatically translates lightly tagged natural language sentences to syntactic queries (query-by-example) thus allowing domain experts to benefit from the advantages of syntactic patterns without a deep understanding of syntax; (ii) our queries run against indexed data, allowing our translated syntactic queries to run at interactive speed; and (iii) our system does not rely on relation schemas and does not make assumptions about the number of arguments involved or their types.

In many respects, our system is similar to the PropMiner system (Akbik et al., 2013) for exploratory relation extraction (Akbik et al., 2014). Both PropMiner and our system support by-example queries in interactive speed. However, the query languages we describe in section 4 are significantly more expressive than PropMiner's language, which supports only binary relations. Furthermore, compared to PropMiner, our annotation pipeline was optimized specifically for the biomedical domain and our system is freely available online.

Technical details. The datasets were annotated for biomedical entities and syntax using a custom SciSpacy pipeline (Neumann et al., 2019)[6], and the syntactic trees were enriched to BART format using pyBART (Tiktinsky et al., 2020). The annotated data is indexed using the Odinson engine (Valenzuela-Escárcega et al., 2020).

---

[6]All abstracts underwent sentence splitting, tokenization, tagging, parsing and NER using all the 4 NER models available in SciSpacy

4 Extractive Query Languages

4.1 Boolean Queries

Boolean queries are the standard in information retrieval (IR): the user provides a set of terms that should, and should not, appear in a document, and the system returns a set of documents that adhere to these constraints. This is a familiar and intuitive model, which can be very effective for initial data exploration as well as for extraction tasks that focus on co-occurrence. We depart from standard boolean queries and extend them by (a) allowing to condition on different linguistic aspects of each token; (b) allowing *capturing* of terms into named variables; and (c) allowing *linguistic expansion* of the captured terms.

The simplest boolean query is a list of terms, where each term is a word, i.e: `infection asymptomatic fatal` The semantics is that all the terms must appear in the query. A term can be made optional by prefixing it with a '?' symbol (`infection asymptomatic ?fatal`). Each term can also specify a list of alternatives: `fatal|deadly|lethal`.

Beyond words. In addition to matching words, terms can also specify linguistic properties: lemmas, parts-of-speech, and domain-specific entity-types: `lemma=infect entity=DISEASE`. Conditions can also be combined: `lemma=cause|reason&tag=NN`. We find that the ability to search for domain-specific types is very effective in boolean queries, as it allows to search for concepts rather than words. In addition to exact match, we also support matching on regular expressions (`lemma=/caus.*/`). The field names word, lemma, entity, tag can be shortened to w,l,e,t.

Captures. Central to our extractive approach is the ability to designate specific search term to be *captured*. Capturing is indicated by prefixing the term with ':' (for an automatically-named capture) or with `name:` (for a named capture). The query `fatal asymptomatic d:e=DISEASE` will look for sentences that contain the terms 'fatal' and 'asymptomatic' as well as a name of a disease, and will capture the disease name under a variable "d". Each query result will be a sentence with a single disease captured. If several diseases appear in the same sentence, each one will be its own result. The user can then focus on the captured entities, and export the entire query result to a CSV file, in which each row contains the sentence, its source, and the captured variables. In the current examples, the result will be a list of disease names that

co-occur with "fatal" and "asymptomatic". We can also issue a query such as `chem:e=SIMPLE_CHEMICAL d:e=DISEASE` to get a list of chemical-disease co-occurrences. Using additional terms, we can narrow down to co-occurrences with specific words, and by using contextual restrictions (§4.3) we can focus on co-occurrences in specific papers or domains.

Expansions. Finally, for captured terms we also support *linguistic expansions*. After the term is matched, we can expand it to a larger linguistic environment based on the underlying syntactic sentence representation. An expansion is expressed by prefixing a term with angle brackets (): `()inf:infection asymptomatic fatal` will capture the word "infection" under the variable "inf" and *expand* it to its surrounding noun-phrase, capturing phrases like "malarialike infection", "asymptomatic infection", "chronic infection" and "a mild subclinical infection 9".

4.2 Sequential (Token) Queries

While boolean queries allow terms to appear in any order, we sometimes care about the exact linear placements of words with respect to each other. The term-specification, capture and expansion syntax is the same as in boolean queries, but here terms must match as in the query. `interspecies transmission` looks for the exact phrase "interspecies transmission" and `tag=NNS transmission` looks for the word transmission immediately preceded by a plural noun. By capturing the noun (`which:tag=NNS transmission`) we obtain a list of terms that includes the words "bacteria", "diseases", "nuclei" and "crossspecies".

Wildcards. sequential queries can also use wildcard symbols: * (matching any single word), '...' (0 or more words), '...2-5...' (2-5 words). The query `interspecies kind:...1-3... transmission` looks for the words "interspecies" and "transmission" with 1 to 3 intervening words, capturing the intervening words under "kind". First results include "host-host", "zoonotic", "virus", "TSE agent", "and interclass".

Repetitions. We also allow to specify repetitions of terms. To do so, the term is enclosed in [ ] and followed by a quantifier. We support the standard list of regular expression quantifiers: *, +, ?, {n,m}. For example, `tag=DT [tag=JJ]* [tag=NN]+`.

4.3 Contextual Restrictions

Each query can be associated with contextual restrictions, which are secondary queries that oper-

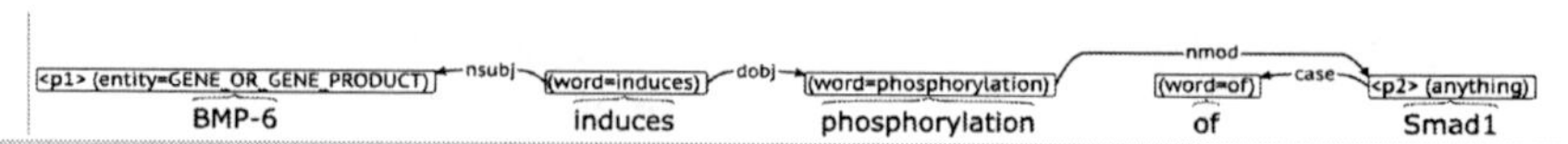

Figure 1: Query Graph of the syntactic query '⟨⟩p1:[e]BMP-6 $induces the $phosphorylation $of ⟨⟩p2:Smad1.'

ate on the same data and restrict the set of sentences that are considered for the main queries. These queries currently have the syntax of the Lucene query language.[7] Our system allows the secondary queries to condition on the paragraph the sentence appears in, and on the title, abstract, authors, publication data, publication venue and MeSH terms of the paper the sentence appears in. Additional sources of information are easy to add. For example, adding the contextual restriction '#d +title:cancer +mesh:"Age Distribution"' restricts a query results to sentences from papers which have the word "cancer" in their title and whose MeSH terms include "Age Distribution". Similarly '#d +title:/corona.*/ +year: [2015 TO 2020]' restricts queries to include sentences from papers published between 2015 and 2020 and have a word starting with *corona* in their title.

These secondary queries greatly increase the power of boolean, sequential and syntactic queries: one could look for interspecies transmissions that relate to certain diseases, or for sentence-level disease-chemical co-occurrences in papers that discuss specific sub-populations.

4.4 Example-based Syntactic Queries

Recent advances in machine learning brought with them accurate syntactic parsing, but parse-trees remain hard to use. We remedy this by employing a novel query language we introduced in (Shlain et al., 2020) which is based on the principle of query-by-example.

The query is specified by starting with a simple natural language sentence that conveys the desired syntactic structure, for example, '*BMP-6 induces the phosphorylation of Smad1*'. Then, words can be marked as anchor words (that need to match exactly) or capture nodes (that are variables). Words can also be neither anchor or capture, in which case they only support the scaffolding of the sentence. The system then translates the sentence with the captures and anchors syntax into a syntactic query graph, which is presented to the user. The user can then restrict capture nodes from "match anything"

to matching specific terms (using the term specification syntax as in boolean or token queries) and can likewise relax the exact-match constraints on anchor words. Like in other query types, capture nodes can be marked for expansion. The syntactic graph is then matched against the pre-parsed and indexed corpus.

This simple markup provides a rich syntax-based query system, while alleviating the user from the need to know linguistic syntax.

For example, consider the query below, the details of which will be discussed shortly:

'⟨⟩p1:[e=GENE_OR_GENE_PRODUCT]BMP-6 $induces the $phosphorylation $of ⟨⟩p2:Smad1'

The words 'induce', 'phosphorylation' and 'of' are anchors (designated by '$'), while 'p1' and 'p2' are captures for 'BMP-6' and 'Smad1'. Both capture nodes are marked for expansion using angle braces ('⟨⟩'). Node p1 is *restricted* to match tokens with the same entity type of BMP-6 (indicated by 'e=GENE_OR...'). The query can be shortened by omitting the entity type and retaining only the entity restriction ('e'):

'⟨⟩p1:[e]BMP-6 $induces the $phosphorylation $of ⟨⟩p2:Smad1'

Here, the entity type is inferred by the system from the entity type of BMP-6.[8] The graph for the query is displayed in Figure 1. It has 5 tokens in a specific syntactic configuration determined by directed labeled edges. The 1st token must have the entity tag of 'GENE_OR...', the 2nd, 3rd, and 4th tokens must be the exact words "induces phosphorylation of", and the 5th is unconstrained.

Sentences whose syntactic graph has a subgraph that aligns to the query and adheres to the constraints will match the query. Example of matching sentences are:
- $\underline{ERK}_{p_1}$ *activation induces phosphorylation of* $\underline{Elk\text{-}1}_{p_2}$.
- $\underline{Thrombopoietin}_{p_1}$ *activates human platelets and induces tyrosine phosphorylation of* $\underline{p80/85\ cortactin}_{p_2}$

---

[7]https://lucene.apache.org/core/6_0_0/queryparser/org/apache/lucene/queryparser/classic/package-summary.html

[8]Similarly, we could specify '$[lemma]induces', resulting in the restriction 'lemma=induce' instead of 'word=induces' for the anchor.

The sentence tokens corresponding to the p1 and p2 graph nodes will be bound to variables with these names: {p1=ERK, p2=Elk-1} for the first sentence and {p1=Thrombopoietin, p2=p80/85 cortactin} for the second.

5   Example Workflow: Risk-factors

We describe a workflow which is based on using our extractive search system over a corpus of all PubMed abstracts. While the described researcher is hypothetical, the results we discuss are real.

Consider a medical researcher who is trying to compile an up to date list of the risk factors for stroke. A PubMed search for "risk factors for stroke" yields 3317 results, and reading through all results is impractical. A Google query for the same phrase brings out an info box from NHLBI[9] listing 16 common risk factors including high blood pressure, diabetes, heart disease, etc. Having a curated list which clearly outlines the risk factors is helpful, but curated lists or survey papers will often not include rare or recent research findings.

The researcher thus turns to extractive search and tries an exploratory boolean query:

'risk factor stroke'

The figure shows the top results for the query and the majority of sentences retrieved indeed specify specific risk factors for stroke. This is an improvement over the PubMed results as the researcher can quickly identify the risk factors discussed without going through the different papers.

Furthermore, the top results contain risk factors like *migrane* or *C804A polymorphism* not listed in the NHLBI knowledge base. However, the full result list is lengthy and extracting all the risk factors from it manually would be tedious. Instead, the researcher notes that many of the top results are variations on the "X is a risk factor for stroke" structure. She thus continues by issuing the following syntactic query, where a capture labeled *r* is used to directly capture the risk factors:

[9]https://www.nhlbi.nih.gov/
health-topics/stroke

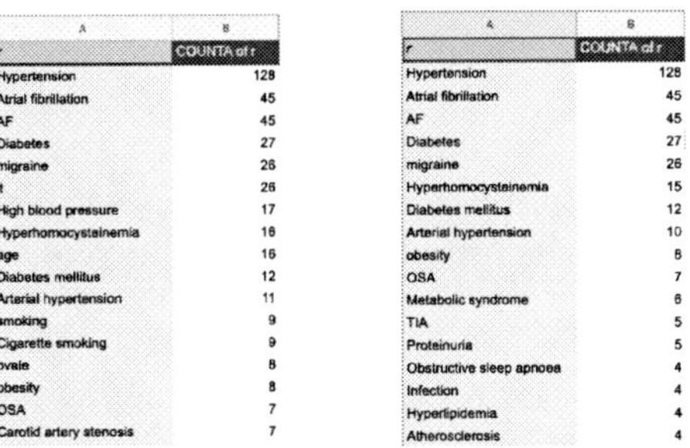

(a) ranked risk factors for stroke (b) ranked disease risk factors

Figure 2: Grouped and ranked results

'r:Diabetes is a $risk $factor for $stroke'.

The figure shows the top results for the query and the risk factors are indeed labeled with *r* as expected. Unfortunately, some of the captured risk factors names are not fully expanded. For example, we capture *syndrome* instead of *metabolic syndrome* and *temperature* instead of *low temperature*. Being interested in capturing the full names, the researcher adds angle brackets '〈〉' to expand the captured elements:

'〈〉r:Diabetes is a $risk $factor for $stroke'.

The full names are now captured as expected.

Now that that researcher has verified that the query yields relevant results, she clicks the download button to download the full result set.

The resulting tab separated file has 1212 rows. Each row includes a result sentence, the captured elements in it (in this case, just the risk factor), and their offsets. Using a spreadsheet to group the rows by risk factor and order the results by frequency, the researcher obtains a list of 640 unique risk factors, 114 of them appearing more than once in the data. Figure 2a lists the top results.

Reviewing the list, the researcher decides that she's not interested in general risk factors, but rather in diseases only. She modifies the query by adding an entity restriction to the 'r' capture:

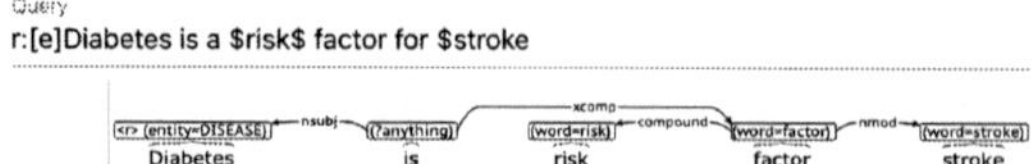

As seen in the query graph, even though the researcher didn't specify the exact entity type, the query parser correctly resolved it to DISEASE. The results now include diseases like sleep apnoea and hypertension but do not include smoking, age and alcohol (see Figure 2b).

Analyzing the results, the researcher now wants to compare the risk factors in the general population to ones listed in research papers dealing with children and infants. Luckily, such papers are indexed with corresponding MeSH terms and the researcher can utilize this fact by appending '#d mesh:Child mesh:Infant -mesh:Adult' to her query. In cases where a desired MeSH term does not exist, an alternative approach is filtering the results based on words in the abstract or title. For example, appending '#d abstract:child abstract:children' to a query will ensure that the result sentences come from abstracts which contain the word *child* or the word *children*.

Happy with the results of the initial query, the researcher can further augment her list by querying for other structures which identify risk factors (e.g. "'r:Diabetes $causes $stroke'", "'$risk $factors for $stroke $include r:Diabetes'", etc.).

Importantly, once the researcher has identified one or more effective queries to extract the risk factors for stroke, the queries can easily be modified in useful ways. For example, with a small modification to our original query we can extract:

risk factors for cancer:

'r:Diabetes is a $risk $factor for $cacner'

diseases which can be caused by smoking:

'$Smoking is a $risk $factor for d:[e]stroke'.

ad-hoc KB of (risk factor, disease) tuples (for self use or as an easily queryable public resource):

'r:Diabetes is a $risk $ factor for d:[e]stroke'.

6 Example Workflow: CORD-19

The COVID-19 Open Research Dataset (Wang et al., 2020) is a collection of 45,000 research papers, including over 33,000 with full text, about COVID-19 and the coronavirus family. The corpus was released by the Allen Institute for AI and associated partners in an attempt to encourage researchers to apply recent advances in NLP to the data to generate insights.

Identifying COVID-19 Aliases Since the CORD-19 corpus includes papers about the entire Coronavirus family of viruses, it's useful to identify papers and sentences dealing specifically with COVID-19. Before converging on the acronym COVID-19 researchers have referred to the virus by many names: nCov-19, SARS-COV-ii, novel coronavirus, etc. Luckily, it's fairly easy to identify many of these aliases using a sequential pattern:

'novel coronavirus (alias:...1-2...)'

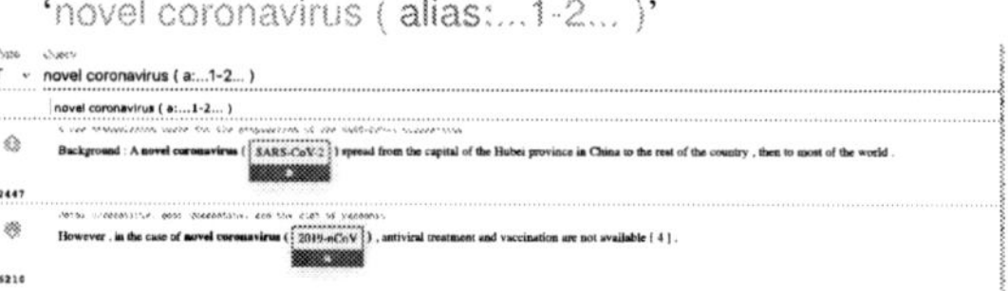

The pattern looks for the words "novel coronavirus" followed by an open parenthesis, one-or-two words which are to be captured under the 'alias' variable, and a closing parenthesis. The query retrieves 52 unique candidate aliases for COVID-19, though some of them refer to older coronaviruses such as "MERS", or non-relevant terms such as "Fig2". After ranking by frequency and validating the results, we can reuse the pattern on newly retrieved aliases to extend the list. Through this iterative process we quickly compile a list of 47 aliases. We marked all occurrences of these terms in the underlying corpus as a new entity type, COVID-19, and re-indexed the dataset with this entity information.

Exploring Drugs and Treatments. To explore drugs and treatments for COVID-19 we search the corpus for chemicals co-occuring with the COVID-19 entity using a boolean query:

'chemical:e=SIMPLE_CHEMICAL|CHEMICAL e=COVID-19'

Table 1(a) shows the top matching chemicals by frequency. While some of the substances listed like *Chloroquine* and *Remdesivir* are drugs being tested for treating COVID-19, others are only hypothesized as useful or appear in other contexts.

To guide the search toward therapeutic substances in different stages of maturity we can add indicative terms to the query. For example, the following query can be used to detect substances at the stage of clinical trials:

'chemical:e=SIMPLE_CHEMICAL|CHEMICAL e=COVID-19 l=trial|experiment', while adding 'l=suggest|hypothesize|candidate' can assist in detecting substances in ideation stage.

Table 1(b,c) shows the frequency distributions of the chemicals resulting from the two queries.

(a) Unrestricted
nucleic acid (171), chloroquine (118), nucleotide (115), NCP (87), CR3022 (47), Ksiazek (46), IgG (45), lopinavir/ritonavir (42), ECMO (40), LPV/r (35), corticosteroids (35), oxygen (32), ribavirin (31), lopinavir (31), Hydroxychloroquine (30), amino acid (30), ritonavir (27), corticosteroid (24), Sofosbuvir (22), amino acids (22), HCQ (19), glucocorticoids (19)

(b) Trial
chloroquine (29), Remdesivir (8), LPV/r (7), lopinavir (6), HCQ (6), ritonavir (4), Arbidol (4), Sofosbuvir (3), nucleotide (3), nucleic acid (3), lopinavir/ritonavir (3), CQ (3), oseltamivir (2), NCT04257656 (2), NCT04252664 (2), Meplazumab (2), Hydroxychloroquine (2), glucocorticoids(2), CEP(2)

(c) Ideation
chloroquine (6), ritonavir (5), S-RBD (4), nucleotide (4), Lopinavir (4), CR3022 (4), Ribavirin (3), nucleic acid (3), logP (3), Li (3), ledipasvir (3), IgG (3), HCQ (3), TGEV (2), teicoplanin (2), nelfinavir (2), NCP (2), HWs (2) glucocorticoids (2), ENPEP (2), ECMO (2), darunavir (2), creatinine (2), creatine (2), CQ (2), corticosteroid (2), CEP (2), ARB (2)

Table 1: Top chemicals co-occuring with the COVID-19 entity and their counts. (a) Unrestricted. (b) with Trial related terms. (c) with Ideation related terms.

While the queries are very basic and include only a few terms for each category, the difference is clearly noticeable: while the Malaria drug Chloroquine tops both lists, the antiviral drug Remdesivir which is currently tested for COVID-19 is second on the list of trial related drugs but does not appear at all as a top result for ideation related drugs.

Importantly, entity co-mention queries like the ones above rely on the availability and accuracy of underlying NER models. As we've seen in Section 5, in cases where the relevant types are not extracted by NER, syntactic queries can be used instead. For example the following query captures sentences including chemicals being used on patients (the abstract or paragraph are required to include COVID-19 related terms).

'he was $treated $with a ⟨⟩chem:treatment
#d paragraph:ncov* paragraph:covid* abstract:ncov* abstract:covid*'

Treatments (via syntactic query)
ribavirin (11), oseltamivir (9), ECMO (6), convalescent plasma (4), TCM (3), LPV/r (3), three fusions of MSCs (2), supportive care (2), protective conditions (2), lopinavir/ritonavir (2), intravenous remdesivir (2), hydroxychloroquine (2), HCQ (2), glucocorticoids (2), FPV (2), effective isolation (2), chloroquine (2), caution (2), bDMARDs (2), azithromycin (2), ARBs (2), antivirals (2), ACE inhibitors (2), 500 mg chloroquine (2), masks (1)

Table 2: Top elements occurring in the syntactic "treated with X" configuration. Note that this query does not rely on NER information.

The top results by frequency are shown in Table 2. The top ranking results show many of the chemicals obtained by equivalent boolean queries[10], but interestingly, they also contain non-chemical treatments like *supportive care*, *isolation* and *masks*. This demonstrates a benefit of using entity agnostic syntactic patterns even in cases where a strong NER model exists.

7 More Examples

While the workflows discussed above pertain mainly to the medical domain, the system is optimized for the broader life science domain. Here are a sample of additional queries, showing different potential use-cases.

Which genes regulate a cell process:
'⟨⟩p1:[e]CD95 v:[l]regulates ⟨⟩p:[e]apoptosis'
Which specie is the natural host of a disease:
'⟨⟩host:[e]bat is a $natural $host of ⟨⟩disease:[e]coronavirus'
Documented LOF mutations in genes:
'$loss $of $function ⟨⟩m:[w]mutation in ⟨⟩gene:[e]PAX8'

8 Conclusion

We presented a search system that targets extracting facts from a biomed corpus and demonstrated its utility in a research and a clinical context over CORD-19 and PubMed. The system works in an Extractive Search paradigm which allows rapid information seeking practices in 3 modes: boolean, sequential and syntactic. The interactive and flexible nature of the system makes it suitable for users in different levels of sophistication.

---

[10]to get a more comprehensive coverage we can issue queries for other syntactic structures like '⟨⟩chem:chemical was used $in $treatment' and combine the results of the different queries.

Acknowledgements The work performed at BIU is supported by funding from the Europoean Research Council (ERC) under the European Union's Horizon 2020 research and innovation programme, grant agreement No. 802774 (iEXTRACT).

References

Alan Akbik, Oresti Konomi, and Michail Melnikov. 2013. Propminer: A workflow for interactive information extraction and exploration using dependency trees. In *Proceedings of the 51st Annual Meeting of the Association for Computational Linguistics: System Demonstrations*, pages 157–162.

Alan Akbik, Thilo Michael, and Christoph Boden. 2014. Exploratory relation extraction in large text corpora. In *Proceedings of COLING 2014, the 25th International Conference on Computational Linguistics: Technical Papers*, pages 2087–2096.

Estelle Chaix, Bertrand Dubreucq, Abdelhak Fatihi, Dialekti Valsamou, Robert Bossy, Mouhamadou Ba, Louise Deléger, Pierre Zweigenbaum, Philippe Bessieres, Loic Lepiniec, et al. 2016. Overview of the regulatory network of plant seed development (seedev) task at the bionlp shared task 2016. In *Proceedings of the 4th bionlp shared task workshop*, pages 1–11.

Louise Deléger, Robert Bossy, Estelle Chaix, Mouhamadou Ba, Arnaud Ferré, Philippe Bessieres, and Claire Nédellec. 2016. Overview of the bacteria biotope task at bionlp shared task 2016. In *Proceedings of the 4th BioNLP shared task workshop*, pages 12–22.

Kim Jin-Dong, Nédellec Claire, Bossy Robert, and Deléger Louise, editors. 2019. *Proceedings of The 5th Workshop on BioNLP Open Shared Tasks*. Association for Computational Linguistics, Hong Kong, China.

Jin-Dong Kim, Tomoko Ohta, Sampo Pyysalo, Yoshinobu Kano, and Jun'ichi Tsujii. 2009. Overview of bionlp'09 shared task on event extraction. In *Proceedings of the BioNLP 2009 workshop companion volume for shared task*, pages 1–9.

Jin-Dong Kim, Yue Wang, Toshihisa Takagi, and Akinori Yonezawa. 2011. Overview of genia event task in bionlp shared task 2011. In *Proceedings of the BioNLP Shared Task 2011 Workshop*, pages 7–15. Association for Computational Linguistics.

Svetlana Kiritchenko, Berry de Bruijn, Simona Carini, Joel Martin, and Ida Sim. 2010. Exact: Automatic extraction of clinical trial characteristics from journal publications. *BMC medical informatics and decision making*, 10:56.

Iain J Marshall, Joël Kuiper, and Byron C Wallace. 2015. RobotReviewer: evaluation of a system for automatically assessing bias in clinical trials. *Journal of the American Medical Informatics Association*, 23(1):193–201.

J.G. Mork, Antonio Jimeno-Yepes, and Alan Aronson. 2013. The nlm medical text indexer system for indexing biomedical literature. *CEUR Workshop Proceedings*, 1094.

Bernd Müller, Christoph Poley, Jana Pössel, Alexandra Hagelstein, and Thomas Gübitz. 2017. LIVIVO : the vertical search engine for life sciences. *Datenbank-Spektrum*, pages 1–6.

Claire Nédellec, Robert Bossy, Jin-Dong Kim, Jung-Jae Kim, Tomoko Ohta, Sampo Pyysalo, and Pierre Zweigenbaum. 2013. Overview of bionlp shared task 2013. In *Proceedings of the BioNLP shared task 2013 workshop*, pages 1–7.

Mark Neumann, Daniel King, Iz Beltagy, and Waleed Ammar. 2019. Scispacy: Fast and robust models for biomedical natural language processing.

Nanyun Peng, Hoifung Poon, Chris Quirk, Kristina Toutanova, and Wen-tau Yih. 2017. Cross-sentence n-ary relation extraction with graph lstms. *Transactions of the Association for Computational Linguistics*, 5:101–115.

Isabel Segura Bedmar, Paloma Martínez, and María Herrero Zazo. 2013. Semeval-2013 task 9: Extraction of drug-drug interactions from biomedical texts (ddiextraction 2013). Association for Computational Linguistics.

Micah Shlain, Hillel Taub-Tabib, Shoval Sadde, and Yoav Goldberg. 2020. Syntactic search by example. In *Proceedings of ACL 2020, System Demonstrations*.

Axel J Soto, Piotr Przybyła, and Sophia Ananiadou. 2018. Thalia: semantic search engine for biomedical abstracts. *Bioinformatics*, 35(10):1799–1801.

Aryeh Tiktinsky, Yoav Goldberg, and Reut Tsarfaty. 2020. pybart: Evidence-based syntactic transformations for ie. In *Proceedings of ACL 2020, System Demonstrations*.

Marco A. Valenzuela-Escárcega, Gus Hahn-Powell, and Dane Bell. 2020. Odinson: A fast rule-based information extraction framework. In *Proceedings of the Twelfth International Conference on Language Resources and Evaluation (LREC 2020)*, Marseille, France. European Language Resources Association (ELRA).

Patrick Verga, Emma Strubell, Ofer Shai, and Andrew McCallum. 2017. Attending to all mention pairs for full abstract biological relation extraction. *arXiv preprint arXiv:1710.08312*.

Lucy Lu Wang, Kyle Lo, Yoganand Chandrasekhar, Russell Reas, Jiangjiang Yang, Darrin Eide, Kathryn Funk, Rodney Michael Kinney, Ziyang Liu, William. Merrill, Paul Mooney, Dewey A. Murdick, Devvret Rishi, Jerry Sheehan, Zhihong Shen, Brandon Stilson, Alex D. Wade, Kuansan Wang, Christopher Wilhelm, Boya Xie, Douglas M. Raymond, Daniel S. Weld, Oren Etzioni, and Sebastian Kohlmeier. 2020. Cord-19: The covid-19 open research dataset. *ArXiv*, abs/2004.10706.

Lucy Lu Wang, Oyvind Tafjord, Sarthak Jain, Arman Cohan, Sam Skjonsberg, Carissa Schoenick, Nick Botner, and Waleed Ammar. 2019. Extracting evidence of supplement-drug interactions from literature. *arXiv preprint arXiv:1909.08135*.

Improving Biomedical Analogical Retrieval with Embedding of Structural Dependencies

Amandalynne Paullada*, **Bethany Percha**[†], **Trevor Cohen**[‡]

*Department of Linguistics, University of Washington, Seattle, WA, USA
[†] Dept. of Medicine and Dept. of Genetics & Genomic Sciences,
Icahn School of Medicine at Mount Sinai, New York, NY, USA
[‡]Dept. of Biomedical Informatics and Medical Education, University of Washington, Seattle, WA, USA
paullada@uw.edu, bethany.percha@mssm.edu, cohenta@uw.edu

Abstract

Inferring the nature of the relationships between biomedical entities from text is an important problem due to the difficulty of maintaining human-curated knowledge bases in rapidly evolving fields. Neural word embeddings have earned attention for an apparent ability to encode relational information. However, word embedding models that disregard syntax during training are limited in their ability to encode the structural relationships fundamental to cognitive theories of analogy. In this paper, we demonstrate the utility of encoding dependency structure in word embeddings in a model we call Embedding of Structural Dependencies (ESD) as a way to represent biomedical relationships in two analogical retrieval tasks: a relationship retrieval (RR) task, and a literature-based discovery (LBD) task meant to hypothesize plausible relationships between pairs of entities unseen in training. We compare our model to skipgram with negative sampling (SGNS), using 19 databases of biomedical relationships as our evaluation data, with improvements in performance on 17 (LBD) and 18 (RR) of these sets. These results suggest embeddings encoding dependency path information are of value for biomedical analogy retrieval.

1 Introduction

Distributed vector space models of language have been shown to be useful as representations of relatedness and can be applied to information retrieval and knowledge base augmentation, including within the biomedical domain (Cohen and Widdows, 2009). A vast amount of knowledge on biomedical relationships of interest, such as therapeutic relationships, drug-drug interactions, and adverse drug events, exists in largely human-curated knowledge bases (Zhu et al., 2019). However, the rate at which new papers are published means new relationships are being discovered faster than human curators can manually update the knowledge bases. Furthermore, it is appealing to automatically generate hypotheses about novel relationships given the information in scientific literature (Swanson, 1986), a process also known as 'literature-based discovery.' A trustworthy model should also be able to reliably represent known relationships that are validated by existing literature.

Neural word embedding techniques such as word2vec[1] and fastText[2] are a widely-used and effective approach to the generation of vector representations of words (Mikolov et al., 2013a) and biomedical concepts (De Vine et al., 2014). An appealing feature of these models is their capacity to solve proportional analogy problems using simple geometric operators over vectors (Mikolov et al., 2013b). In this way, it is possible to find analogical relationships between words and concepts without the need to specify the relationship type explicitly, a capacity that has recently been used to identify therapeutically-important drug/gene relationships for precision oncology (Fathiamini et al., 2019). However, neural embeddings are trained to predict co-occurrence events without consideration of syntax, limiting their ability to encode information about relational structure, which is an essential component of cognitive theories of analogical reasoning (Gentner and Markman, 1997). Additionally, recent work (Peters et al., 2018) has found that contextualized word embeddings from language models such as ELMo, when evaluated on analogy tasks, perform worse on semantic relation tasks than static embedding models.

The present work explores the utility of encoding syntactic structure in the form of dependency paths into neural word embeddings for analogical

---

[1]https://github.com/tmikolov/word2vec
[2]https://fasttext.cc/

Proceedings of the BioNLP 2020 workshop, pages 38–48
Online, July 9, 2020 ©2020 Association for Computational Linguistics

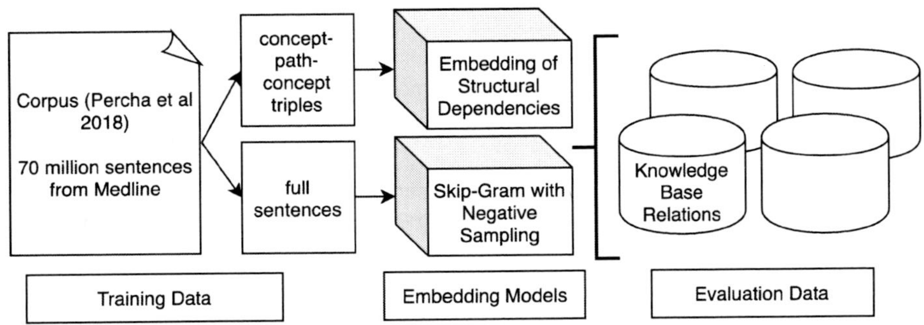

Figure 1: Overview of training and evaluation pipeline. Two embedding models, Embedding of Structural Dependencies (ESD) and Skip-gram with Negative Sampling (SGNS), are trained on data from a corpus of $\approx$70 million sentences from Medline. The resulting representations are then evaluated on data collected from biomedical knowledge bases.

retrieval of biomedical relations. To this end, we build and evaluate vector space models for representing biomedical relationships, using a corpus of dependency-parsed sentences from biomedical literature as a source of grammatical representations of relationships between concepts.

We compare two methods for learning biomedical concept embeddings, the skip-gram with negative sampling (SGNS) algorithm (Mikolov et al., 2013a) and Embedding of Semantic Predications (ESP) (Cohen and Widdows, 2017), which adapts SGNS to encode concept-predicate-concept triples. In the current work, we adapt ESP to encode dependency paths, an approach we call Embedding of Structural Dependencies (ESD). We train ESD and SGNS on a corpus of approximately 70 million sentences from biomedical research paper abstracts from Medline, and evaluate each model's ability to solve analogical retrieval problems derived from various biomedical knowledge bases. We train ESD on concept-path-concept triples extracted from these sentences, and SGNS on full sentences that have been minimally preprocessed with named entities (see §3). Figure 1 shows the pipeline from training to evaluation.

From an applications perspective, we aim to evaluate the utility of these representations of relationships for two tasks. The first involves correctly identifying a concept that is related in a particular way to another concept, when this relationship has already been described explicitly in the biomedical literature. This task is related to the NLP task of relationship extraction, but rather than considering one sentence at a time, distributional models represent information from across all of the instances in which this pair have co-occurred, as well as

information about relationships between similar concepts. We refer to this task as *relationship retrieval (RR)*. The second task involves identifying concepts that are related in a particular way to one another, where this relationship has not been described in the literature previously. We refer to this task as *literature-based discovery (LBD)*, as identifying such implicit knowledge is the main goal of this field (Swanson, 1986).

We evaluate on four kinds of biomedical relationships, characterized by the semantic types of the entity pairs involved, namely *chemical-gene*, *chemical-disease*, *gene-gene*, and *gene-disease* relationships.

The following paper is structured as follows. §2 describes vector space models of language as they are evaluated for their ability to solve proportional analogy problems, as well as prior work in encoding dependency paths for downstream applications in relation extraction. §3 presents the dependency path corpus from Percha and Altman (2018). §4 summarizes the knowledge bases from which we develop our evaluation data sets. §5 describes the training details for each vector space model. §6 and §7 describe the methods and results for the *RR* and *LBD* evaluation paradigms. §8 and §9 offer discussion and conclude the paper. Code and evaluation data will be made available at `https://github.com/amandalynne/ESD`.

2 Background

We look to prior work in using proportional analogies as a test of relationship representation in the general domain with existing studies on vector space models trained on generic English. While our biomedical data is largely in English, we constrain our evaluation to specific biomedical concepts and relationships as we apply and extend established methods.

Vector space models of language and analogical reasoning

Vector space models of semantics have been applied in information retrieval, cognitive science and computational linguistics for decades (Turney and Pantel, 2010), with a resurgence of interest in recent years. Mikolov et al. (2013a) and Mikolov et al. (2013b) introduce the skip-gram architecture. This work demonstrated the use of a continuous vector space model of language that could be used for analogical reasoning when vector offset methods are applied, providing the following canonical example: if x_i is the vector corresponding to word i, $x_{king} - x_{man} + x_{woman}$ yields a vector that is close in proximity to x_{queen}. This result suggests that the model has learned something about semantic gender. They identified some other linguistic patterns recoverable from the vector space model, such as pluralization: $x_{apple} - x_{apples} \approx x_{car} - x_{cars}$, and developed evaluation sets of proportional analogy problems that have since been widely used as benchmarks for distributional models (see for example (Levy et al., 2015)).

However, work soon followed that pointed out some of the shortcomings of attributing these results to the models' analogical reasoning capacity. For example, Linzen (2016) showed that the vector for 'queen' is itself one of the nearest neighbors to the vector for 'woman,' and so it can be argued that the model does not actually learn relational information that can be applied to analogical reasoning, but rather, can rely on the direct similarity between the target terms in the analogy to produce desirable results.

Furthermore, Gladkova et al. (2016) introduce the Better Analogy Test Set (BATS) to provide an evaluation set for analogical reasoning that includes a broader set of semantic and syntactic relationships between words. This set proved far more challenging for embedding-based approaches. Newman-Griffis et al. (2017) provide results of vector offset methods applied to a dataset of biomedical analogies derived from UMLS triples, showing that certain biomedical relationships are more difficult to learn with analogical reasoning than others.

Because the aim of this project is to robustly learn a handful of biomedical relationships, we are less concerned about the linguistic generalizability of these particular representations, but future work will examine the application of these vector space models to analogies in the general domain.

Dependency embeddings

Levy and Goldberg (2014a) adapt the SGNS model to encode direct dependency relationships, rather than dependency paths. In this approach, a dependency-type/relative pair is treated as a target for prediction when the head of a phrase is observed (e.g. $P(scientist/nsubj|discovers)$). The dependency-based skipgram embeddings were shown to better reflect the functional roles of words than those trained on narrative text, which tended to emphasize topical associations. Recent work (Zhang et al. (2018), Zhou et al. (2018), Li et al. (2019)) has also integrated dependency path representations in neural architectures for biomedical relation extraction, framing it as a classification task rather than an analogical reasoning task. The work of Washio and Kato (2018) is perhaps the most closely related to our approach, in that neural embeddings are trained on word-path-word triples. Aside from our application of domain-specific Named Entity Recognition (NER), a key methodological difference between this work and the current work is that their approach represents word pairs as a linear transformation of the concatenation of their embeddings, while we use XOR as a binding operator (following the approach of Kanerva (1996)), which was first used to model biomedical analogical retrieval with semantic predications extracted from the literature by Cohen et al. (2011)[3]. On account of the use of a binding operator, individual entities, pairs of entities and dependency paths are all represented in a common vector space.

3 Text Data

We train both the ESD and SGNS models on data released by Percha and Altman (2018). This corpus[4] consists of about 70 million sentences from a subset of **MEDLINE** (approximately 16.5 million abstracts) which have `PubTator` (Wei et al., 2013) annotations applied to identify phrases that denote names of *chemicals* (including drugs and other chemicals of interest), *genes* (and the proteins they code for), and *diseases* (including side effects

---

[3]For related work, see Widdows and Cohen (2014)

[4]Version 7 of the corpus retrieved at `https://zenodo.org/record/3459420`

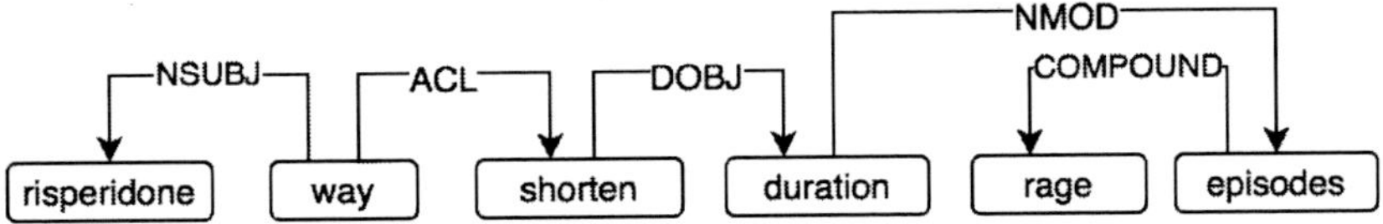

"Liquid **risperidone** may be a safe and effective way to shorten the duration of **rage** episodes."

```
way_nsubj_START_ENTITY way_acl_shorten shorten_dobj_duration
duration_nmod_episodes episodes_compound_END_ENTITY
```

Figure 2: Example of a path of dependencies between two entities of interest. The full parse is not shown, but rather, the minimum path of dependency relations between the two entities given the sentence.

and other phenotypes). Throughout this paper, we use these shorthand names for each of these categories, following the convention established in Wei et al. (2013) and followed by Percha and Altman (2018).

The following example sentence from an article processed by `PubTator` shows how multi-word phrases that denote biomedical entities of interest, in this case *atypical depression* and *seasonal affective disorder*, are concatenated by underscores to constitute single tokens:

Chromium has a beneficial effect on eating-related atypical symptoms of depression, and may be a valuable agent in treating atypical_depression and seasonal_affective_disorder.

Percha and Altman (2018) also provide pruned Stanford dependency (De Marneffe and Manning, 2008) parses for the sentences in the corpus, consisting, for each sentence, of the minimal path of dependency relations connecting pairs of biomedical named entities identified by `PubTator`. Specifically, they extract dependency paths that connect chemicals to genes, chemicals to diseases, genes to diseases, and genes to genes. Figure 2 shows an example of a dependency path of relations between two terms, *risperidone* and *rage*. We use these dependency paths as representations for predicates that denote biomedical relationships of interest by concatenating the string representations of each path element, which are shown below the sentence in Figure 2. Following Percha and Altman (2018), we exclude paths that denote a coordinating conjunction between elements and paths that denote an appositive construction, both of which are highly common in the set. In this corpus of 70 million sentences, there are about 44 million unique dependency paths that connect concepts of interest, the vast majority (around 40 million) of which appear just once in the corpus. 540,011 of these paths appear at least 5 times in the corpus.

4 Knowledge Bases

We construct our evaluation data sets with exemplars from knowledge bases for four primary kinds of biomedical relationships, characterized by the interactions between pairs of entities of the following types: *chemical-gene*, *chemical-disease*, *gene-disease*, and *gene-gene*.

We evaluate on pairs of entities from the following knowledge bases: DrugBank (Wishart et al., 2018), Online Mendelian Inheritance in Man (OMIM) (Hamosh et al., 2005), PharmGKB (PGKB) (Whirl-Carrillo et al., 2012), Reactome (Fabregat et al., 2016), Side Effect Resource (SIDER) (Kuhn et al., 2016), and Therapeutic Target Database (TTD) Wang et al. (2020).

Each knowledge base consists of pairs of entities that relate in a specific way. For example, SIDER Side Effects consists of *chemical-disease*-typed pairs such that the chemical is known to have the disease as a side effect, e.g. (*sertraline, insomnia*). Meanwhile, another *chemical-disease* pair from a different database, Therapeutic Target Database (TTD) indications, is such that the chemical is indicated as a treatment for the disease, e.g. (*carphenazine, schizophrenia*). In constructing our evaluation sets, we process all terms such that they are lower-cased, and multi-word terms are concatenated by underscores. Furthermore, we eliminate from our evaluation sets any knowledge base terms that do not appear in the training corpus described in §3 at least 5 times. It should be noted that across these sets, a single biomedical entity may appear with numerous spellings and naming conventions.

Table 2 shows the corresponding relationship type for each of the knowledge bases we use, as well as the number of pairs from each that are used in our evaluation data. The relationship retrieval data consists of knowledge base pairs that appear in our training corpus connected by a dependency path at least once, while the literature-based discovery targets are those knowledge base pairs that do not appear connected by a dependency path in the corpus.

5   Training Details

SGNS With SGNS, a shallow neural network is trained to estimate the probability of encountering a context term, t_c, within a sliding window centered on an observed term, t_o. The training objective involves maximizing this probability for true context terms $P(t_c|t_o)$, and minimizing it for randomly drawn counterexamples $t_{\neg c}$, $P(t_{\neg c}|t_o)$, with probability estimated as the sigmoid function of the scalar product between the input weight vector for the observed term and the output weight vector of the context term, $\sigma(\overrightarrow{t_o}.\overrightarrow{t_{c|\neg c}})$. We used the `Semantic Vectors`[5] implementation of SGNS (which performs similarly to the `fastText` implementation across a range of analogical retrieval benchmarks (Cohen and Widdows, 2018)) to train 250-dimensional embeddings, with a sliding window radius of two, on the complete set of full sentences from the corpus described in §3 as the training corpus. As previously mentioned, multi-word phrases corresponding to named entities recognized by the `PubTator` system in these sentences are concatenated by underscores, and consequently receive a single vector representation.

ESD With ESD, a shallow neural network is trained to estimate the probability of encountering the object, o, of a subject-predicate-object triple sPo. The training objective involves maximizing this probability for true objects $P(o|s, P)$ and minimizing it for randomly drawn counterexamples, $\neg o$, $P(\neg o|s, P)$. We adapted the `Semantic Vectors`[5] implementation of ESP to encode dependency paths, with binary vectors as representational basis (Widdows and Cohen, 2012) and the non-negative normalized Hamming distance ($NNHD$) to estimate the similarity between them.

$$NNHD = \max\left(0, 1 - \frac{2 \times \text{Hamming distance}}{\text{dimensionality}}\right)$$

With this representational paradigm, probability can be estimated as $NNHD(o, s \otimes P)$, where $\otimes$ represents the use of pairwise exclusive OR as a *binding operator*, in accordance with the Binary Spatter Code (Kanerva, 1996). While ESP

was originally developed to encode knowledge extracted from the literature using a small set of predefined predicates (e.g. TREATS), we adapt it here to encode a large variety (n=546,085) of dependency paths. For training, we concatenate the dependency relations (the underscored parts in Figure 2) into a single predicate token for which a vector is learned. Some examples of path tokens (concatenated dependency relations) can be seen in Table 1. Unlike the original ESP implementation where predicate vectors were held constant, we permit dependency path vectors to evolve during training[6]. Further details on ESP can be found in (Cohen and Widdows, 2017). For the current work, we set the dimensionality at 8000 bits (as this is equivalent in representational capacity to 250-dimensional single precision real vectors). For ESD, Table 1 shows the nearest neighboring dependency path vectors to the bound product $I(metformin) \otimes O(diabetes)$, illustrating paths that indicate the relationship between these terms, and ESD's capability to learn similar representations for paths with similar meaning.

Both SGNS and ESD were trained over five epochs, with a subsampling threshold of 10^{-5}, a minimum term frequency threshold of 5 (which includes concatenated dependency paths for ESD), and a maximum frequency threshold of 10^6.

6   Evaluation Methods

We use a proportional analogy ranked retrieval task for both the RR and LBD tasks, following prior work as described in §2. Figure 3 visualizes this process. From a set of (X, Y) entity pairs from a knowledge base, given a term C and all terms D such that (C, D) is a pair in the set, we select n random (A, B) cue pairs from a disjoint set of pairs. We refer to (C, D) pairs as 'target pairs,' correct D completions as 'targets,' and (A, B) pairs as 'cues.' The vectors for the cue terms (A, B) and the term C are summed in the following fashion to produce the resulting vector v. Given an analogical pair $A:B::C:D$, where A and C, B and D are of the same semantic type, respectively, we develop cue vectors for the target D in each model as follows:

$$SGNS : \vec{v} = \overrightarrow{B} - \overrightarrow{A} + \overrightarrow{C}$$
$$ESD : \vec{v} = \overrightarrow{I(A)} \otimes \overrightarrow{O(B)} \otimes \overrightarrow{I(C)}$$

---

[6]This capability has been used to to predict drug interactions, with performance exceeding that of models with orders of magnitude more parameters (Burkhardt et al., 2019).

SCORE	PATH
0.974	controlled_nmod_start_entity_end_entity_amod_controlled
0.935	add-on_nmod_start_entity_end_entity_amod_add-on
0.565	reduces_nsubj_start_entity_reduces_dobj_requirement_requirement_nmod_end_entity
0.537	associated_compound_start_entity_end_entity_nsubj_associated
0.516	start_entity_conj_efficacy_efficacy_acl_treating_treating_dobj_end_entity
0.438	treatment_amod_start_entity_treatment_nmod_end_entity

Table 1: Nearest neighboring dependency path embeddings to $I(metformin) \otimes O(diabetes)$ where I and O indicate input and output weight vectors respectively.

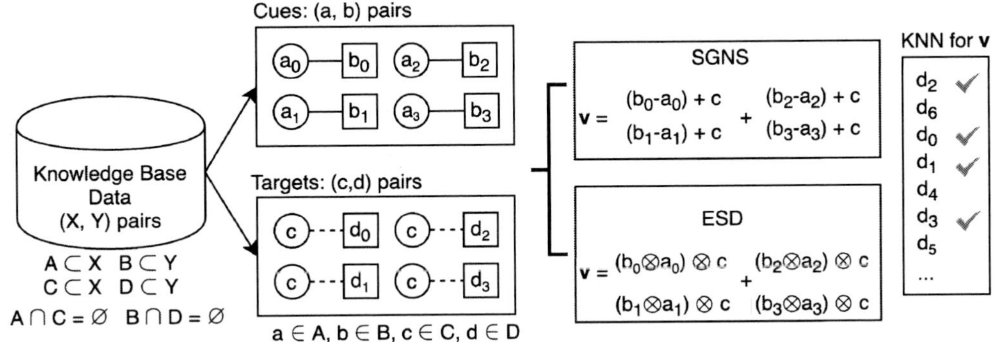

Figure 3: Overview of analogical ranked retrieval paradigm.

where I and O represent the input and output weight vectors of the ESD model, respectively. The SGNS method is the same as the 3COSADD method as described in Levy and Goldberg (2014b).

A K-nearest neighbor search is performed for $\mathbf{v}$ (using cosine distance for SGNS, NNHD for ESD) over the search space, and we record the ranks for each correct D target. The search space is constrained such that it consists of those terms from our training corpus that have a vector in both ESD and SGNS, a total of about 300,000 terms overall. For ESD, this space consists of the output weight vectors for each concept. For the proportional analogy task using K-nearest neighbors to rank completions to the analogy, the desired outcome is for the correct targets to be highly similar to the analogy cue vector v, such that the highest ranks are assigned to the correct target terms D in a search over the entire vector space. In this fashion, we perform this KNN search for every (X, Y) pair in the knowledge base and record the ranks for correct targets. We then compare the ranks of terms D across both vector spaces; the higher the ranks, the better the model is at capturing relational similarity.

Table 2 shows, for each knowledge base, how many total unique X terms and total (X, Y) pairs are used for each task. Additionally, we show the average number of correct Y terms per X and the maximum number of correct Y terms per X. For the relationship retrieval task, we consider those (X, Y) pairs which are connected by at least one dependency path in our corpus. Meanwhile, (X, Y) pairs for the LBD task must *not* be connected by a dependency path in the corpus (we treat these held-out pairs as a proxy for estimating the quality of novel hypotheses). We know from the (X, Y) pair's presence in the knowledge base that it is a gold standard pair for the given relationship type, but from the models' perspective this information is not available from the text alone. Thus, we believe it is a good test of the models' ability to generate plausible hypotheses. To reiterate, the methodology for both the relationship retrieval and literature-based discovery evaluations is the same; the only difference is in which pairs of terms from each knowledge base are used for evaluation data.

We examine the role of increasing the number of cues in improving retrieval. For example, for a given (C, D) target pair, we can combine vectors

		Relationship Retrieval				Literature-based Discovery			
		Total X	Total Pairs	Mean Y / X	Max Y / X	Total X	Total Pairs	Mean Y / X	Max Y / X
Chem-Gene	Gene Targets (DrugBank)	1626	6290	4	107	3569	37162	10	420
	PGKB	535	2089	4	48	1563	28053	18	144
	Agonists (TTD)	148	172	1	3	307	462	2	7
	Antagonists (TTD)	188	200	1	2	508	620	1	5
	Gene Targets (TTD)	1179	1436	1	7	4088	6430	2	15
	Inhibitors (TTD)	522	669	1	7	1273	2082	2	15
Chem-Disease	Side Effects (SIDER)	334	1289	4	31	892	6591	7	46
	Drug Indication (SIDER)	1077	2737	3	22	2160	8356	4	45
	Biomarker-Disease (TTD)	298	417	1	11	253	321	1	6
	Drug Indication (TTD)	1749	1958	1	6	2664	2999	1	10
	Disease Targets (TTD)	710	1502	2	22	1085	3088	3	27
Gene-Disease	OMIM	2197	2870	1	9	3461	5545	2	11
	PGKB	600	1693	3	34	1609	12605	8	73
Gene-Gene	Enzymes (DrugBank)	966	3622	4	33	1781	16242	9	71
	Carriers (DrugBank)	203	345	2	27	444	1174	3	18
	Transporters (DrugBank)	510	2357	5	44	1140	13889	12	94
	PGKB	497	2595	5	50	940	14142	15	89
	Complex (Reactome)	1757	3061	2	9	2550	6593	3	31
	Reaction (Reactome)	579	1031	2	9	1274	4024	3	29

Table 2: Total unique X terms, total (X, Y) pairs, average number of correct Y terms per X, and maximum number of correct Y terms per X for each knowledge base.

for multiple (A, B) pairs with the C term vector to produce a final cue vector that is closer to the target D. When multiple cues are used, we superpose the cue vector for each of the cues, and normalize the resulting vector, with normalization of real vectors to unit length in SGNS, and normalization of binary vectors using the majority rule with ties split at random with ESD. Cues are always selected from the subset of knowledge base pairs that co-occur in our training corpus. We ensure that none of the (A, B) cue terms overlap with each other, nor with the (C, D) target terms, to assure that self-similarity does not inflate performance. We produced results for a range of 1, 5, 10, 25, and 50 cues, finding that the best results come from using 25 cues; we only report these resulting scores in §7.

As a baseline inspired partly by Linzen (2016), we compute the similarity of vectors for B and D terms and C and D terms compared directly to each other, omitting the analogical task. The intuition here is that C and D terms are potentially close together in the vector space merely due to frequent co-occurrence in the corpus, and any analogical reasoning performance is merely relying on that fact. Meanwhile, terms B and D can be close together in the vector space simply because they are the same semantic type, and thus occur in similar contexts. In this case, relational analogy might not explain the performance, but mere distributional similarity. In the $B{:}D$ comparison setting, cues B are added together to create a single cue vector with which to perform the KNN ranking over terms in which to find the target term D. These cue terms

B are extracted from the same A, B cue pairs as those used for the full analogy setting to ensure a reasonable comparison across methods. In the $C{:}D$ comparison setting, no cues are aggregated.

7 Results

We present qualitative and quantitative results for each vector space model's ability to represent and retrieve relational information.

Qualitative Results Table 3 shows a side-by-side comparison of the top 10 retrieved terms given the vector for the term *risperidone* composed with 25 randomly selected (drug, indication) cues from SIDER. The goal is to complete the proportional analogy corresponding to the treatment relationship. Of the top 10 terms retrieved in the ESD vector space, 4 are correct completions to the analogy, while 3 more are plausible completions based on literature. 'Tardive oromandibular dystonia,' while of the correct semantic type targeted by this analogy, is actually a side effect of risperidone. A majority of the retrieved results, however, are known or plausible treatment targets. Meanwhile, most of the top 10 terms retrieved by SGNS are names of other drugs that are similar to risperidone. Additionally, 'psychiatric and visual disturbances' and 'tardive dyskinesia' are side effects of risperidone, not treatment targets. Notably, all of the results retrieved with ESD are of the correct semantic type, i.e., they are disorders, while SGNS retrieves a mix of drugs and side effects.

Quantitative Results For each C term in each evaluation set, we record the ranks of all D tar-

rank	ESD (ours)	SGNS
1	separation anxiety	risperidone ×
2	**schizophrenia**	olanzapine ×
3	depressed state	quetiapine ×
4	**bipolar mania**	aripiprazole ×
5	tardive oromanibular dystonia	clozapine ×
6	treatment of trichotillomania ∗	psychiatric and visual disturbances
7	pervasive developmental disorder (NOS) ∗	ziprasidone ×
8	borderline personality disorder	amisulpride ×
9	**psychotic disorders**	paliperidone ×
10	**mania**	tardive dyskinesia

Table 3: Top 10 results for a K-nearest neighbor search over terms for treatment targets for the drug risperidone (an antipsychotic drug), using 25 (drug, indication) pairs from SIDER as cues. **Bolded** terms are correct targets, i.e., they are listed as treatment targets for risperidone in SIDER. ∗: a disorder that risperidone treats or might treat, based on external literature or a synonym for a target from SIDER; ×: a chemical, i.e., something that could not be a treatment target for a drug.

get terms resulting from the K-nearest neighbor search. For ease of comparison, we normalize all raw ranks by the length of the full search space (324363 terms in total), and then subtract this value from 1 so that lower ranks (i.e., better results) are displayed as higher numbers, for ease of interpretation. For a baseline score, we ran a simulation in which the entire search space was shuffled randomly 100 times, and recorded the median ranks of multiple target D terms, given some C. We find that the median rank for D terms in a randomly shuffled space tended toward the middle of the ranked list. Thus, the baseline score is established as 0.5; any score lower than this means the model performed worse than a random shuffle at retrieving target terms. In Table 4, 1 is the highest possible score, and 0 is the lowest.

We report results at 25 (A, B) cues, the setting for which performance was best for both ESD and SGNS. 'Full' in Table 4 refers to evaluation with a full $A{:}B{::}C{:}D$ analogy, while 'B:D' refers to the baseline that compares vectors for terms directly, rather than using relational information. We do not report $C{:}D$ comparison results, as they were categorically worse than both Full and $B{:}D$ results.

8 Discussion

The results in Table 4 show that ESD outperforms SGNS on the *RR* task for 18 of 19 databases, and for 17 of 19 databases on the *LBD* task. It is clear that literature-based discovery is harder than relationship retrieval, as the scores are generally lower across the board for this task. We discuss the results

for each task separately.

8.1 Relationship retrieval

For a total of 12 out of 19 sets, ESD on full analogies outperforms ESD on direct *B:D* comparisons, suggesting that the model has learned generalizable relationship information for these types of relations rather than relying on distributional term similarity. Because *gene-gene* pairs consist of entities of the same semantic type, it can be argued that *B:D* similarity should be very high, and yet scores are higher for the full analogy over the *B:D* baseline for most of these sets, for both ESD and SGNS. For SIDER side effects, the *B:D* baseline for ESD shows higher scores than the full analogy for both *LBD* and *RR*; one reason for this could be that there is a high degree of side effect overlap between drugs, and so the side effect terms themselves are highly similar to each other.

8.2 Literature-based discovery

The best performance on a majority of the sets comes from the ESD *B:D* model, suggesting that the model relies on term similarity over relational information for performance. Although SGNS doesn't perform the best overall, the full analogy model tends to outperform its B:D counterpart, suggesting that SGNS has managed to extrapolate relational information to the retrieval of held-out targets. As previously mentioned, performance on this task is made difficult due to the lack of normalization of concepts across our datasets. Additionally, as Table 4 shows, several top ranked terms are plausible analogy completions, but do not appear as

| | | Relationship retrieval | | | | LBD | | | |
| | | ESD (ours) | | SGNS | | ESD (ours) | | SGNS | |
		Full	B:D	Full	B:D	Full	B:D	Full	B:D
Chem-Gene	Gene Targets (DrugBank)	**0.912**	0.897	0.839	0.212	0.715	**0.806**	0.496	0.250
	PGKB	0.969	**0.994**	0.705	0.361	0.737	**0.918**	0.366	0.317
	Agonists (TTD)	0.997	0.907	**0.998**	0.647	0.802	0.781	**0.924**	0.708
	Antagonists (TTD)	**1.000**	0.900	0.999	0.732	0.802	0.703	**0.831**	0.750
	Gene Targets (TTD)	**0.998**	0.867	0.994	0.387	0.746	**0.760**	0.625	0.479
	Inhibitors (TTD)	**0.998**	0.874	0.993	0.415	**0.773**	0.759	0.682	0.392
Chem-Disease	Side Effects (SIDER)	0.997	**0.999**	0.967	0.942	0.952	**0.994**	0.799	0.932
	Drug Indication (SIDER)	**1.000**	0.995	0.949	0.588	0.969	**0.988**	0.663	0.605
	Biomarker-Disease (TTD)	0.996	**0.997**	0.944	0.781	0.932	**0.977**	0.799	0.726
	Drug Indication (TTD)	**1.000**	0.994	0.981	0.675	0.977	**0.992**	0.722	0.661
	Disease Targets (TTD)	0.990	**0.997**	0.900	0.711	0.887	**0.989**	0.663	0.648
Gene-Disease	OMIM	**0.997**	0.911	0.950	0.599	0.668	**0.792**	0.578	0.578
	PGKB	0.982	**0.996**	0.781	0.624	0.836	**0.969**	0.592	0.618
Gene-Gene	Enzymes (DrugBank)	**1.000**	**1.000**	0.987	0.981	0.979	**0.999**	0.900	0.975
	Carriers (DrugBank)	0.987	**1.000**	0.636	0.555	0.841	**0.962**	0.360	0.487
	Transporters (DrugBank)	**1.000**	**1.000**	0.974	0.947	0.996	**0.999**	0.870	0.951
	PGKB	**0.999**	0.995	0.899	0.471	0.907	**0.956**	0.479	0.425
	Complex (Reactome)	**1.000**	0.819	**1.000**	0.206	**0.866**	0.731	0.838	0.399
	Reaction (Reactome)	**1.000**	0.917	0.996	0.273	**0.878**	0.826	0.699	0.366

Table 4: Results for relationship retrieval (RR) and literature-based discovery (LBD) for full analogy (A:B::C:D) and B:D retrieval. Scores are displayed here as the median of scores (1 - normalized rank) for all D terms in a knowledge base evaluation set.

gold-standard targets in the databases. Considering the case of SIDER, which is built from automatically extracted information (not human-curated) the plausible results here are missing from the database but are supported by evidence from published papers (e.g. Oravecz and Štuhec (2014)).

9   Conclusion

We have compared two vector space models of language, Embedding of Structural Dependencies and Skip-gram with Negative Sampling, for their ability to represent biomedical relationships from literature in an analogical retrieval task. Our results suggest that encoding structural information in the form of dependency paths connecting biomedical entities of interest can improve performance on two analogical retrieval tasks, relationship retrieval and literature-based discovery. In future work, we would like to compare our methods with knowledge base completion techniques using contextualized vectors from language models as in Bosselut et al. (2019) as another method applicable to literature-based discovery.

Acknowledgements

This research was supported by U.S. National Library of Medicine Grant No. R01 LM011563. The authors would like to thank the anonymous reviewers for their feedback.

References

Antoine Bosselut, Hannah Rashkin, Maarten Sap, Chaitanya Malaviya, Asli Celikyilmaz, and Yejin Choi. 2019. COMET: Commonsense transformers for automatic knowledge graph construction. In *Proceedings of the 57th Annual Meeting of the Association for Computational Linguistics*, pages 4762–4779, Florence, Italy. Association for Computational Linguistics.

Hannah A Burkhardt, Devika Subramanian, Justin Mower, and Trevor Cohen. 2019. Predicting adverse drug-drug interactions with neural embedding of semantic predications. In *AMIA Annual Symposium Proceedings*, volume 2019, page 992. American Medical Informatics Association.

Trevor Cohen and Dominic Widdows. 2009. Empirical distributional semantics: methods and biomedical applications. *Journal of biomedical informatics*, 42(2):390–405.

Trevor Cohen and Dominic Widdows. 2017. Embedding of semantic predications. *Journal of biomedical informatics*, 68:150–166.

Trevor Cohen and Dominic Widdows. 2018. Bringing order to neural word embeddings with embeddings augmented by random permutations (earp). In *Proceedings of the 22nd Conference on Computational Natural Language Learning*, pages 465–475.

Trevor Cohen, Dominic Widdows, Roger Schvaneveldt, and Thomas C Rindflesch. 2011. Finding schizophrenia's prozac emergent relational similarity in predication space. In *International Symposium on Quantum Interaction*, pages 48–59. Springer.

Marie-Catherine De Marneffe and Christopher D Manning. 2008. The stanford typed dependencies representation. In *Coling 2008: proceedings of the workshop on cross-framework and cross-domain parser evaluation*, pages 1–8. Association for Computational Linguistics.

Lance De Vine, Guido Zuccon, Bevan Koopman, Laurianne Sitbon, and Peter Bruza. 2014. Medical semantic similarity with a neural language model. In *Proceedings of the 23rd ACM international conference on conference on information and knowledge management*, pages 1819–1822.

Antonio Fabregat, Konstantinos Sidiropoulos, Phani Garapati, Marc Gillespie, Kerstin Hausmann, Robin Haw, Bijay Jassal, Steven Jupe, Florian Korninger, Sheldon McKay, et al. 2016. The reactome pathway knowledgebase. *Nucleic acids research*, 44(D1):D481–D487.

Safa Fathiamini, Amber M Johnson, Jia Zeng, Vijaykumar Holla, Nora S Sanchez, Funda Meric-Bernstam, Elmer V Bernstam, and Trevor Cohen. 2019. Rapamycin- mtor+ braf=? using relational similarity to find therapeutically relevant drug-gene relationships in unstructured text. *Journal of biomedical informatics*, 90:103094.

Dedre Gentner and Arthur B Markman. 1997. Structure mapping in analogy and similarity. *American psychologist*, 52(1):45.

Anna Gladkova, Aleksandr Drozd, and Satoshi Matsuoka. 2016. Analogy-based detection of morphological and semantic relations with word embeddings: What works and what doesn't. In *Proceedings of the NAACL-HLT SRW*, pages 47–54, San Diego, California, June 12-17, 2016. ACL.

Ada Hamosh, Alan F Scott, Joanna S Amberger, Carol A Bocchini, and Victor A McKusick. 2005. Online mendelian inheritance in man (omim), a knowledgebase of human genes and genetic disorders. *Nucleic acids research*, 33(suppl_1):D514–D517.

Pentti Kanerva. 1996. Binary spatter-coding of ordered k-tuples. In *International Conference on Artificial Neural Networks*, pages 869–873. Springer.

Michael Kuhn, Ivica Letunic, Lars Juhl Jensen, and Peer Bork. 2016. The sider database of drugs and side effects. *Nucleic acids research*, 44(D1):D1075–D1079.

Omer Levy and Yoav Goldberg. 2014a. Dependency-based word embeddings. In *Proceedings of the 52nd Annual Meeting of the Association for Computational Linguistics (Volume 2: Short Papers)*, pages 302–308.

Omer Levy and Yoav Goldberg. 2014b. Linguistic regularities in sparse and explicit word representations. In *Proceedings of the eighteenth conference on computational natural language learning*, pages 171–180.

Omer Levy, Yoav Goldberg, and Ido Dagan. 2015. Improving distributional similarity with lessons learned from word embeddings. *Transactions of the Association for Computational Linguistics*, 3:211–225.

Zhiheng Li, Zhihao Yang, Chen Shen, Jun Xu, Yaoyun Zhang, and Hua Xu. 2019. Integrating shortest dependency path and sentence sequence into a deep learning framework for relation extraction in clinical text. *BMC medical informatics and decision making*, 19(1):22.

Tal Linzen. 2016. Issues in evaluating semantic spaces using word analogies. In *Proceedings of the 1st Workshop on Evaluating Vector-Space Representations for NLP*, pages 13–18, Berlin, Germany. Association for Computational Linguistics.

Tomas Mikolov, Kai Chen, Greg Corrado, and Jeffrey Dean. 2013a. Efficient estimation of word representations in vector space. In *Proceedings of International Conference on Learning Representations (ICLR)*.

Tomáš Mikolov, Wen-tau Yih, and Geoffrey Zweig. 2013b. Linguistic regularities in continuous space word representations. In *Proceedings of the 2013 Conference of the North American Chapter of the Association for Computational Linguistics: Human Language Technologies*, pages 746–751, Atlanta, Georgia. Association for Computational Linguistics.

Denis Newman-Griffis, Albert Lai, and Eric Fosler-Lussier. 2017. Insights into analogy completion from the biomedical domain. In *BioNLP 2017*, pages 19–28, Vancouver, Canada,. Association for Computational Linguistics.

Robert Oravecz and Matej Štuhec. 2014. Trichotillomania successfully treated with risperidone and naltrexone: a geriatric case report. *Journal of the American Medical Directors Association*, 15(4):301–302.

Bethany Percha and Russ B Altman. 2018. A global network of biomedical relationships derived from text. *Bioinformatics*, 34(15):2614–2624.

Matthew Peters, Mark Neumann, Luke Zettlemoyer, and Wen-tau Yih. 2018. Dissecting contextual word embeddings: Architecture and representation. In *Proceedings of the 2018 Conference on Empirical Methods in Natural Language Processing*, pages 1499–1509, Brussels, Belgium. Association for Computational Linguistics.

Don R Swanson. 1986. Fish oil, raynaud's syndrome, and undiscovered public knowledge. *Perspectives in biology and medicine*, 30(1):7–18.

Peter D Turney and Patrick Pantel. 2010. From frequency to meaning: Vector space models of semantics. *Journal of artificial intelligence research*, 37:141–188.

Yunxia Wang, Song Zhang, Fengcheng Li, Ying Zhou, Ying Zhang, Zhengwen Wang, Runyuan Zhang, Jiang Zhu, Yuxiang Ren, Ying Tan, et al. 2020. Therapeutic target database 2020: enriched resource for facilitating research and early development of targeted therapeutics. *Nucleic acids research*, 48(D1):D1031–D1041.

Koki Washio and Tsuneaki Kato. 2018. Filling missing paths: Modeling co-occurrences of word pairs and dependency paths for recognizing lexical semantic relations. *arXiv preprint arXiv:1809.03411*.

Chih-Hsuan Wei, Hung-Yu Kao, and Zhiyong Lu. 2013. Pubtator: a web-based text mining tool for assisting biocuration. *Nucleic acids research*, 41(W1):W518–W522.

Michelle Whirl-Carrillo, Ellen M McDonagh, JM Hebert, Li Gong, K Sangkuhl, CF Thorn, Russ B Altman, and Teri E Klein. 2012. Pharmacogenomics knowledge for personalized medicine. *Clinical Pharmacology & Therapeutics*, 92(4):414–417.

Dominic Widdows and Trevor Cohen. 2012. Real, complex, and binary semantic vectors. In *International Symposium on Quantum Interaction*, pages 24–35. Springer.

Dominic Widdows and Trevor Cohen. 2014. Reasoning with vectors: A continuous model for fast robust inference. *Logic Journal of the IGPL*, 23(2):141–173.

David S Wishart, Yannick D Feunang, An C Guo, Elvis J Lo, Ana Marcu, Jason R Grant, Tanvir Sajed, Daniel Johnson, Carin Li, Zinat Sayeeda, et al. 2018. Drugbank 5.0: a major update to the drugbank database for 2018. *Nucleic acids research*, 46(D1):D1074–D1082.

Yijia Zhang, Hongfei Lin, Zhihao Yang, Jian Wang, Shaowu Zhang, Yuanyuan Sun, and Liang Yang. 2018. A hybrid model based on neural networks for biomedical relation extraction. *Journal of biomedical informatics*, 81:83–92.

Huiwei Zhou, Shixian Ning, Yunlong Yang, Zhuang Liu, Chengkun Lang, and Yingyu Lin. 2018. Chemical-induced disease relation extraction with dependency information and prior knowledge. *Journal of biomedical informatics*, 84:171–178.

Yongjun Zhu, Olivier Elemento, Jyotishman Pathak, and Fei Wang. 2019. Drug knowledge bases and their applications in biomedical informatics research. *Briefings in bioinformatics*, 20(4):1308–1321.

DeSpin: a prototype system for detecting spin in biomedical publications

Anna Koroleva
Zurich University of Applied Sciences (ZHAW),
Waedenswil, Switzerland
Swiss Institute of Bioinformatics (SIB),
Lausanne, Switzerland
aakorolyova@gmail.com

Sanjay Kamath
Total,
France
sanjay@lri.fr

Patrick M.M. Bossuyt
Academic Medical Center,
University of Amsterdam,
Amsterdam, Netherlands
p.m.bossuyt@amsterdamumc.nl

Patrick Paroubek
LIMSI, CNRS, Université Paris-Saclay,
Orsay, France
pap@limsi.fr

Abstract

Improving the quality of medical research reporting is crucial to reduce avoidable waste in research and to improve the quality of health care. Despite various initiatives aiming at improving research reporting – guidelines, checklists, authoring aids, peer review procedures, etc. – overinterpretation of research results, also known as distorted reporting or spin, is still a serious issue in research reporting.

In this paper, we propose a Natural Language Processing (NLP) system for detecting several types of spin in biomedical articles reporting randomized controlled trials (RCTs). We use a combination of rule-based and machine learning approaches to extract important information on trial design and to detect potential spin.

The proposed spin detection system includes algorithms for text structure analysis, sentence classification, entity and relation extraction, semantic similarity assessment. Our algorithms achieved operational performance for the these tasks, F-measure ranging from 79,42 to 97.86% for different tasks. The most difficult task is extracting reported outcomes.

Our tool is intended to be used as a semi-automated aid tool for assisting both authors and peer reviewers to detect potential spin. The tool incorporates a simple interface that allows to run the algorithms and visualize their output. It can also be used for manual annotation and correction of the errors in the outputs.

The proposed tool is the first tool for spin detection. The tool and the annotated dataset are freely available.

1 Background

It is widely acknowledged nowadays that that the quality of reporting of research results in the clinical domain is suboptimal. As a consequence, research findings can often not be replicated, and billions of euros may be wasted yearly (Ioannidis, 2005).

Numerous initiatives aim at improving the quality of research reporting. Guidelines and checklists have been developed for every type of clinical research. Still, the quality of reporting remains low: authors fail to choose and follow a correct guideline/checklist (Samaan et al., 2013). Automated tools, such as Penelope[1], are introduced to facilitate the use of guidelines/checklists. It was proved that authoring aids improve the completeness of reporting (Barnes et al., 2015).

Enhancing the quality of peer reviewing is another step to improve research reporting. Peer reviewing requires assessing a large number of information items. Nowadays, Natural Language Processing (NLP) is applied to facilitate laborious manual tasks such as indexing of medical literature (Huang et al., 2011) and systematic review process (Ananiadou et al., 2009). Similarly, the peer reviewing process can be partially automated with the help of NLP.

Our project tackles a specific issue of research reporting that, to our knowledge, has not been addressed by the NLP community: spin, also referred to as overinterpretation of research results. In the context of clinical trials assessing a new (experi-

---

At the time of reported work, Anna Koroleva was a PhD student at LIMSI-CNRS in Orsay, France and at the Academic Medical Center, University of Amsterdam in Amsterdam, the Netherlands. Sanjay Kamath was a PhD student at LIMSI-CNRS and LRI Univ. Paris-Sud in Orsay, France.

[1]https://www.penelope.ai/

Proceedings of the BioNLP 2020 workshop, pages 49–59
Online, July 9, 2020 ©2020 Association for Computational Linguistics

mental) intervention, spin consists in exaggerating the beneficial effects of the studied intervention (Boutron et al., 2010).

Spin is common in articles reporting randomized controlled trials (RCTs) - clinical trials comparing health interventions, to which participants are allocated randomly to avoid biases - with non-significant primary outcome. Abstracts are more prone to spin than full texts. Spin is found in a high percentage of abstracts of articles in surgical research (40%) (Fleming, 2016), cardiovascular diseases (57%) (Khan et al., 2019), cancer (47%) (Vera-Badillo et al., 2016), obesity (46.7%) (Austin et al., 2018), otolaryngology (70%) (Cooper et al., 2018), anaesthesiology (32,2%) (Kinder et al., 2018), and wound care (71%) (Lockyer et al., 2013). Although the problem of spin has started to attract attention in the medical community in the recent years, the shown prevalence of spin proves that it often remains unnoticed by editors and peer reviewers.

Abstracts are often the only part of the article available to readers, and spin in abstracts of RCTs poses a serious threat to the quality of health care by causing overestimation of the intervention by clinicians (Boutron et al., 2014), which may lead to the use of an ineffective of unsafe intervention in clinical practice. Besides, spin in research articles is linked to spin in press releases and health news (Haneef et al., 2015; Yavchitz et al., 2012), which has the negative impact of raising false expectations regarding the intervention among the public.

The importance of the problem of spin motivated our work. We aimed at developing NLP algorithms to aid authors and readers in detecting spin. We focused on randomized controlled trials (RCTs) as they are the most important source of evidence for Evidence-based medicine, and spin in RCTs has high negative impact.

Our work lies within the scope of the Methods in Research on Research (MiRoR) project[2], an international project devoted to improving the planning, conduct, reporting and peer reviewing of health care research. For the design and development of our toolkit, we benefited from advice from the MiRoR consortium members.

In this paper, we introduce a prototype of a system, called DeSpin (Detector of Spin), that automatically detects potential spin in abstracts of RCTs and relevant supporting information. This

___
[2]http://miror-ejd.eu/

prototype comprises a set of spin-detecting algorithms and a simple interface to run the algorithms and display their output.

This paper is organized as follows: first, we provide an overview of some existing semi-automated aid systems for authors, reviewers and readers of biomedical articles. Second, we introduce in more detail the notion of spin, the types of spin that we address, and the information that is required to assess an article for spin. After that, we describe our current algorithms, methods employed and provide their evaluation. Finally, we discuss the potential future development of the prototype.

2 Related work

Although there has been no attempt to automate spin detection in biomedical articles, a number of works addressed developing automated aid tools to assist authors and readers of scientific articles in performing various other tasks. Some of these tools were tested and were shown to reduce the workload and improve the performance of human experts on the corresponding task.

2.1 Authoring aid tools

Barnes et al. (2015) assessed the impact of a writing aid tool based on the CONSORT statement (Schulz et al., 2010) on the completeness of reporting of RCTs. The tools was developed for six domains of the Methods section (trial design, randomization, blinding, participants, interventions, and outcomes) and consisted of reminders of the corresponding CONSORT item(s), bullet points enumerating the key elements to report, and good reporting examples. The tool was assessed in an RCT in which the participants were asked to write a Methods section of an article based on a trial protocol, either using the aid tool ('intervention' group) or without using the tool ('control' group). The results of 41 participants showed that the mean global score for reporting completeness was higher with the use of the tool than without it.

2.2 Aid tools for readers and reviewers

Kiritchenko et al. (2010) developed a system called ExaCT to automatically extract 21 key characteristics of clinical trial design, such as treatment names, eligibility criteria, outcomes, etc. ExaCT consists of an information extraction algorithm that looks for text fragments corresponding to the target information elements, a web-based user interface through which human experts can view and correct the suggested fragments.

The National Library of Medicine's Medical Text Indexer (MTI) is a system providing automatic recommendations based on the Medical Subject Headings (MeSH) terms for indexing medical articles (Mork et al., 2013). MTI is used to assist human indexers, catalogers, and NLM's History of Medicine Division in their work. Its use by indexers was shown to grow over years (used to index 15.75% of the articles 2002 vs 62.44% in 2014) and to improve the performance (precision, recall and F-measure) of indexers (Mork et al., 2017).

Marshall et al. (2015) addressed the task of automating assessment of risk of bias in clinical trials. Bias is phenomenon related to spin: it is a systematic error or a deviation from the truth in the results or conclusions that can cause an under- or overestimation of the effect of the examined treatment (Higgins and Green, 2008). The authors developed a system called RobotReviewer that used machine learning to assess an article for the risk of different types of bias and to extract text fragments that support these judgements. These works showed that automated risk of bias assessment can be achieve reasonable performance, and the extraction of supporting text fragments reached similar quality to that of human experts. Marshall et al. (2017) further developed RobotReviewer, adding functionality for extracting the PICO (Population, Interventions/Comparators, Outcomes) elements from articles and detecting study design (RCT), for the purpose of automated evidence synthesis. Soboczenski et al. (2019) assessed RobotReviewer in a user study involving 41 participants, evaluating time spent for bias assessment, text fragment suggestions by machine learning, and usability of the tool. Semi-automation in this study was shown to be quicker than manual assessment; 91% of the automated risk of bias judgments and 62% of supporting text suggestions were accepted by the human reviewers.

The cited works demonstrate that semi-automated aid tools can prove useful for both authors and readers/reviewers of medical articles and has a potential to improve the quality of the articles and facilitate the analysis of the texts.

3 Spin: definition and types

We adopt the definition and classification of spin introduced by Boutron et al. (2010) and Lazarus et al. (2015), who divided instances of spin into several types and subtypes.

We addressed the following types of spin:

1. Outcome switching – unjustified change of the pre-defined trial outcomes, leading to reporting only the favourable outcomes that support the hypothesis of the researchers (Goldacre et al., 2019). Outcome switching is one of the most common types of spin. It can consist in omitting the primary outcome in the results / conclusions of the abstract, or in the focus on significant secondary outcomes, e.g.:

   *The primary end point of this trial was **overall survival**. <...> This trial showed a significantly increased **R0 resection rate** although it failed to demonstrate a **survival** benefit.*

   In this example, the primary outcome ("overall survival"), the results for which were not favourable, is mentioned in the conclusion, but it is not reported in the first place and occurs within a concessive clause (starting by "although"). This way of reporting puts the focus on the other, favourable, outcome ("R0 resection rate").

2. Interpreting non-significant outcome as a proof of equivalence of the treatments, e.g.:

   *The median PFS was 10.3 months in the XELIRI and 9.3 months in the FOLFIRI arm (p = 0.78). Conclusion: The XELIRI regimen showed **similar PFS** compared to the FOLFIRI regimen.*

   The results for the outcome "median PFS" are not significant, which is often erroneously interpreted as a proof of similarity of the treatments. However, a non-significant result means that the null hypothesis of a difference could not be rejected, which is not equivalent to a demonstration of similarity of the treatments. This would require the rejection of the null hypothesis of a difference, or a substantial difference, in outcomes between treatments.

3. Focus on within-group comparisons, e.g.:

   ***Both groups showed robust improvement** in both symptoms and functioning.*

   The goal of randomized controlled trials is to compare two treatments with regard to some outcomes. If the superiority of the experimental treatment over the control treatment was not shown, within-group comparisons (reporting the changes within a group of patients receiving a treatment, instead of comparing patients receiving different treatments) can be used to persuade the reader of beneficial effects of the experimental treatment.

Two concepts are vital for spin detection and play a key role in our algorithms:

1. The primary outcome of a trial – the most important variable monitored during the trial to assess how the studied treatment impacts it. Primary outcomes are recorded in trial registries (open online databases storing the information about registered clinical trials), and should be defined in the text of clinical articles, e.g.:

   *The primary end point was **a difference of $> 20\%$ in the microvascular flow index of small vessels among groups**.*

2. Statistical significance of the primary outcome. Statistical hypothesis testing is used to check for a significant difference in outcomes between two patient groups, one receiving the experimental treatment and the other receiving the control treatment. Statistical significance is often reported as a P-value compared to predefined threshold, usually set to 0.05. Spin most often occurs when the results for the primary outcome are not significant (Boutron et al., 2010; Fleming, 2016; Khan et al., 2019; Vera-Badillo et al., 2016; Austin et al., 2018; Cooper et al., 2018; Kinder et al., 2018; Lockyer et al., 2013), although trials with significant effect on the primary outcome may also be prone to spin (Beijers et al., 2017).

Trial results are commonly reported as an effect on the (primary) outcome[3], along with the p-value.

***Microcirculatory flow indices of small and medium vessels** were **significantly** higher in the levosimendan group as compared to the control group ($p < 0.05$).*

Statistical significance levels of trial outcomes are vital for spin detection, as spin is commonly related to non-significant results for

---

[3]It is important to distinguish between the notions of outcome, effect and result in this context: an outcome is a measure/variable monitored during a clinical trial; effect refers to the change in an outcome observed during a trial; trial results refer to the set of effects for all measured outcomes.

the primary outcome, or to selective reporting of significant outcomes only.

4 Algorithms

Spin is a complex notion and thus detecting spin cannot be seen as a binary classification problem. We believe that the most viable approach to spin detection is to assess each (sub)type of spin separately. We aimed at developing algorithms to extract and analyse pieces of information relevant to the addressed types of spin. The extracted information and its analysis, provided by our tool, can help human experts in making the conclusion on presence or absence of spin of the given (sub)type.

Detection of spin and related information is a complex task which cannot by fully automated. Our system is designed as a semi-automated tool that finds potential instances of the addressed types of spin and extracts the supporting information that can help the user to make the final decision on the presence of spin. In this section, we present the algorithms currently included in the system, according to the types of spin that they are used to detect.

As we aim at detecting spin in the Results and Conclusions sections of articles' abstracts, we first need an algorithm analyzing the given article to detect its abstract and the Results and Conclusions sections within the abstract. We will not mention this algorithm in the list of algorithms for each spin type to avoid repetition. If we talk about extracting some information from the abstract, it implies that the text structure analysis algorithm was applied.

4.1 Outcome switching

We focus on the switching (change/omission) of the primary outcome. Primary outcome switching can occur at several points:

- the primary outcome(s) recorded in the trial registry can differ from the primary outcome(s) declared in the article;

- the primary outcome(s) declared in the abstract can differ from the primary outcome(s) declared in the body of the article;

- the primary outcome(s) recorded in the trial registry can be omitted when reporting the results for the outcomes in the abstract;

- the primary outcome(s) recorded in the article can be omitted when reporting the results for the outcomes in the abstract.

Primary outcome switching detection involves the following algorithms:

1. Identification of primary outcomes in trial registries and in the article's text.

2. Identification of reported outcomes from sentences reporting the results, e.g. (reported outcomes are in bold):

   *The results of this study showed that **symptom Scores** in massage group were improved significantly compared with control group, and the rate of **dyspnea**, **cough** and **wheeze** in the experimental group than the control group were reduced by approximately 45%, 56% and 52%.*

3. Assessment of semantic similarity of pairs of outcomes extracted by the above algorithms to check for missing outcomes. We perform the assessment for the following sets of outcomes:

   - The primary outcome extracted from the registry is compared to the primary outcome(s) declared in the article;
   - The primary outcome extracted from the abstract is compared to the primary outcome(s) declared in the body of the article;
   - The primary outcome extracted from the article is compared to the outcomes reported in the abstract;
   - The primary outcome extracted from the registry is compared to the outcomes reported in the abstract.

These assessments allow to detect switching of the primary outcome at all the possible stages. If the primary outcome in the registry and in the article, or in the abstract and body of the article differ, we conclude that there is potential outcome switching, which is reported to the user. Similarly, if the primary outcome (from the article or from the registry) is missing from the list of the reported outcomes, we suspect selective reporting of outcomes, and the system reports it to the user.

In the example on the page 3, the system should extract "overall survival" as the primary outcome, and "R0 resection rate" and "survival" as reported outcomes. The similarity between "overall survival" and "R0 resection rate" is low, while the similarity between "overall survival" and "survival" is high, thus, we conclude that the primary outcome "overall survival" is reported as "survival".

As semantic similarity often depends on the context, the conclusions of the system are presented to the user, who can check them to make the conclusions on correctness of the analysis.

4. Assessing the discourse prominence of the reported primary outcome (detected by the previous algorithms) by checking if it is reported the first place among all the outcomes; if it is reported in a concessive clause.

In the example above, the system will detect that the primary outcome "survival" is reported within a concessive clause (starting by "although") and will flag the sentence as potentially focusing on secondary outcomes.

4.2 Interpreting non-significant outcome as a proof of equivalence of the treatments

As we stated above, conclusions on the similarity/equivalence of the studies treatments are justified only if the trial was of non-inferiority or equivalence type. Thus, we employ two algorithms to detect this type of spin:

1. Identification of statements of similarity between treatments, e.g.:

   *Both products caused **similar** leukocyte counts diminution and had **similar** safety profiles.*

2. Identifying the markers of non-inferiority or equivalence trial design, e.g.:

   *ONCEMRK is a phase 3, multicenter, double-blind, **noninferiority** trial comparing raltegravir 1200mg QD with raltegravir 400mg BID in treatment-naive HIV-1–infected adults.*

If there is a statement of similarity of treatments while no markers of non-inferiority / equivalence design are found, we conclude the presence of spin and report it to the user.

4.3 Focus on within-group comparisons

Any statement in the results and conclusions of the abstract that presents a comparison of two states of a patient group without comparing it to another group is a within-group comparison. This type of spin is detected by a single algorithm that identifies within-group comparisons that are further reported to the user:

*Young Mania Rating Scale total scores **improved with ritanserin**.*

4.4 Other algorithms

We support extraction of some information that is not directly involved in the detection of spin, but that can help user in spin assessment and that can be used in the future when new spin types are added. The algorithms include:

1. Extraction of measures of statistical significance, both numerical and verbal (in bold):

   *Study group patients had a **significant** lower reintubation rate than did controls; six patients (17%) versus 19 patients (48%), **P<0.05**; respectively.*

2. Extraction of the relation between the reported outcomes and their statistical significance, extracted at the previous stages. For the example above, we extract pairs (”reintubation rate”, ”significant”) and (”reintubation rate”, ”P<0.05”).

   These algorithms, in combination with the assessment of semantic similarity of extracted outcomes, allows to identify the significance level for the primary outcome.

5 Methods

In this section, we briefly outline the methods used in our algorithms, the datasets used for evaluation, and the current performance of the algorithms. Our approach is based on some previous works for the related tasks. As the details on development of the algorithms, annotating the data and testing different approaches are described in detail in the corresponding articles, we limit ourselves here to only a brief description of the best-performing method that we selected for each task.

The methods we employ can be divided into two groups: machine learning, including deep learning, used for the core tasks for which we have sufficient training data, and rule-based methods, used for the simpler tasks or for tasks where we do not have enough data for machine learning.

5.1 Rule-based methods

We developed rules for the following tasks:

- To find the abstract, we use regular expressions rules that are evaluated on the set of 3938 PubMed Central (PMC)[4] articles in XML format with a specific tag for the abstract, used as the gold standard. To evaluate our algorithm, we applied it to the raw texts extracted from the XML files and compared the extracted abstracts to those obtained using the XML tag.

- To extract outcomes from trial registries, we use regular expressions to extract the trial registration number from the article; using it, we find on the web, download and parse the registry entry corresponding to the trial.

- To extract significance levels, we use rules based on regular expressions and token, lemma and pos-tag information.

- To assess the discourse prominence of an outcome, to detect statements of similarity between treatments, within-group comparisons and markers of non-inferiority design, we employ rules based on token, lemma and pos-tag information.

We annotated abstracts of 180 articles (2402 sentences) for similarity statements and within-group comparisons (Koroleva, 2020). The proportion of these types of statements in our corpus is low: we identified only 72 similarity statements and 127 within-group comparisons. The evaluation of statements of similarity between treatments and within-group comparisons was performed with two settings: 1) using the whole text of abstracts; 2) using only the Results and Conclusions sections of the abstract, which raised the precision, as expected (Table 1).

5.2 Machine learning methods

For the core tasks of our system, we either used an existing annotated corpus or annotated our own corpora. Our corpora were annotated by a single annotator (AK), consulted by consulted our medical advisors from the MiRoR network (Isabelle Boutron, Patrick Bossuyt and Liz Wager).

We tested several approaches for each task, including rule-based and machine-learning approaches (see details below). Overall, we found that the best performance on our tasks was shown

---

[4]https://www.ncbi.nlm.nih.gov/pmc/

Algorithm	Method	Annotated dataset	Precision	Recall	F1
Primary outcomes extraction	Deep learning	2,000 sentences / 1,694 outcomes	86.99	90.07	88.42
Reported outcomes extraction	Deep learning	1,940 sentences / 2,251 outcomes	81.17	78.09	79.42
Outcome similarity assessment	Deep learning	3,043 pairs of outcomes	88.93	90.76	89.75
Similarity statements extraction	Rules	180 abstracts / 2402 sentences whole abstract results and conclusions	 77.8 85.1	 87.5 87.5	 82.4 86.3
Within-group comparisons	Rules	180 abstracts / 2402 sentences whole abstract results and conclusions	 53.2 71.9	 90.6 90.6	 67.1 80.1
Abstract extraction	Rules	3938 abstracts	94.7	94	94.3
Text structure analysis: sections of abstract	Deep learning	PubMed200k	97.82	95.81	96.8
Extraction of significance levels	Rules	664 sentences / 1,188 significance level markers	99.18	96.58	97.86
Outcome - significance level relation extraction	Deep learning	2,678 pairs of outcomes and significance level markers	94.3	94	94

Table 1: Overview of algorithms, methods, results and annotated datasets

by a deep learning approach that was recently proved to be highly successful in many NLP applications. It employs language representations pretrained on large unannotated data and fine-tuned on a relatively small amount of annotated data for a specific downstream task. The language representations that we tested include: BERT (Bidirectional Encoder Representations from Transformers) models (Devlin et al., 2018), trained on a general-domain corpus of 3.3B words; BioBERT model (Lee et al., 2019), trained on the BERT corpus and a biomedical corpus of 18B words; and SciBERT models (Beltagy et al., 2019), trained on the BERT corpus and a scientific corpus of 3.1B words. For each task, we chose the best-performing model.

Details about the annotated datasets that we used and the tested approaches can be found below. The best results for each task are summarised in Table 1.

5.2.1 Identification of sections in the abstract

For identifying sections within the abstract (in particular, Results and Conclusions), we used the PubMed 200k dataset introduced in Dernoncourt and Lee (2017). This dataset contains approximately 200,000 abstracts of RCTs with 2.3 million sentences. Each sentence is annotated with one of the following classes, corresponding to the sections of the abstract: background, objective, method, result, or conclusion. We used the train-dev-test split provided by the developers of the dataset.

We compared a rule-based approach and BERT, SciBERT and BioBERT models, fine-tuned for the sentence classification task on the PubMed 200k dataset. The best performance was shown by the fine-tuned BioBERT model.

5.2.2 Outcome extraction

The outcome extraction task includes two subtasks: extracting primary and reported outcomes. For each subtask, we annotated a separate corpus. For primary outcome extraction, we annotated a corpus of 2,000 sentences, coming from 1,672 articles. The sentences were selected randomly, from both abstracts and full texts, without restriction to a particular medical domain. A total of 1,694 primary outcomes was annotated (Koroleva, 2019a). For reported outcome extraction, we annotated reported outcomes in the abstracts of articles for which we annotated the primary outcomes. The corpus contains 1,940 sentences from 402 articles, with a total of 2,251 reported outcomes (Koroleva, 2019a).

We compared a rule-based system and several machine learning algorithms for primary and reported outcome extraction. Details about the annotated datasets and the methods that we tested can be found in Koroleva et al. (EasyChair, 2020). We selected the best performing approach to be included in our tool

For primary outcomes extraction, the best performance was demonstrated by the BioBERT model fine-tuned for named entity recognition task on our corpus of 2,000 sentences annotated for primary outcomes. For reported outcomes extraction, the best performance was achieved by the SciBERT model fine-tuned for named entity recognition task on our corpus of 1,940 sentences annotated with reported outcomes.

5.2.3 Assessment of semantic similarity of outcomes

To annotate semantic similarity between outcomes, we used pairs of sentences from our corpora of outcomes: the first sentence in each pair comes from the corpus of primary outcomes, the second sentence comes from the corpus of reported outcomes, and both sentences are from the same article. We assigned a binary label of similarity (similar/dissimilar) to each pair of outcomes in each sentence pair. The corpus contains 3,043 pairs of outcomes (Koroleva, 2019b).

We tested several semantic similarity measures (string-based, lexical, vector-based) and the BERT, SciBERT and BioBERT models, fine-tuned for sentence pair classification task on the corpus of outcome pairs. Details on the corpus annotation and on the methods tested can be found in Koroleva et al. (2019). The best performance was shown by the fine-tuned BioBERT model.

5.2.4 Extraction of the relation between reported outcomes and statistical significance levels

To annotate the relation between reported outcomes and statistical significance levels, we selected sentences containing markers of statistical significance from the corpus annotated with reported outcomes. We annotated the pairs of outcomes and significance levels with a binary label ("positive": the significance level is related to the outcome; "negative": the significance level is not related to the outcome). The final corpus contains 663 sentences with 2,552 annotated relations (Koroleva, 2019c).

We tested several machine learning algorithms and the BERT, SciBERT and BioBERT model fine-tuned for the relation extraction task on the annotated corpus. The details on the corpus and the method can be found in Koroleva and Paroubek (2019). The best result for this task was achieved by the fine-tuned BioBERT model.

6 Interface

Our prototype system allows the user to load a text (with or without annotations), run algorithms, visualize their output, correct, add or remove annotations. The expected input is an article reporting an RCT in the text format, including the abstract.

Figure 1 shows the interface with an example of a processed text.

The main items of the drop-down menu on the top of the page are **Annotations**, allowing to visualize and manage the annotations, and **Algorithms**, allowing to run the described algorithms to detect potential spin and the related information. The text fragments identified by the algorithms can be highlighted in the text. When running the algorithms, a report is generated that contains the extracted information and its analysis by the tool (e.g. a mismatch between the outcomes in the text and in the trial registry; absence of the declared primary outcome among the reported outcomes in the abstract). The report is saved into the Metadata section of Annotations menu, which can be accessed through the interface, and can be exported to a file via the **Generate report** item of the Algorithms menu. Human experts can use this report to check the extracted information and the analysis performed by the tool, and to make a final decision on the presence/absence of a given type of spin.

7 Results and conclusions

The current functionality, methods in use, annotated datasets and the best achieved results are outlined in Table 1. Performance is assessed per-token for outcome and significance level extraction and per-unit for other tasks.

In this paper, we presented a first prototype tool for assisting authors and reviewers to detect spin and related information in abstracts of articles reporting RCTs. The employed algorithms show operational performance in complex semantic tasks, even with relatively low volume of available annotated data. We envisage two possible applications of our system: as an authoring aid or as peer-reviewing tool. The authoring aid version can be further developed into an educational tool, explaining the notion of spin and its types to the user.

Possible directions for future work include: improving the implementation and interface (adding prompts for interaction with the user; facilitating installation process), algorithms (improving current performance, adding detection of new spin

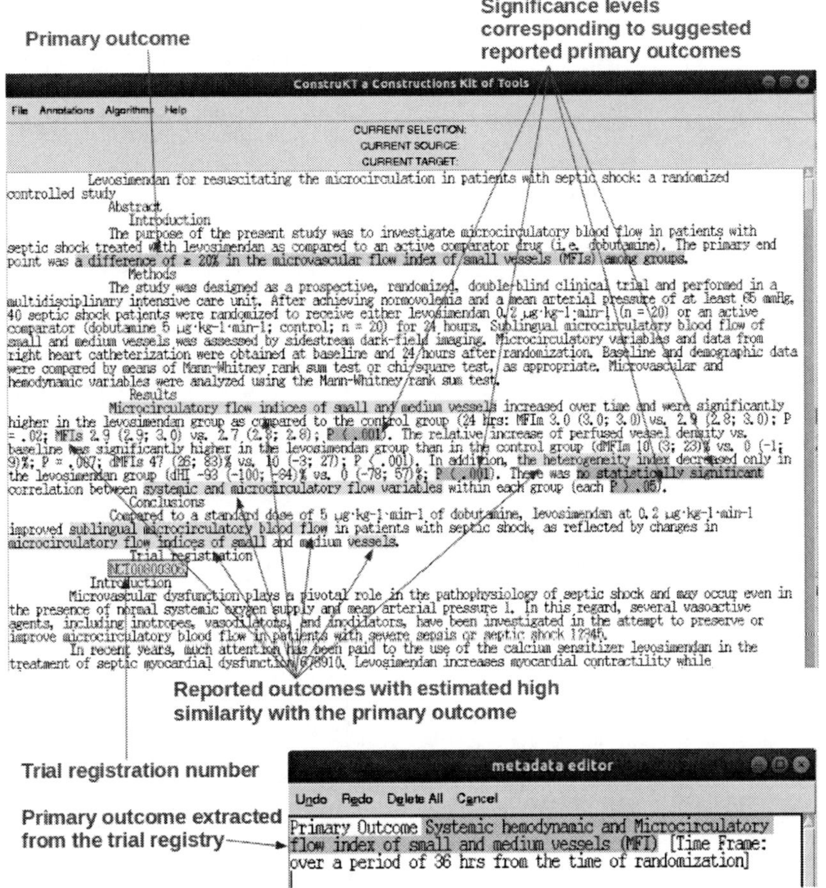

Figure 1: Example of a processed text

types), application (promoting the tool among the target audience; encouraging users to submit their manually annotated data, to be used to improve the algorithms), and optimization (parallel processing of multiple input text files). Our system can be easily incorporated into other text processing tools.

Another interesting yet challenging direction for the future work is detecting spin/distorted reporting in texts belonging to scientific domains other than biomedicine. First of all, a qualitative study of spin is needed to define and classify spin in each scientific domain (similar to the work of Boutron et al. (2010) and Lazarus et al. (2015) for clinical trials). To our best knowledge, there have been no attempts to conduct such a study for non-biomedical texts. It is therefore difficult to hypothesise whether spin-detection algorithms developed for texts reporting clinical trials could be applicable for other domains. It appears that the definition and the types of spin are domain-specific (e.g. outcome-related types of spin, prevalent in the biomedical domain, would not be relevant in domains that do not use the notion of outcome). Hence, we suppose that spin-detection algorithms are domain-specific as well and cannot be applied to other domains.

8 Availability

The proposed prototype tool and associated models are available at:
https://github.com/aakorolyova/DeSpin.

Acknowledgments

This project has received funding from the European Union's Horizon 2020 research and innovation programme under the Marie Sklodowska-Curie grant agreement No 676207.

References

Sophia Ananiadou, Brian Rea, Naoaki Okazaki, Rob Procter, and James Thomas. 2009. Supporting systematic reviews using text mining. *Social Science Computer Review - SOC SCI COMPUT REV*, 27:509–523.

Jennifer Austin, Christopher Smith, Kavita Natarajan, Mousumi Som, Cole Wayant, and Matt Vassar. 2018. Evaluation of spin within abstracts in obesity randomized clinical trials: A cross-sectional review: Spin in obesity clinical trials. *Clinical Obesity*, 9:e12292.

Caroline Barnes, Isabelle Boutron, Bruno Giraudeau, Raphael Porcher, Douglas Altman, and Philippe Ravaud. 2015. Impact of an online writing aid tool for writing a randomized trial report: The cobweb (consort-based web tool) randomized controlled trial. *BMC medicine*, 13:221.

Lian Beijers, Bertus F. Jeronimus, Erick H. Turner, Peter de Jonge, and Annelieke M. Roest. 2017. Spin in rcts of anxiety medication with a positive primary outcome: a comparison of concerns expressed by the us fda and in the published literature. *BMJ Open*, 7(3).

Iz Beltagy, Arman Cohan, and Kyle Lo. 2019. Scibert: Pretrained contextualized embeddings for scientific text. *arXiv preprint arXiv:1903.10676*.

Isabelle Boutron, Douglas Altman, Sally Hopewell, Francisco Vera-Badillo, Ian Tannock, and Philippe Ravaud. 2014. Impact of spin in the abstracts of articles reporting results of randomized controlled trials in the field of cancer: the SPIIN randomized controlled trial. *Journal of Clinical Oncology*.

Isabelle Boutron, Susan Dutton, Philippe Ravaud, and Douglas Altman. 2010. Reporting and interpretation of randomized controlled trials with statistically nonsignificant results for primary outcomes. *JAMA*.

Craig M. Cooper, Harrison M. Gray, Andrew E. Ross, Tom A. Hamilton, Jaye B. Downs, Cole Wayant, and Matt Vassar. 2018. Evaluation of spin in the abstracts of otolaryngology randomized controlled trials: Spin found in majority of clinical trials. *The Laryngoscope*.

Franck Dernoncourt and Ji Young Lee. 2017. PubMed 200k RCT: a dataset for sequential sentence classification in medical abstracts. In *8th IJCNLP (Volume 2: Short Papers)*, pages 308–313, Taipei, Taiwan. Asian Federation of NLP.

Jacob Devlin, Ming-Wei Chang, Kenton Lee, and Kristina Toutanova. 2018. BERT: pre-training of deep bidirectional transformers for language understanding. *CoRR*, abs/1810.04805.

Padhraig S. Fleming. 2016. Evidence of spin in clinical trials in the surgical literature. *Ann Transl Med.*, 4,19(385).

Ben Goldacre, Henry Drysdale, Aaron Dale, Ioan Milosevic, Eirion Slade, Philip Hartley, Cicely Marston, Anna Powell-Smith, Carl Heneghan, and Kamal R. Mahtani. 2019. Compare: a prospective cohort study correcting and monitoring 58 misreported trials in real time. *Trials*, 20(1):118.

Romana Haneef, Clement Lazarus, Philippe Ravaud, Amelie Yavchitz, and Isabelle Boutron. 2015. Interpretation of results of studies evaluating an intervention highlighted in google health news: a cross-sectional study of news. *PLoS ONE*.

Julian P. Higgins and Sally Green, editors. 2008. *Cochrane handbook for systematic reviews of interventions*. Wiley & Sons Ltd., West Sussex.

Minlie Huang, Aurélie Névéol, and Zhiyong Lu. 2011. Recommending mesh terms for annotating biomedical articles. *Journal of the American Medical Informatics Association : JAMIA*, 18:660–7.

John Ioannidis. 2005. Why most published research findings are false. *PLoS medicine*, 2:e124.

Muhammad Khan, Noman Lateef, Tariq Siddiqi, Karim Abdur Rehman, Saed Alnaimat, Safi Khan, Haris Riaz, M Hassan Murad, John Mandrola, Rami Doukky, and Richard Krasuski. 2019. Level and prevalence of spin in published cardiovascular randomized clinical trial reports with statistically nonsignificant primary outcomes: A systematic review. *JAMA Network Open*, 2:e192622.

N.C. Kinder, M.D. Weaver, Cole Wayant, and Matt Vassar. 2018. Presence of 'spin' in the abstracts and titles of anaesthesiology randomised controlled trials. *British Journal of Anaesthesia*, 122.

Svetlana Kiritchenko, Berry de Bruijn, Simona Carini, Joel D. Martin, and Ida Sim. 2010. Exact: automatic extraction of clinical trial characteristics from journal publications. *BMC Med Inform Decis Mak.*

Anna Koroleva. 2019a. MiRoR11 - P2 - Annotated corpus for primary and reported outcomes extraction.

Anna Koroleva. 2019b. MiRoR11 - P2 - Annotated corpus for semantic similarity of clinical trial outcomes.

Anna Koroleva. 2019c. MiRoR11 - P2 - Annotated corpus for the relation between reported outcomes and their significance levels.

Anna Koroleva. 2020. MiRoR11 - P2 - Annotated dataset for spin-related types of statements (statements of similarity and within-group comparisons).

Anna Koroleva, Sanjay Kamath, and Patrick Paroubek. 2019. Measuring semantic similarity of clinical trial outcomes using deep pre-trained language representations. *Journal of Biomedical Informatics: X*, 4:100058.

Anna Koroleva, Sanjay Kamath, and Patrick Paroubek. EasyChair, 2020. Extracting outcomes from articles reporting randomized controlled trials using pretrained deep language representations. EasyChair Preprint no. 2940.

Anna Koroleva and Patrick Paroubek. 2019. Extracting relations between outcomes and significance levels in randomized controlled trials (RCTs) publications. In *Proceedings of the 18th BioNLP Workshop and Shared Task*, pages 359–369, Florence, Italy. Association for Computational Linguistics.

Clément Lazarus, Romana Haneef, Philippe Ravaud, and Isabelle Boutron. 2015. Classification and prevalence of spin in abstracts of non-randomized studies evaluating an intervention. *BMC Med Res Methodol.*

Jinhyuk Lee, Wonjin Yoon, Sungdong Kim, Donghyeon Kim, Sunkyu Kim, Chan Ho So, and Jaewoo Kang. 2019. Biobert: a pre-trained biomedical language representation model for biomedical text mining. *arXiv preprint arXiv:1901.08746.*

Suzanne Lockyer, Robert Willard Hodgson, Jo C. Dumville, and Nicky Cullum. 2013. ”Spin” in wound care research: the reporting and interpretation of randomized controlled trials with statistically non-significant primary outcome results or unspecified primary outcomes. In *Trials.*

Iain Marshall, Joël Kuiper, Edward Banner, and Byron C. Wallace. 2017. Automating biomedical evidence synthesis: RobotReviewer. In *Proceedings of ACL 2017, System Demonstrations*, pages 7–12, Vancouver, Canada. Association for Computational Linguistics.

Iain James Marshall, Joël Kuiper, and Byron C. Wallace. 2015. Robotreviewer: Evaluation of a system for automatically assessing bias in clinical trials. *Journal of the American Medical Informatics Association : JAMIA*, 23.

James Mork, Alan Aronson, and Dina Demner-Fushman. 2017. 12 years on – is the NLM medical text indexer still useful and relevant? *Journal of Biomedical Semantics*, 8(1).

James G. Mork, Antonio Jimeno-Yepes, and Alan R. Aronson. 2013. The NLM medical text indexer system for indexing biomedical literature. *CEUR Workshop Proceedings*, 1094.

Zainab Samaan, Lawrence Mbuagbaw, Daisy Kosa, Victoria Borg Debono, Rejane Dillenburg, Shiyuan Zhang, Vincent Fruci, Brittany Dennis, Monica Bawor, and Lehana Thabane. 2013. A systematic scoping review of adherence to reporting guidelines in health care literature. *Journal of multidisciplinary healthcare*, 6:169–88.

Kenneth F. Schulz, Douglas G. Altman, and David Moher. 2010. Consort 2010 statement: updated guidelines for reporting parallel group randomised trials. *BMJ*, 340.

Frank Soboczenski, Thomas Trikalinos, Joël Kuiper, Randolph G. Bias, Byron Wallace, and Iain J. Marshall. 2019. Machine learning to help researchers evaluate biases in clinical trials: A prospective, randomized user study. *BMC Medical Informatics and Decision Making*, 19.

Francisco E. Vera-Badillo, Marc Napoleone, Monika K. Krzyzanowska, Shabbir M.H. Alibhai, An-Wen Chan, Alberto Ocana, Bostjan Seruga, Arnoud J. Templeton, Eitan Amir, and Ian F. Tannock. 2016. Bias in reporting of randomised clinical trials in oncology. *European Journal of Cancer*, 61:29 – 35.

Amélie Yavchitz, Isabelle Boutron, Aida Bafeta, Ibrahim Marroun, Pierre Charles, Jean J. C. Mantz, and Philippe Ravaud. 2012. Misrepresentation of randomized controlled trials in press releases and news coverage: a cohort study. *PLoS Med.*

Towards Visual Dialog for Radiology

Olga Kovaleva‡*, Chaitanya Shivade†*, Satyananda Kashyap⋆, Karina Kanjaria⋆,
Adam Coy⋆, Deddeh Ballah⋆, Joy Wu⋆, Yufan Guo⋆, Alexandros Karargyris⋆,
David Beymer⋆, Anna Rumshisky‡, Vandana Mukherjee⋆
‡ University of Massachussets, Lowell † Amazon
⋆ IBM Almaden Research Center.

Abstract

Current research in machine learning for radiology is focused mostly on images. There exists limited work in investigating intelligent interactive systems for radiology. To address this limitation, we introduce a realistic and information-rich task of Visual Dialog in radiology, specific to chest X-ray images. Using MIMIC-CXR, an openly available database of chest X-ray images, we construct both a synthetic and a real-world dataset and provide baseline scores achieved by state-of-the-art models. We show that incorporating medical history of the patient leads to better performance in answering questions as opposed to conventional visual question answering model which looks only at the image. While our experiments show promising results, they indicate that the task is extremely challenging with significant scope for improvement. We make both the datasets (synthetic and gold standard) and the associated code publicly available to the research community.

1 Introduction

Answering questions about an image is a complex multi-modal task demonstrating an important capability of artificial intelligence. A well-defined task evaluating such capabilities is Visual Question Answering (VQA) (Antol et al., 2015) where a system answers free-form questions reasoning about an image. VQA demands careful understanding of elements in an image along with intricacies of the language used in framing a question about it. Visual Dialog (VisDial) (Das et al., 2017; de Vries et al., 2016) is an extension to the VQA problem, where a system is required to engage in a dialog about the image. This adds significant complexity to VQA where a system should now be able to associate the question in the image, and reason over additional information gathered from previous question answers in the dialog.

Although limited work exploring VQA in radiology exists, VisDial in radiology remains an unexplored problem. With the healthcare setting increasingly requiring efficiency, evaluation of physicians is now based on both the quality and the timeliness of patient care. Clinicians often depend on official reports of imaging exam findings from radiologists to determine the appropriate next step. However, radiologists generally have a long queue of imaging studies to interpret and report, causing subsequent delay in patient care (Bhargavan et al., 2009; Siewert et al., 2016). Furthermore, it is common practice for clinicians to call radiologists asking follow-up questions on the official reporting, leading to further inefficiencies and disruptions in the workflow (Mangano et al., 2014).

Visual dialog is a useful imaging adjunct that can help expedite patient care. It can potentially answer a physician's questions regarding official interpretations without interrupting the radiologist's workflow, allowing the radiologist to concentrate their efforts on interpreting more studies in a timely manner. Additionally, visual dialog could provide clinicians with a preliminary radiology exam interpretation prior to receiving the formal dictation from the radiologist. Clinicians could use the information to start planning patient care and decrease the time from the completion of the radiology exam to subsequent medical management (Halsted and Froehle, 2008).

In this paper, we address these gaps and make the following contributions: 1) we introduce construction of RadVisDial - the first publicly available dataset for visual dialog in radiology, derived from the MIMIC-CXR (Johnson et al., 2019) dataset, 2) we compare several state-of-the-art models for VQA and VisDial applied to these images, and 3) we conduct a comprehensive set of experiments

---

* Equal contribution, Work done at IBM Research

Proceedings of the BioNLP 2020 workshop, pages 60–69
Online, July 9, 2020 ©2020 Association for Computational Linguistics

highlighting different challenges of the problem and propose solutions to overcome them.

2 Related Work

Most of the large publicly available datasets (Kaggle, 2017; Rajpurkar et al., 2017) for radiology consist of images associated with a limited amount of structured information. For example, Irvin et al. (2019); Johnson et al. (2019) make images available along with the output of a text extraction module that produces labels for 13 abnormalities in a chest X-ray. Of note recently, the task of generating reports from radiology images has become popular in the research community (Jing et al., 2018; Wang et al., 2018). Two recent shared tasks at Image-CLEF explored the VQA problem with radiology images (Hasan et al., 2018; Abacha et al., 2019). Lau et al. (2018) also released a small dataset VQA-RAD for the specific task.

The first VQA shared task at ImageCLEF (Hasan et al., 2018) used images from articles at PubMed Central. While Abacha et al. (2019) and Lau et al. (2018) use clinical images, the sizes of these datasets are limited. They are a mix of several modalities including 2D modalities such as X-rays, and 3D modalities such as ultrasound, MRI, and CT scans. They also cover several anatomic locations from the brain to the limbs. This makes a multimodal task with such images overly challenging, with shared task participants developing separate models (Al-Sadi et al., 2019; Abacha et al., 2018; Kornuta et al., 2019) to first address these subtasks (such as modality detection) before actually solving the problem of VQA.

We address these limitations and build up on MIMIC-CXR (Johnson et al., 2019) the largest publicly available dataset of chest X-rays and corresponding reports. We focus on the problem of visual dialog for a single modality and anatomy in the form of 2D chest X-rays. We restrict the number of questions and generate answers for them automatically which allows us to report results on a large set of images.

3 Data

3.1 MIMIC-CXR

The MIMIC-CXR dataset[1] consists of 371,920 chest X-ray images in the Digital Imaging and Communications (DICOM) format along with 206,576 reports. Each report is well structured and typically consists of sections such as `Medical Condition`, `Comparison`, `Findings`, and `Impression`. Each report can map to one or more images and each patient can have one or more reports. The images consist of both frontal and lateral views. The frontal views are either anterior-posterior (AP) or posterior-anterior (PA). The initial release of data also consists of annotations for 14 labels (13 abnormalities and one `No Findings` label) for each image. These annotations are obtained by running the CheXpert labeler (Irvin et al., 2019); a rule-based NLP pipeline against the associated report. The labeler output assigns one of four possibilities for each of the 13 abnormalities: {*yes, no, maybe, not mentioned in the report*}.

3.2 Visual Dialog dataset construction

Every training record of the original VisDial dataset (Das et al., 2017) consists of three elements: an image I, a caption for the image C, and a dialog history H consisting of a sequence of ten question-answer pairs. Given the image I, the caption C, a possibly empty dialog history H, and a follow-up question q, the task is to generate an answer a where $\{q, a\} \in H$. Following the original formulation, we synthetically create our dataset using the plain text reports associated with each image (this synthetic dataset will be considered to be silver-standard data for the experiments described in section 5). The `Medical Condition` section of the radiology report is a single sentence describing the medical history of the patient. We treat this sentence from the `Medical Condition` section as the *caption* of the image. We use NegBio (Peng et al., 2018) for extracting sections within a report.

We discard all images that do not have a medical condition in their report. Further, each CheXpert label is formulated as a question probing the presence of a disorder, and the output from the labeler is treated as the corresponding answer. Thus, ignoring the `No Findings` label, there are 52 possible question-answer pairs as a result of 13 questions and 4 possible answers.

We decided to focus on PA images for most of our experiments as this is the most informative view for chest X-rays, according to our team radiologists. The original VisDial dataset (Das et al., 2017) consists of ten questions per dialog and one dialog per image. Since we only have a set of 13

---

[1] https://physionet.org/content/mimic-cxr/1.0.0/

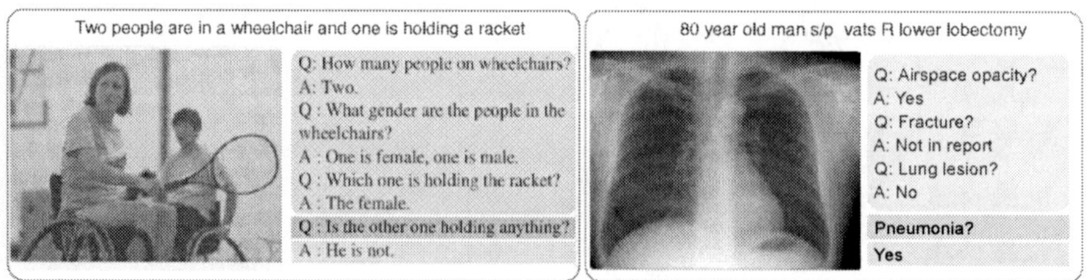

Figure 1: Comparison of VisDial 1.0 (left) with our synthetically constructed dataset (right).

possible questions, we limit the length of the dialog to 5 randomly sampled questions. The resulting dataset has 91060 images in the PA view (with train/validation/test splits containing 77205, 7340 and 6515 images, respectively). This synthetic data will be made available through the MIMIC Derived Data Repository.[2] Thus any individual with access to MIMIC-CXR will have access to our data. Figure 1 shows an example from our dataset and how it compares with one from VisDial 1.0.

3.3 Evaluation

The questions in our dataset are limited to probing the presence of an abnormality in a chest X-ray. Similarly, the answers are limited to one of the four choices. Owing to the restricted nature of the problem, we deviate from the evaluation protocol outlined in (Das et al., 2017) and instead calculate the F1-score for each of the four answers. We also report a macro-averaged F1 score across the four answers to make model comparisons easier.

4 Models

For our experiments, we selected a set of models designed for image-based question answering tasks. Namely, we experimented with three architectures: Stacked Attention Network (SAN) (Yang et al., 2016), Late Fusion Network (LF) (Das et al., 2017), and Recursive Visual Attention Network (RVA) (Niu et al., 2019). Following the original VisDial study (Das et al., 2017), we use an encoder-decoder structure with a discriminative decoder for each of the models. Below we give an overview of all the three algorithms.

4.1 Stacked Attention Network

The original configuration of SAN was introduced for the general-domain VQA task. The model performs multi-step reasoning by refining question-

<hr>

[2]https://physionet.org/physiotools/
mimic-code/HEADER.shtml

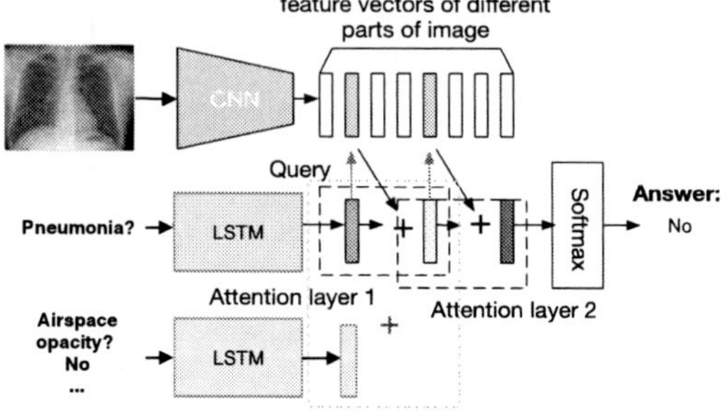

Figure 2: The modified architecture of the SAN model (image taken from (Yang et al., 2016)). The proposed modification shown in orange incorporates the history of dialog turns in the same way as the question through an LSTM. In our ablation experiments the changed part either reduces to encoding an image caption only or gets cut completely.

guided attention over image features in an iterative manner. The attended image features are then combined with the question features for answer prediction. SAN has been successfully adapted for medical VQA tasks such as VQA-RAD (Lau et al., 2018) and VQA-Med task of the ImageCLEF 2018 challenge (Ionescu et al., 2018). In our setup, we use a stack of two image attention layers and an LSTM-based question representation.

To take the dialog history into account and therefore adjust the SAN model for the needs of the Visual Dialog task, we modify the first image attention layer of the network by adding a term for LSTM representation of the history. This modification forces the image attention weights to become both question- and history-guided (see Figure 2).

4.2 Late Fusion Network

Proposed by (Das et al., 2017) as a baseline model for the Visual Dialog task, Late Fusion Network encodes the question and the dialog history through two separate RNNs, and the image through a CNN. The resulting representations are simply concatenated in a single vector, which is then used by a decoder for predicting the answer. We use this model unchanged, as released in the original Visual Dialog challenge.

4.3 Recursive Visual Attention

This model is the winner of the 2019 Visual Dialog challenge[3]. It recursively browses the past history of dialog turns until the current question is paired with the turn containing the most relevant information. This strategy is particularly useful for resolving co-references, naturally occurring in general-domain dialog questions. As previously, we do not modify the architecture of the model.

5 Experiments

This section presents our down-sampling strategy, gives details about conducted ablation studies, and describes experiments with various representations of images and texts.

5.1 Downsampling

A closer analysis of our data showed that the majority of the reports processed by the CheXpert labeler resulted in no mention of most of the 13 pathologies. This presented a heavily skewed dataset that would lead to a biased model instead of true visual understanding. This issue is not unique to radiology; it is observed even in the current benchmarks for VQA, and attempts have been made to mitigate the resulting problems (Hudson and Manning, 2019; Zhang et al., 2016; Agrawal et al., 2018).

In order to dissuade the answer biases, we performed data balancing, specifically by downsampling major labels in our dataset. As mentioned above, the CheXpert labeler outputs four possible answers for 13 labels. To investigate the skew in the data, we plotted a distribution of the 52 question-answer pairs (Figure 3). Further, we downsampled the question-answer pairs to fit a smoother answer distribution with the method presented in GQA based on the Earth Mover's Distance method (Hudson and Manning, 2019; Rubner et al., 2000). We iterated over the 52 pairs in decreasing frequency order and downsampled the categories belonging to the skewed head of the distribution. The relative label ranks by frequency remained the same for the balanced sets as with the unbalanced sets. For example, the pairs { *'Other pleural findings'* → *'Not*

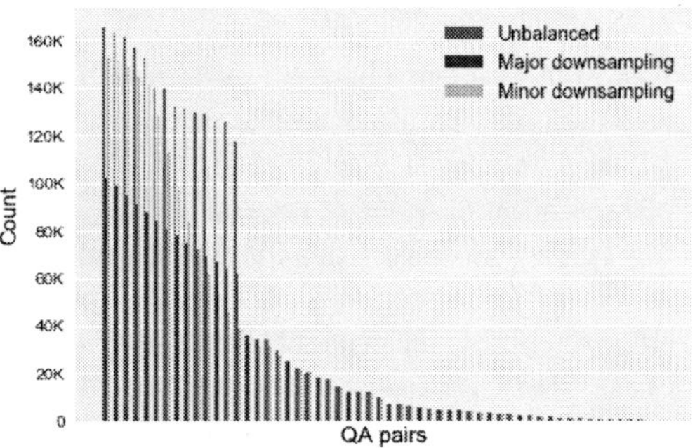

Figure 3: Downsampling strategies. Every bar along the X axis represents a single question-answer pair, where questions (13 in total) and answers (4 in total) are obtained through CheXpert.

in report' } and { *'Fracture'* → *'Not in report'* } remained the first and second largest counts in both the unbalanced and downsampled versions of the datasets. To reduce the disparity between dominant and underrepresented categories, we tuned the parameters outlined in (Hudson and Manning, 2019). We experimented with two different sets of parameter values and obtained two datasets with more balanced question-answer distributions. We further refer to them as "minor" and "major" downsampling, reflecting the total amount of data reduced (shown in blue and gray in Figure 3).

5.2 Evaluating importance of context

To assess the importance of the dialog context for question answering, we compare the performance of different variations of the Stacked Attention Network, selected as the best-performing model in the previous experiment (see subsection 6.1). In particular, we examine three scenarios: (a) the model makes a prediction based solely on a given image (essentially solving the VQA task rather than the Visual Dialog task), (b) the model makes its prediction given an image and its caption, and (c) the model makes its prediction given an image, a caption, and a history of question-answer pairs. Similar to the model modifications described in subsection 4.1 and Figure 2, we achieve the goal through experimenting with the SAN model by changing its first image attention layer to accordingly take in (a) question and image features, (b) question, image, and caption features, and (c) question, image, and full dialog history features.

---

[3]https://visualdialog.org/challenge/2019

5.3 Image representations

We test three approaches for pre-trained image representations. The first approach uses a ResNet-101 architecture (He et al., 2016) for multiclass classification of input X-ray images into 14 finding labels extracted from the associated reports (as described in section 3.2). Our second method aims to replicate the original CheXpert study (Irvin et al., 2019). Here we use a DenseNet-121 image classifier trained for prediction of five pre-selected and clinically important labels, namely, atelectasis, cardiomegaly, consolidation, edema, and pleural effusion. In both ResNet and DenseNet-based approaches we take the features obtained from the last pooling layer.

Finally, we adopted a bottom-up mechanism for image region proposal introduced by Anderson et al. (2018). More specifically, we first trained a neural network predicting bounding boxes for the image regions, corresponding to a set of 11 handcrafted clinical annotations adopted from an existing chest X-ray dataset[4]. We then represented every region as a latent feature vector of a trained patch-wise convolution autoencoder, and (3) concatenated all the obtained vectors to represent the entire image.

Based on the results of the experiment (subsection 6.3), we found that ResNet-101 image vectors yielded the best performance, so we used them in other experiments.

5.4 Effect of incorporating a lateral view

One of the crucial aspects of X-ray radiography exams is to capture the subject from multiple views. Typically, in case of chest X-rays, radiologists order an additional lateral view to confirm and locate findings that are not clearly visible from a frontal (PA or AP) view. We test whether the VisDial models are able to leverage the additional visual information offered by a lateral (LAT) view. We filter the data down to the patients whose chest X-ray exams had both a frontal and lateral views and re-sample the resulting data-set into train (52952 PA and 8086 AP images), validation (6614 PA and 964 AP images), and test (6508 PA and 1035 AP images). We train a separate ResNet-101 model for each of the three views on this re-sampled data using the method described in the previous section. The vector representations of a frontal view and the corresponding lateral view are concatenated as an aggregate image representation.

5.5 Text representations

Finally, we investigate the best way for representing the textual data by incorporating different pre-trained word vectors. More specifically, we measure the performance of our best-performing SAN model reached with (a) randomly initialized word embeddings trained jointly with the rest of the models, (b) domain-independent GloVe Common Crawl embeddings (Pennington et al., 2014), and (c) domain-specific fastText embeddings trained by (Romanov and Shivade, 2018). The latter are initialized with GloVe embeddings trained on Common Crawl, followed by training on 12M PubMed abstracts, and finally on 2M clinical notes from MIMIC-III database (Johnson et al., 2016). In all the experiments, we use 300-dimensional word vectors. We also experimented with transformer-based contextual vectors using BERT (Devlin et al., 2019). More specifically, instead of using LSTM representations of the textual data, we extracted the last layer vectors from ClinicalBERT (Alsentzer et al., 2019) pre-trained on MIMIC notes, and averaged them over input sequence tokens.

5.6 Question order

In a visual dialog setting, a model is conditioned on the image vector, the image caption, and the dialog history to predict the answer to a new question. We hypothesized that a model should be able to answer later questions in a dialog better since it has more information from the previous questions and their answers. As described in Section 3.2, we randomly sample 5 questions out of 13 possible choices to construct a dialog. We re-ordered the question-answer pairs in the dialog to reflect the order in which the corresponding abnormality label mentions occurred in the report. However, results for questions ordered based on their occurrence in the narrative did not vary from the setup with a random order of questions.

6 Results

We report macro-averaged F1-scores achieved on the same unbalanced validation set for each of the experiments. When experimenting with different configurations of the same model, we also break down the aggregate score to the F1 scores for individual answer options.

---

[4]https://www.kaggle.com/c/
rsna-pneumonia-detection-challenge

Model	'Yes'	'No'	'Maybe'	'Not in report'	Macro F1
SAN (VQA)	0.24	0	0.09	0.84	0.29
SAN (caption only)	0.30	0.09	0.09	0.81	0.33
SAN (full history)	0.22	0.26	0.04	0.83	**0.34**

Table 1: Ablation experiments. Per-answer F1-scores along with the macro F1-score are shown for tested SAN configurations.

6.1 Downsampling

Our results show (Table 2) consistent improvement of the scores across all the models as the training data becomes more balanced. All the models yielded comparable scores, with SAN being slightly better than other models (0.34 against 0.33 macro F1-score). Later in our experiments, we used the major down sampled version of the data-set.

Model	Unbalanced	Downsampled	
		Minor	Major
SAN	0.25	0.28	0.34
LF	0.28	0.31	0.33
RvA	0.24	0.33	0.33

Table 2: Data balancing experiments. Macro F1 scores are reported for every tested model.

6.2 Evaluating importance of context

One of the main findings of our study revealed the importance of contextual information for answering questions about a given image. As shown in Table 1, adding the image caption and the history of turns results in incremental increases of macro F1-scores. Notably, the VQA setup in which the model relies on the image only, it fails to detect the 'No' answer, whereas the history-aware configuration leads to a significant performance gain for this particular label. As expected and due to the skewed nature of the data-set, the highest and the lowest per-label scores were achieved for the most and the least frequent labels ('Not in report' and 'Maybe'), respectively.

6.3 Image representation

Out of the tested image representations, ResNet-derived vectors perform consistently better than the other approaches (see Table 3). Although in our DenseNet-121 image classification pre-training we were able to replicate the performance of (Irvin et al., 2019), the Visual Dialog scores for the corresponding vectors turned out to be lower. We believe this might be due to the fact that, by design, the network uses a limited set of pre-training classes not sufficient to generalize well to a full set of diseases used in the Visual Dialog task.

Model	DenseNet-121	Region Proposal	ResNet
SAN	0.27	0.29	**0.34**
LF	**0.33**	0.31	**0.33**
RvA	0.29	0.32	**0.33**

Table 3: Comparative performance (macro-F1) of Visual Dialog models on the test set with different image representations.

6.4 Effect of incorporating a lateral view

As expected, for both variations of the frontal view (i.e. AP and PA) appending lateral image vectors enhanced the performance of the tested SAN model (see Table 4). This suggests that lateral and frontal image vectors complement each other, and the models can benefit from using both. However, in our data-set only a subset of reports has both views available, which significantly reduces the amount of training data.

6.5 Word embeddings

Another observation from our experiments is that domain-specific pre-trained word embeddings contribute to better scores (see Table 5). This is due to the fact that domain-specific embeddings contain medical knowledge that helps the model make more justified predictions.

When using BERT, we did not notice gains in performance, which most likely means that the last-layer averaging strategy is not optimal and more sophisticated approaches such as (Xiao, 2018) are required . Alternatively, the final representation of the CLS can be used to represent input text.

View		'Yes'	'No'	'Maybe'	'Not in report'	Macro F1
AP+LAT	AP	0.40	0.21	0.12	0.79	0.381
	LAT	0.41	0.23	0.13	0.75	0.379
	AP + LAT	0.41	0.22	0.12	0.79	**0.385**
PA+LAT	PA	0.30	0.30	0.08	0.88	0.392
	LAT	0.32	0.32	0.07	0.86	0.391
	PA + LAT	0.32	0.34	0.06	0.87	**0.396**

Table 4: Effect of adding the lateral view to a frontal view (AP and PA).

Embedding	'Yes'	'No'	'Maybe'	'Not in report'	Macro F1
Random	0.26	0.22	0.04	0.73	0.31
GloVe (common crawl)	0.27	0	0.09	0.80	0.29
fastText (MedNLI)	0.24	0.22	0.07	0.84	**0.33**

Table 5: Comparative performance of the SAN model with different word embeddings.

7 Comparison with the gold-standard data

To complement our experiments with the silver data and investigate the applicability of the trained models to real-world scenarios, we also collected a set of gold standard data which consisted of two expert radiologists having a dialog about a particular chest X-ray. These X-ray images were randomly sampled PA views from the test our data. In this section, we present the data collection workflow, outline the associated challenges, compare the resulting data-set with the silver-standard, and report the performance of trained models.

7.1 Gold Standard Data Collection

We laid the foundations for our data collection in a manner similar to that of the general visual dialog challenge (Das et al., 2017). Two radiologists, designated as a "questioner" and an "answerer", conversed with each other following a detailed annotation guideline created to ensure consistency. The "answerer" in each scenario was provided with an image and a caption (medical condition). The "questioner" was provided with only the caption, and tasked with asking follow-up questions about the image, visible only to the "answerer". In order to make the gold data-set comparable to the silver-standard one, we restricted the beginning of each answer to contain a direct response of *'Yes'*, *'No'*, *'Maybe'*, or *'Not mentioned'*. In our annotation guidelines *'Not mentioned'* referred to the lack of evidence of the given medical condition that was asked by the "questioner" radiologist. The answer was elaborated with additional information if the radiologists found it necessary. The whole data collection procedure resulted in 100 annotated dialogs.

7.2 Gold standard results

Following the gold standard data collection, we performed some preliminary analyses with the best silver standard SAN model. Our gold standard data was split into train (70), validation (20), and test (10) sets. We experimented with three setups: (a) evaluating the silver-data trained networks on the gold standard data, (b) training and evaluating the models on the gold data, and (c) fine-tuning the silver-data trained networks on the gold standard data. Table 6 shows the results of these experiments. We found the best macro-F1 score of 0.47 was achieved by the silver data-trained SAN network fine-tuned on the gold standard data. We observed that the model could not directly predict any of the classes if directly evaluated on the gold data-set, suggesting that it was trained to fit the data patterns significantly different from those present in the collected data-set. However, pre-training on the silver data serves as a good starting point for further model fine-tuning. The obtained scores in general imply that there are many differences between the gold and silver data, including their vocabularies, answer distributions, and level of question detail.

7.3 Comparison of gold and silver data

To provide a meaningful analysis of the sources of difference between the gold and silver datasets,

Train data	'Yes'	'No'	Macro F1
Silver	0.00	0.00	0.00
Gold	0.27	0.77	0.35
Silver+gold	0.60	0.82	**0.47**

Table 6: Comparative performance of the SAN model trained on different combinations of silver and gold data, and evaluated on the test subset of gold data. Note that the gold annotations did not contain *'Not in report'* and *'Maybe'* options.

we grouped the gold questions semantically by using the CheXpert vocabulary for the 13 labels used for the construction of the silver dataset. The gold questions that are unable to be grouped via CheXpert were mapped manually using expert clinical knowledge. We systematically compared the gold and silver dialogs on the same 100 chest X-rays and noted the following differences.

- **Frequency of semantically equivalent questions**. Just under half of the gold question types were semantically covered by the questions in the silver dataset.

- **Granularity of questions**. We observed that the silver dataset tends to ask highly granular questions about specific findings (e.g. "consolidation") as expected. The radiology experts, however, asked a range of low (e.g. "Are there any bone abnormalities?), medium (e.g. "Are the lungs clear?") and high (e.g. "Is there evidence of pneumonia?") granularity questions. The gold dialogs tend to start with broader (low granularity) questions and narrow the differential diagnosis down as the dialogs progress.

- **Question sense.** The radiologists also asked questions in the form of whether some structure is "normal" (e.g. *"Is the soft tissue normal?"*). Whereas, the silver questions only asked whether an abnormality is present. Since chest X-rays are screening exams where a good proportion of the images may be "normal", having more questions asking whether different anatomies are normal would, therefore, yield more *'Yes'* answers.

- **Answer distributions** The answer distributions of the gold and silver data differ greatly. Specifically, while the gold data was com-

posed heavily of *'Yes'* or *'No'* answers, the silver comprised mostly of *'Not in report'*.

8   Discussion

Our main finding is that the introduced task of visual dialog in radiology presents a lot of challenges from the machine learning perspective, including a skewed distribution of classes and a required ability to reason over both visual and textual input data. The best of our baseline models achieved 0.34 macro-averaged F1-score, indicating on a significant scope for potential improvements. Our comparison of gold and silver standard data shows some trends are in line with medical doctors' strategies in medical history taking, starting with broader, general questions and then narrowing the scope of their questions to more specific findings (Talbot et al.; Campillos-Llanos et al., 2020).

Despite the difficulty and the practical usefulness of the task, it is important to list the limitations of our study. The questions were limited to presence of 13 abnormalities extracted by CheXpert and the answers were limited to 4 options. The studies used in this work (from MIMIC-CXR) originate from a single tertiary hospital in the United States. Moreover, they correspond to a specific group of patients, namely those admitted to the Emergency Department (ED) from 2012 to 2014. Therefore, the data and hence the model reflect multiple real-world biases. It should also be noted that chest X-rays are mostly used for screening than diagnostic purposes. A radiology image is only one of the many data points (e.g. labs, demographics, medications) used while making a diagnosis. Therefore, although predicting presence of abnormalities (e.g. pneumonia) based on brief knowledge of the patient's medical history and the chest X-ray might be a good exercise and a promising first step in evaluating machine learning models, it is clinically limited.

There are plenty of directions for future work that we intend to pursue. To make the synthetic data more realistic and expressive, both questions and answers should be diversified with the help of clinicians' expertise and external knowledge bases such as UMLS(Bodenreider, 2004). We plan to enrich the data with more question types, addressing, for example, the location or the size of a given lung abnormality. We plan to collect more real life dialog between radiologists and augment the two datasets to get a richer set of more expressive dialog. We anticipate that bridging the gap between the silver- and the gold-standard data in terms of natural language formulations would significantly reduce the difference in model performance for the two setups.

Another direction is to develop a strategy to manage the uncertain labels such as *'Maybe'* and *'Not in report'* to make the dataset more balanced.

9   Conclusion

We explored the task of Visual Dialog for radiology using chest X-rays and released the first publicly available silver- and gold-standard datasets for this task. Having conducted a set of rigorous experiments with state-of-the-art machine learning models used for the combination of visual and language reasoning, we demonstrated the complexity of the task and outlined the promising directions for further research.

Acknowledgments

We would like to thank Mousumi Roy for her help in this project. We are also thankful to Mehdi Moradi for helpful discussions.

References

AB Abacha, SA Hasan, VV Datla, J Liu, D Demner-Fushman, and H Müller. 2019. Vqa-med: Overview of the medical visual question answering task at image-clef 2019. In CLEF2019 Working Notes. CEUR Workshop Proceedings (CEURWS. org), ISSN, pages 1613–0073.

Asma Ben Abacha, Soumya Gayen, Jason J Lau, Sivaramakrishnan Rajaraman, and Dina Demner-Fushman. 2018. Nlm at imageclef 2018 visual question answering in the medical domain. In CLEF (Working Notes).

Aishwarya Agrawal, Dhruv Batra, Devi Parikh, and Aniruddha Kembhavi. 2018. Don't just assume; look and answer: Overcoming priors for visual question answering. In Proceedings of the IEEE Conference on Computer Vision and Pattern Recognition, pages 4971–4980.

Aisha Al-Sadi, Bashar Talafha, Mahmoud Al-Ayyoub, Yaser Jararweh, and Fumie Costen. 2019. Just at imageclef 2019 visual question answering in the medical domain. In CLEF (Working Notes).

Emily Alsentzer, John Murphy, William Boag, Wei-Hung Weng, Di Jin, Tristan Naumann, and Matthew McDermott. 2019. Publicly available clinical BERT embeddings. In Proceedings of the 2nd Clinical Natural Language Processing Workshop, pages 72–78, Minneapolis, Minnesota, USA. Association for Computational Linguistics.

Peter Anderson, Xiaodong He, Chris Buehler, Damien Teney, Mark Johnson, Stephen Gould, and Lei Zhang. 2018. Bottom-up and top-down attention for image captioning and visual question answering. In Proceedings of the IEEE Conference on Computer Vision and Pattern Recognition, pages 6077–6086.

Stanislaw Antol, Aishwarya Agrawal, Jiasen Lu, Margaret Mitchell, Dhruv Batra, C Lawrence Zitnick, and Devi Parikh. 2015. Vqaa: Visual question answering. In Proceedings of the IEEE International Conference on Computer Vision, pages 2425–2433.

Mythreyi Bhargavan, Adam H Kaye, Howard P Forman, and Jonathan H Sunshine. 2009. Workload of radiologists in united states in 2006–2007 and trends since 1991–1992. Radiology, 252(2):458–467.

Olivier Bodenreider. 2004. The unified medical language system (umls): integrating biomedical terminology. Nucleic acids research, 32(suppl_1):D267–D270.

Leonardo Campillos-Llanos, Catherine Thomas, Éric Bilinski, Pierre Zweigenbaum, and Sophie Rosset. 2020. Designing a virtual patient dialogue system based on terminology-rich resources: Challenges and evaluation. Natural Language Engineering, 26(2):183–220.

Abhishek Das, Satwik Kottur, Khushi Gupta, Avi Singh, Deshraj Yadav, José MF Moura, Devi Parikh, and Dhruv Batra. 2017. Visual dialog. In Proceedings of the IEEE Conference on Computer Vision and Pattern Recognition, pages 326–335.

Jacob Devlin, Ming-Wei Chang, Kenton Lee, and Kristina Toutanova. 2019. Bert: Pre-training of deep bidirectional transformers for language understanding. In Proceedings of the 2019 Conference of the North American Chapter of the Association for Computational Linguistics: Human Language Technologies, Volume 1 (Long and Short Papers), pages 4171–4186.

Mark J Halsted and Craig M Froehle. 2008. Design, implementation, and assessment of a radiology workflow management system. American Journal of Roentgenology, 191(2):321–327.

Sadid A Hasan, Yuan Ling, Oladimeji Farri, Joey Liu, Henning Müller, and Matthew Lungren. 2018. Overview of imageclef 2018 medical domain visual question answering task. In CLEF (Working Notes).

Kaiming He, Xiangyu Zhang, Shaoqing Ren, and Jian Sun. 2016. Deep residual learning for image recognition. In Proceedings of the IEEE conference on computer vision and pattern recognition, pages 770–778.

Drew A Hudson and Christopher D Manning. 2019. Gqa: A new dataset for real-world visual reasoning and compositional question answering. In Proceedings of the IEEE Conference on Computer Vision and Pattern Recognition, pages 6700–6709.

Bogdan Ionescu, Henning Müller, Mauricio Villegas, Alba García Seco de Herrera, Carsten Eickhoff, Vincent Andrearczyk, Yashin Dicente Cid, Vitali Liauchuk, Vassili Kovalev, Sadid A Hasan, et al. 2018. Overview of imageclef 2018: Challenges, datasets and evaluation. In International Conference of the Cross-Language Evaluation Forum for European Languages, pages 309–334. Springer.

Jeremy Irvin, Pranav Rajpurkar, Michael Ko, Yifan Yu, Silviana Ciurea-Ilcus, Chris Chute, Henrik Marklund, Behzad Haghgoo, Robyn Ball, Katie Shpanskaya, et al. 2019. Chexpert: A large chest radiograph dataset with uncertainty labels and expert comparison. In Proceedings of AAAI.

Baoyu Jing, Pengtao Xie, and Eric Xing. 2018. On the automatic generation of medical imaging reports. In Proceedings of the 56th Annual Meeting of the Association for Computational Linguistics (Volume 1: Long Papers), pages 2577–2586.

Alistair EW Johnson, Tom J Pollard, Seth Berkowitz, Nathaniel R Greenbaum, Matthew P Lungren, Chihying Deng, Roger G Mark, and Steven Horng. 2019. Mimic-cxr: A large publicly available database of labeled chest radiographs. arXiv preprint arXiv:1901.07042.

Alistair EW Johnson, Tom J Pollard, Lu Shen, H Lehman Li-wei, Mengling Feng, Mohammad Ghassemi, Benjamin Moody, Peter Szolovits, Leo Anthony Celi, and Roger G Mark. 2016. Mimic-iii, a freely accessible critical care database. Scientific data, 3:160035.

Kaggle. 2017. Data science bowl. https://www.kaggle.com/c/data-science-bowl-2017.

Tomasz Kornuta, Deepta Rajan, Chaitanya Shivade, Alexis Asseman, and Ahmet S Ozcan. 2019. Leveraging medical visual question answering with supporting facts. In CLEF (Working Notes).

Jason J Lau, Soumya Gayen, Asma Ben Abacha, and Dina Demner-Fushman. 2018. A dataset of clinically generated visual questions and answers about radiology images. Scientific Data, 5:180251.

Mark D Mangano, Arifeen Rahman, Garry Choy, Dushyant V Sahani, Giles W Boland, and Andrew J Gunn. 2014. Radiologists' role in the communication of imaging examination results to patients: perceptions and preferences of patients. American Journal of Roentgenology, 203(5):1034–1039.

Yulei Niu, Hanwang Zhang, Manli Zhang, Jianhong Zhang, Zhiwu Lu, and Ji-Rong Wen. 2019. Recursive visual attention in visual dialog. In Proceedings of the IEEE Conference on Computer Vision and Pattern Recognition, pages 6679–6688.

Yifan Peng, Xiaosong Wang, Le Lu, Mohammadhadi Bagheri, Ronald Summers, and Zhiyong Lu. 2018. Negbio: a high-performance tool for negation and uncertainty detection in radiology reports. AMIA Summits on Translational Science Proceedings, 2018:188.

Jeffrey Pennington, Richard Socher, and Christopher Manning. 2014. Glove: Global vectors for word representation. In Proceedings of the 2014 conference on empirical methods in natural language processing (EMNLP), pages 1532–1543.

Pranav Rajpurkar, Jeremy Irvin, Aarti Bagul, Daisy Ding, Tony Duan, Hershel Mehta, Brandon Yang, Kaylie Zhu, Dillon Laird, Robyn L Ball, et al. 2017. Mura: Large dataset for abnormality detection in musculoskeletal radiographs. arXiv preprint arXiv:1712.06957.

Alexey Romanov and Chaitanya Shivade. 2018. Lessons from natural language inference in the clinical domain. In Proceedings of the 2018 Conference on Empirical Methods in Natural Language Processing, pages 1586–1596.

Yossi Rubner, Carlo Tomasi, and Leonidas J Guibas. 2000. The earth mover's distance as a metric for image retrieval. International journal of computer vision, 40(2):99–121.

Bettina Siewert, Olga R Brook, Mary Hochman, and Ronald L Eisenberg. 2016. Impact of communication errors in radiology on patient care, customer satisfaction, and work-flow efficiency. American Journal of Roentgenology, 206(3):573–579.

Thomas B Talbot, Kenji Sagae, Bruce John, and Albert A Rizzo. Designing useful virtual standardized patient encounters.

Harm de Vries, Florian Strub, A. P. Sarath Chandar, Olivier Pietquin, Hugo Larochelle, and Aaron C. Courville. 2016. Guesswhat?! visual object discovery through multi-modal dialogue. In 2017 IEEE Conference on Computer Vision and Pattern Recognition, pages 4466–4475.

Xiaosong Wang, Yifan Peng, Le Lu, Zhiyong Lu, and Ronald M Summers. 2018. Tienet: Text-image embedding network for common thorax disease classification and reporting in chest x-rays. In Proceedings of the IEEE Conference on Computer Vision and Pattern Recognition, pages 9049–9058.

Han Xiao. 2018. bert-as-service. https://github.com/hanxiao/bert-as-service.

Zichao Yang, Xiaodong He, Jianfeng Gao, Li Deng, and Alex Smola. 2016. Stacked attention networks for image question answering. In Proceedings of the IEEE Conference on Computer Vision and Pattern Recognition, pages 21–29.

Peng Zhang, Yash Goyal, Douglas Summers-Stay, Dhruv Batra, and Devi Parikh. 2016. Yin and yang: Balancing and answering binary visual questions. In Proceedings of the IEEE Conference on Computer Vision and Pattern Recognition, pages 5014–5022.

A BERT-based One-Pass Multi-Task Model for Clinical Temporal Relation Extraction

Chen Lin[1], Timothy Miller[1*], Dmitriy Dligach[2], Farig Sadeque[1], Steven Bethard[3] and Guergana Savova[1]

[*]Co-first author
[1]Boston Children's Hospital and Harvard Medical School
[2]Loyola University Chicago
[3]University of Arizona
[1]{first.last}@childrens.harvard.edu
[2]ddligach@luc.edu
[3]bethard@email.arizona.edu

Abstract

Recently BERT has achieved a state-of-the-art performance in temporal relation extraction from clinical Electronic Medical Records text. However, the current approach is inefficient as it requires multiple passes through each input sequence. We extend a recently-proposed one-pass model for relation classification to a one-pass model for relation extraction. We augment this framework by introducing global embeddings to help with long-distance relation inference, and by multi-task learning to increase model performance and generalizability. Our proposed model produces results on par with the state-of-the-art in temporal relation extraction on the THYME corpus and is much "greener" in computational cost.

1 Introduction

The analysis of many medical phenomena (e.g., disease progression, longitudinal effects of medications, treatment regimen and outcomes) heavily depends on temporal relation extraction from the clinical free text embedded in the Electronic Medical Records (EMRs). At a coarse level, a clinical event can be linked to the document creation time *(DCT)* as Document Time Relations *(DocTimeRel)*, with possible values of *BEFORE, AFTER, OVERLAP, and BEFORE_OVERLAP* (Styler IV et al., 2014). At a finer level, a narrative container (Pustejovsky and Stubbs, 2011) can temporally subsume an event as a *contains* relation. The THYME corpus (Styler IV et al., 2014) consists of EMR clinical text and is annotated with time expressions (TIMEX3), events (EVENT), and temporal relations (TLINK) using an extension of TimeML (Pustejovsky et al., 2003; Pustejovsky and Stubbs, 2011). It was used in the Clinical Temp-Eval series (Bethard et al., 2015, 2016, 2017).

While the performance of DocTimeRel models has reached above 0.8 F1 on the THYME corpus,

the CONTAINS task remains a challenge for both conventional learning approaches (Sun et al., 2013; Bethard et al., 2015, 2016, 2017) and neural models (structured perceptrons (Leeuwenberg and Moens, 2017), convolutional neural networks (CNNs) (Dligach et al., 2017; Lin et al., 2017), and Long Short-Term memory (LSTM) networks (Tourille et al., 2017; Dligach et al., 2017; Lin et al., 2018; Galvan et al., 2018)). The difficulty is that the limited labeled data is insufficient for training deep neural models for complex linguistic phenomena. Some recent work (Lin et al., 2019) has used massive pre-trained language models (BERT; Devlin et al., 2018) and their variations (Lee et al., 2019) for this task and significantly increased the CONTAINS score by taking advantage of the rich BERT representations. However, that approach has an input representation that is highly wasteful – the same sentence must be processed multiple times, once for each candidate relation pair.

Inspired by recent work in Green AI (Schwartz et al., 2019; Strubell et al., 2019), and one-pass encodings for multiple relations extraction (Wang et al., 2019), we propose a one-pass encoding mechanism for the CONTAINS relation extraction task, which can significantly increase the efficiency and scalability. The architecture is shown in Figure 1. The three novel modifications to the original one-pass relational model of Wang et al. (2019) are: (1) Unlike Wang et al. (2019), our model operates in the relation extraction setting, meaning it must distinguish between relations and non-relations, as well as classifying by relation type. (2) We introduce a pooled embedding for relational classification across long distances. Wang et al. (2019) focused on short-distance relations, but clinical CONTAINS relations often span multiple sentences, so a sequence-level embedding is necessary for such long-distance inference. (3) We use the same BERT encoding of the input instance for both

Proceedings of the BioNLP 2020 workshop, pages 70–75
Online, July 9, 2020 ©2020 Association for Computational Linguistics

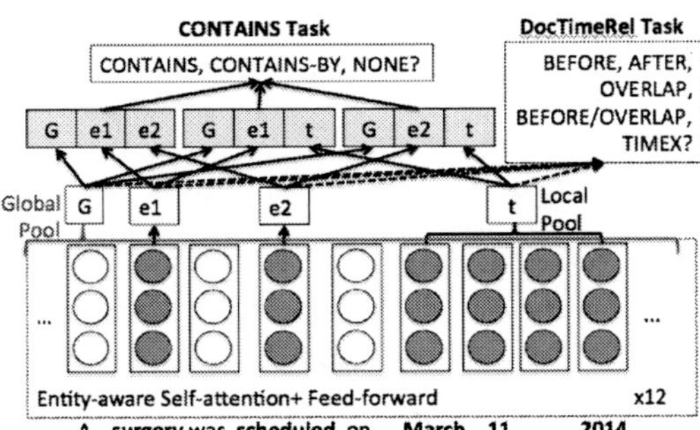

Figure 1: Model Architecture. e1, e2, and t represent entity-embeddings for "surgery", "scheduled", and "March 11, 2014" respectively. G is the pooled embedding for the entire input instance.

DocTimeRel and CONTAINS tasks, i.e. adding multi-task learning (MTL) on top of one-pass encoding. DocTimeRel and CONTAINS are related tasks. For example, if a medical event A happens BEFORE the DCT, while event B happens AFTER the DCT, it is unlikely that there is a CONTAINS relation between A and B. MTL provides an effective way to leverage useful knowledge learned in one task to benefit other tasks. What is more, MTL can potentially employ a regularization effect that alleviates overfitting to a specific task.

2 Methodology

2.1 Twin Tasks

Apache cTAKES (Savova et al., 2010)(http://ctakes.apache.org) is used for segmenting and tokenizing the THYME corpus in order to generate instances. Each instance is a sequence of tokens with the gold standard event and time expression annotations marked in the token sequences by logging their positional information. Using the entity-aware self-attention based on relative distance (Wang et al., 2019), we can encode every entity, E_i, by its BERT embedding, e_i. If an entity e_i consists of multiple tokens (many time expressions are multi-token), it is average-pooled (local pool in Figure 1) over the embedding of the corresponding tokens in the last BERT layer.

For the CONTAINS task, we create relation candidates from all pairs of entities within an input sequence. Each candidate is represented by the concatenation of three embeddings, e_i, e_j, and G, as $[G{:}e_i{:}e_j]$, where G is an average-pooled embedding over the entire sequence, and is different from the embedding of [CLS] token. The [CLS] token is

the conventional token BERT inserts at the start of every input sequence and its embedding is viewed as the representation of the entire sequence. The concatenated embedding is passed to a linear classifier to predict the CONTAINS, CONTAINED-BY, or NONE relation, $\hat{r_{ij}}$, as in eq. (1).

$$P(\hat{r_{ij}}|\mathbf{x}, E_i, E_j)=softmax(W^L[G:e_i:e_j] + b) \tag{1}$$

where $W^L \in \mathbb{R}^{3d_z \times l_r}$, d_z is the dimension of the BERT embedding, $l_r = 3$ for the CONTAINS labels, b is the bias, and $\mathbf{x}$ is the input sequence.

Similarly, for the DocTimeRel (dtr) task we feed each entity's embedding, e_i, together with the global pooling G, to another linear classifier to predict the entity's five "temporal statuses": TIMEX if the entity is a time expression or the dtr type (BEFORE, AFTER, etc.) if the entity is an event:

$$P(\hat{dtr_i}|\mathbf{x}, E_i) = softmax(W^D[G:e_i] + b) \tag{2}$$

where $W^D \in \mathbb{R}^{2d_z \times l_d}$, and $l_d = 5$.

For the combined task, we define loss as:

$$L(\hat{r_{ij}}, r_{ij}) + \alpha(L(\hat{dtr_i}, dtr_i) + L(\hat{dtr_j}, dtr_j)) \tag{3}$$

where $\hat{r_{ij}}$ is the predicted relation type, $\hat{dtr_i}$ and $\hat{dtr_j}$ are the predicted temporal statuses for E_i and E_j respectively, r_{ij} is the gold relation type, and dtr_i and dtr_j are the gold temporal statuses. α is a weight to balance CONTAINS loss and dtr loss.

2.2 Window-based token sequence processing

Following Lin et al. (2019), we use a set window of tokens (Token-Window) disregarding natural sentence boundaries for generating instances. BERT may still take punctuation tokens into account. Each token sequence is limited by a set number of entities (Entity-Window) to be processed. We apply a sliding token window (windows may overlap), thus every entity gets processed. Positional information for each entity is output along the token sequence and is propagated through different layers via the entity-aware self-attention mechanism (Wang et al., 2019).

3 Experiments

3.1 Data and Settings

We adopt the THYME corpus (Styler IV et al., 2014) for model fine-tuning and evaluation. The

Model	P	R	F1
Multi-pass	0.735	0.613	0.669
Multi-pass+Silver	0.674	0.695	0.684
One-pass	0.647	0.671	0.659
One-pass+[CLS]	0.665	0.673	0.669
One-pass+Pooling	0.670	0.689	0.680
One-pass+Pooling+MTL	0.686	0.687	**0.686**

Table 1: Model performance of *CONTAINS* relation on colon cancer test set. Multi-pass baselines are from Lin et al. (2019)'s system without and with self-training using silver instances (system predictions on a unlabeled colon cancer set). We tested a one pass system with just argument embeddings; with the [CLS] token as the global context vector ([CLS]); with argument embeddings plus a globally pooled context vector (Pooling); and with global pooling as well as multi-task learning (MTL) with DocTimeRel.

one-pass multi-task model is fine-tuned on the THYME Colon Cancer training set with uncased BERT base model, using the code released by Wang et al. (2019)[1] as a base. The batch size is set to 4, the learning rate is selected from (1e-5, 2e-5, 3e-5, 5e-5), the Token-Window size is selected from (60, 70, 100), the Entity-Window size is selected from (8, 10, 16), the training epochs are selected from (2, 3, 4, 5), the clipping distance k (the maximum relative position to consider) is selected from (3, 4, 5), and α is selected from (0.01, 0.05). A single NVIDIA GTX Titan Xp GPU is used for the computation. The best model is selected on the Colon cancer development set and tested on the Colon cancer test set, and on THYME Brain cancer test set for portability assessment.

3.2 Results on THYME

Table 1 shows performance of our one-pass models for the CONTAINS task on the Clinical TempEval colon cancer test set. The one-pass (OP) model alone obtains an F1 score of 0.659. Adding the [CLS] token as the global context vector increases the F1 score to 0.669. Using a globally average-pooled context vectors G instead of [CLS] improves performance to 0.680, better than the multi-pass model without silver instances (Lin et al., 2019). Applying the MTL setting, the one-pass twin-task (CONTAINS and DocTimeRel) model without any silver data reaches 0.686 F1, which is on par with the multi-pass model trained with additional silver instances on the CONTAINS task,

Model	Single	MTL
AFTER	0.86	0.83
BEFORE	0.88	0.89
BEFORE/OVERLAP	0.63	0.56
OVERLAP	0.89	0.85
TIMEX	0.98	0.98
OVERALL	0.88	0.86

Table 2: Model performance in F1-scores of temporal statuses on colon cancer test set. Single: One-pass+Pooling for a single dtr Task; MTL: One-pass+Pooling for twin tasks: CONTAINS and dtr.

Model	P	R	F1
Lin et al. (2019)	0.473	0.700	0.565
One-pass+Pooling	0.506	0.643	0.566
One-pass+Pooling+MTL	0.545	0.624	**0.582**

Table 3: Model performance of *CONTAINS* relation on brain cancer test set.

0.684 F1 (Lin et al., 2019).

Table 2 shows the performance of our one-pass models for the DocTimeRel task on the Clinical TempEval colon cancer test set. The single-task model achieves 0.88 weighted average F1, while the MTL model compromises the performance to 0.86 F1. Of note, this result is not directly comparable to Bethard et al. (2016) results because the Clinical TempEval evaluation script does not take into account if an entity is correctly recognized as a time expression (TIMEX). There are two types of entities in the THYME annotation: events and time expressions (TIMEX). The Bethard et al. (2016) evaluation on DocTimeRel was focused on all events, and classified an event into four DocTimeRel types. Our evaluation was for all entities. For a given entity, we classify it as a TIMEX or an event; if it is an event, we classify it into four DocTimeRel types, for a total of five classes.

Table 3 shows the portability of our one-pass models on the THYME brain cancer test set. Without any tuning on brain cancer data, the MTL model with global pooling performs at 0.582 F1, which is better than the multi-pass model trained with additional silver instances (0.565 F1) reported in Lin et al. (2019), trading roughly equal amounts of precision for recall to obtain a better balance. Without MTL, the one-pass CONTAINS model with global context embeddings (One-pass+Pooling) achieves 0.566 F1 on the brain cancer test set, significantly lower than the MTL

Model	flops/inst	inst#	Ratio
OP	218,767,889	20k	1
OP+MTL	218,783,260	20k	1
Multi-pass	218,724,880	427k	23
Multi-pass+Silver	218,724,880	497k	25

Table 4: Computational complexity in flops per instance (flops/inst)×total number of instances (inst#).

model (using a Wilcoxon Signed-rank test over document-by-document comparisons, as in (Cherry et al., 2013), p-value=0.01962).

3.3 Computational Efficiency

Table 4 shows the computational burden for different models in terms of floating point operations (flops). The flops are derived from TensorFlow's profiling tool on saved model graphs. The second column is the flops per one training instance, the third column lists the number of instances for different model settings. The total computational complexity for one training epoch is thus the multiplication between column 2 and 3. The *Ratio* column is the relative ratio of total complexity using the OP total flops as the comparator.

For relation extraction, all entities within a sequence must be paired. If there are n entities in a token sequence, there are $n \times (n-1)/2$ ways to combine those entities for relational candidates. The multi-pass model would encode the same sequence $n \times (n-1)/2$ times, while the one-pass model would only encode it once and add the pairing computation on top of the BERT encoding represented in Figure 1 with very minor increase in computation per one instance (about 43K flops); and the MTL model adds another 15k flops; but they are of the same magnitude, 219K flops. The one-pass models save a lot of passes on the training instances, 20k vs. 497k, which results in a significant difference in computational load, 1 vs. 25, which could be several hours to several days difference in GPU hours. The exact number of training instances processed by the one-pass model is affected by the Token-Window and Entity-Window hyper-parameters. However, even in the worst case scenario, when the Token-Window is set to 100, and the Entity-Window is set to 8, there are 108K training instances for the one-pass model, which is still substantially fewer training instances than what are used for the multi-pass model. In addition, since the one-pass models do not run the extra steps used for generating silver instances (Lin et al., 2019), the time savings is even greater.

4 Discussion

Through table 1 row 3-5, we can see that sequence-wise embedding, either global pooling G or [CLS], is important for clinical temporal relation extraction which involves long-distance relations that may go across multiple natural sentences. Entity embeddings are good for tasks that focus on short-distance relations (such as (Gábor et al., 2018)), but may not be sufficient for picking enough context for long-distance relations.

Combining MTL with a one-pass mechanism produces a more efficient and generalizable model. With merely additional 15k flops (table 4 row 1 and 2), the model achieves high performance for both tasks. However, we found that it is hard for both tasks to get top performance. If the weight for dtr loss is increased, the dtr F1 increases at the cost of the CONTAINS scores. Even though the majority of entities in CONTAINS relations have aligned dtr values (e.g., in Figure 2(#1), both entities have matching dtr value, AFTER), some relations do have conflicted dtr values. For example, in Figure 2(#2), the dtr for *screening* is BEFORE, while *test* is a BEFORE_OVERLAP (the present perfect tense signifies *tests* happened in the past but lasts through present, hence BEFORE_OVERLAP). Even though it is a gold CONTAINS annotation, the model may be confused by an event that happened in the past (*screening*) to contain another event (*test*) that is longer than its temporal scope. Due to these conflicts, we thus pick the more challenging CONTAINS task as our priority and set α relatively low (0.01) in order to optimize the model towards the CONTAINS task, ignoring some of the dtr errors or conflicts. In the meantime, the MTL setting does help prevent the model from overfitting to one specific task, thus achieving some level of generalization. The significant 1.6% increase in F1-score on the Brain test set in table 3 demonstrates the improved generalizability.

In conclusion, we built a "green" model for a challenging problem. Deployed on a single gpu with 25 times better efficiency, it succeeded in both temporal tasks, achieved better generalizability, and suited to other pre-trained models (Liu et al., 2019; Alsentzer et al., 2019; Beltagy et al., 2019; Lan et al., 2019; Yang et al., 2019, etc.)

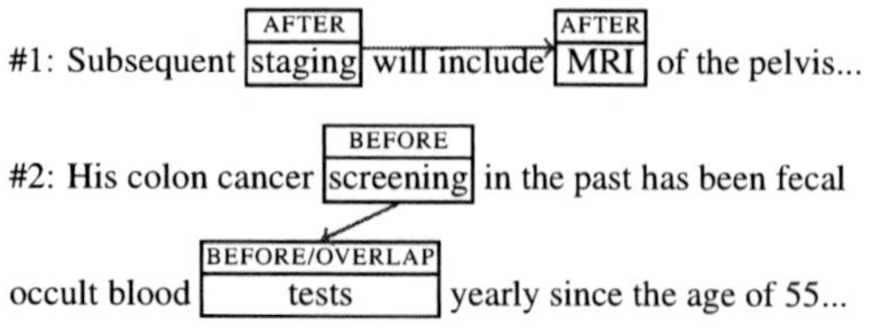

Figure 2: CONTAINS Relations with matching(#1)/conflicting(#2) DocTimeRel values.

Acknowledgments

The study was funded by R01LM10090, R01GM114355 and UG3CA243120 from the Unites States National Institutes of Health. The content is solely the responsibility of the authors and does not necessarily represent the official views of the National Institutes of Health. We thank the anonymous reviewers for their valuable suggestions and criticism. The Titan Xp GPU used for this research was donated by the NVIDIA Corporation.

References

Emily Alsentzer, John R Murphy, Willie Boag, Wei-Hung Weng, Di Jin, Tristan Naumann, and Matthew McDermott. 2019. Publicly available clinical bert embeddings. *arXiv preprint arXiv:1904.03323*.

Iz Beltagy, Kyle Lo, and Arman Cohan. 2019. Scibert: A pretrained language model for scientific text. In *Proceedings of the 2019 Conference on Empirical Methods in Natural Language Processing and the 9th International Joint Conference on Natural Language Processing (EMNLP-IJCNLP)*, pages 3606–3611.

Steven Bethard, Leon Derczynski, Guergana Savova, Guergana Savova, James Pustejovsky, and Marc Verhagen. 2015. Semeval-2015 task 6: Clinical tempeval. In *Proceedings of the 9th International Workshop on Semantic Evaluation (SemEval 2015)*, pages 806–814.

Steven Bethard, Guergana Savova, Wei-Te Chen, Leon Derczynski, James Pustejovsky, and Marc Verhagen. 2016. Semeval-2016 task 12: Clinical tempeval. *Proceedings of SemEval*, pages 1052–1062.

Steven Bethard, Guergana Savova, Martha Palmer, James Pustejovsky, and Marc Verhagen. 2017. Semeval-2017 task 12: Clinical tempeval. *Proceedings of the 11th International Workshop on Semantic Evaluation (SemEval-2017)*, pages 563–570.

Colin Cherry, Xiaodan Zhu, Joel Martin, and Berry de Bruijn. 2013. la recherche du temps perdu: extracting temporal relations from medical text in the 2012 i2b2 nlp challenge. *Journal of the American Medical Informatics Association*, 20(5):843–848.

Jacob Devlin, Ming-Wei Chang, Kenton Lee, and Kristina Toutanova. 2018. Bert: Pre-training of deep bidirectional transformers for language understanding. *arXiv preprint arXiv:1810.04805*.

Dmitriy Dligach, Timothy Miller, Chen Lin, Steven Bethard, and Guergana Savova. 2017. Neural temporal relation extraction. *EACL 2017*, page 746.

Kata Gábor, Davide Buscaldi, Anne-Kathrin Schumann, Behrang QasemiZadeh, Haïfa Zargayouna, and Thierry Charnois. 2018. SemEval-2018 task 7: Semantic relation extraction and classification in scientific papers. In *Proceedings of The 12th International Workshop on Semantic Evaluation*, pages 679–688, New Orleans, Louisiana. Association for Computational Linguistics.

Diana Galvan, Naoaki Okazaki, Koji Matsuda, and Kentaro Inui. 2018. Investigating the challenges of temporal relation extraction from clinical text. In *Proceedings of the Ninth International Workshop on Health Text Mining and Information Analysis*, pages 55–64.

Zhenzhong Lan, Mingda Chen, Sebastian Goodman, Kevin Gimpel, Piyush Sharma, and Radu Soricut. 2019. Albert: A lite bert for self-supervised learning of language representations. *arXiv preprint arXiv:1909.11942*.

Jinhyuk Lee, Wonjin Yoon, Sungdong Kim, Donghyeon Kim, Sunkyu Kim, Chan Ho So, and Jaewoo Kang. 2019. BioBERT: a pre-trained biomedical language representation model for biomedical text mining. *Bioinformatics*.

Tuur Leeuwenberg and Marie-Francine Moens. 2017. Structured learning for temporal relation extraction from clinical records. In *Proceedings of the 15th Conference of the European Chapter of the Association for Computational Linguistics*.

Chen Lin, Timothy Miller, Dmitriy Dligach, Hadi Amiri, Steven Bethard, and Guergana Savova. 2018. Self-training improves recurrent neural networks performance for temporal relation extraction. In *Proceedings of the Ninth International Workshop on Health Text Mining and Information Analysis*, pages 165–176.

Chen Lin, Timothy Miller, Dmitriy Dligach, Steven Bethard, and Guergana Savova. 2017. Representations of time expressions for temporal relation extraction with convolutional neural networks. *BioNLP 2017*, pages 322–327.

Chen Lin, Timothy Miller, Dmitriy Dligach, Steven Bethard, and Guergana Savova. 2019. A bert-based universal model for both within-and cross-sentence clinical temporal relation extraction. In *Proceedings of the 2nd Clinical Natural Language Processing Workshop*, pages 65–71.

Yinhan Liu, Myle Ott, Naman Goyal, Jingfei Du, Mandar Joshi, Danqi Chen, Omer Levy, Mike Lewis, Luke Zettlemoyer, and Veselin Stoyanov. 2019. Roberta: A robustly optimized bert pretraining approach. *arXiv preprint arXiv:1907.11692*.

James Pustejovsky, José M Castano, Robert Ingria, Roser Sauri, Robert J Gaizauskas, Andrea Setzer, Graham Katz, and Dragomir R Radev. 2003. Timeml: Robust specification of event and temporal expressions in text. *New directions in question answering*, 3:28–34.

James Pustejovsky and Amber Stubbs. 2011. Increasing informativeness in temporal annotation. In *Proceedings of the 5th Linguistic Annotation Workshop*, pages 152–160. Association for Computational Linguistics.

Guergana K Savova, James J Masanz, Philip V Ogren, Jiaping Zheng, Sunghwan Sohn, Karin C Kipper-Schuler, and Christopher G Chute. 2010. Mayo clinical text analysis and knowledge extraction system (ctakes): architecture, component evaluation and applications. *Journal of the American Medical Informatics Association*, 17(5):507–513.

Roy Schwartz, Jesse Dodge, Noah A Smith, and Oren Etzioni. 2019. Green ai. *arXiv preprint arXiv:1907.10597*.

Emma Strubell, Ananya Ganesh, and Andrew McCallum. 2019. Energy and policy considerations for deep learning in nlp. *Proceedings of the 57th Annual Meeting of the Association for Computational Linguistics*, pages 3645–3650.

William F Styler IV, Steven Bethard, Sean Finan, Martha Palmer, Sameer Pradhan, Piet C de Groen, Brad Erickson, Timothy Miller, Chen Lin, Guergana Savova, et al. 2014. Temporal annotation in the clinical domain. *Transactions of the Association for Computational Linguistics*, 2:143–154.

Weiyi Sun, Anna Rumshisky, and Ozlem Uzuner. 2013. Evaluating temporal relations in clinical text: 2012 i2b2 challenge. *Journal of the American Medical Informatics Association*, 20(5):806–813.

Julien Tourille, Olivier Ferret, Aurelie Neveol, and Xavier Tannier. 2017. Neural architecture for temporal relation extraction: A bi-lstm approach for detecting narrative containers. In *Proceedings of the 55th Annual Meeting of the Association for Computational Linguistics (Volume 2: Short Papers)*, volume 2, pages 224–230.

Haoyu Wang, Ming Tan, Mo Yu, Shiyu Chang, Dakuo Wang, Kun Xu, Xiaoxiao Guo, and Saloni Potdar. 2019. Extracting multiple-relations in one-pass with pre-trained transformers. *Proceedings of the 57th Annual Meeting of the Association for Computational Linguistics*, pages 1371–1377.

Zhilin Yang, Zihang Dai, Yiming Yang, Jaime Carbonell, Ruslan Salakhutdinov, and Quoc V Le. 2019. Xlnet: Generalized autoregressive pretraining for language understanding. *Advances in Neural Information Processing Systems 32*, pages 5754–5764.

Experimental Evaluation and Development of a Silver-Standard for the MIMIC-III Clinical Coding Dataset

Thomas Searle[1], Zina Ibrahim[1], Richard JB Dobson[1,2]
[1]Department of Biostatistics and Health Informatics,
Institute of Psychiatry, Psychology and Neuroscience,
King's College London, London, U.K.
[2]Institute of Health Informatics, University College London,
London, London, U.K.
{firstname.lastname}@kcl.ac.uk

Abstract

Clinical coding is currently a labour-intensive, error-prone, but critical administrative process whereby hospital patient episodes are manually assigned codes by qualified staff from large, standardised taxonomic hierarchies of codes. Automating clinical coding has a long history in NLP research and has recently seen novel developments setting new state of the art results. A popular dataset used in this task is MIMIC-III, a large intensive care database that includes clinical free text notes and associated codes. We argue for the reconsideration of the validity MIMIC-III's assigned codes that are often treated as gold-standard, especially when MIMIC-III has not undergone secondary validation. This work presents an open-source, reproducible experimental methodology for assessing the validity of codes derived from EHR discharge summaries. We exemplify the methodology with MIMIC-III discharge summaries and show the most frequently assigned codes in MIMIC-III are under-coded up to 35%.

1 Introduction

Clinical coding is the process of translating statements written by clinicians in natural language to describe a patient's complaint, problem, diagnosis and treatment, into an internationally-recognised coded format (World Health Organisation, 2011). Coding is an integral component of healthcare and provides standardised means for reimbursement, care administration, and for enabling epidemiological studies using electronic health record (EHR) data (Henderson et al., 2006).

Manual clinical coding is a complex, labour-intensive, and specialised process. It is also error-prone due to the subtleties and ambiguities common in clinical text and often strict timelines imposed on coding encounters. The annual cost of clinical coding is estimated to be $25 billion in the US alone (Farkas and Szarvas, 2008).

To alleviate the burden of the status quo of manual coding, several Machine learning (ML) automated coding models have been developed (Larkey and Croft, 1996; Aronson et al., 2007; Farkas and Szarvas, 2008; Perotte et al., 2014; Ayyar et al., 2016; Baumel et al., 2018; Mullenbach et al., 2018; Falis et al., 2019). However, despite continued interest, translation of ML systems into real-world deployments has been limited. An important factor contributing to the limited translation is the fluctuating quality of the manually-coded real hospital data used to train and evaluate such systems, where large margins of error are a direct consequence of the difficulty and error-prone nature of manual coding. To our knowledge, the literature contains only two systematic evaluations of the quality of clinically-coded data, both based on UK trusts and showing accuracy to range between 50 to 98% Burns et al. (2012) and error rates between 1%-45.8% CHKS Ltd (2014) respectively. In Burns et al. (2012), the actual accuracy is likely to be lower because the reviewed trusts used varying statistical evaluation methods, validation sources (clinical text vs clinical registries), sampling modes for accuracy estimation (random vs non-random), and the quality of validators (qualified clinical coders vs lay people). CHKS Ltd (2014) highlight that 48% of the reviewed trusts used discharge summaries alone or as the primary source for coding an encounter, to minimise the amount of raw text used for code assignment. However, further portions of the documented encounter are often needed to assign codes accurately.

The Medical Information Mart for Intensive Care (MIMIC-III) database (Johnson et al., 2016) is the largest free resource of hospital data and constitutes a substantial portion of the training of automated coding models. Nevertheless, MIMIC-III is significantly under-coded for specific conditions (Kokotailo and Hill, 2005), and has been shown to exhibit reproducibility issues in the problem of

Proceedings of the BioNLP 2020 workshop, pages 76–85
Online, July 9, 2020 ©2020 Association for Computational Linguistics

mortality prediction (Johnson et al., 2017). Therefore, serious consideration is needed when using MIMIC-III to train automated coding solutions.

In this work, we seek to understand the limitations of using MIMIC-III to train automated coding systems. To our knowledge, no work has attempted to validate the MIMIC-III clinical coding dataset for all admissions and codes, due to the time-consuming and costly nature of the endeavour. To illustrate the burden, having two clinical coders, working 38 hours a week re-coding all 52,726 admission notes at a rate of 5 minutes and $3 per document, would amount to ∼$316,000 and ∼115 weeks work for a 'gold standard' dataset. Even then, documents with a low inter-annotator agreement would undergo a final coding round by a third coder, further raising the approximate cost to ∼$316,000 and stretching the 70 weeks.

In this work, we present an experimental evaluation of coding coverage in the MIMIC-III discharge summaries. The evaluation uses text extraction rules and a validated biomedical named entity recognition and linking (NER+L) tool, Med-CAT (Kraljevic et al., 2019) to extract ICD-9 codes, reconciling them with those already assigned in MIMIC-III. The training and experimental setup yield a reproducible open-source procedure for building silver-standard coding datasets from clinical notes. Using the approach, we produce a silver-standard dataset for ICD-9 coding based on MIMIC-III discharge summaries.

This paper is structured as follows: Section 2 reviews essential background and related work in automated clinical coding, with a particular focus on MIMIC-III. Section 3 presents our experimental setup and the semi-supervised development of a silver standard dataset of clinical codes derived from unstructured EHR data. The results are presented in Section 4, while Section 5 discusses the wider impact of the results and future work.

2 Background

2.1 Clinical Coding Overview

The International Statistical Classification of Diseases and Health Related Problems (ICD) provides a hierarchical taxonomic structure of clinical terminology to classify morbidity data (World Health Organisation, 2011). The framework provides consistent definitions across global health care services to describe adverse health events including illness, injury and disability. Broadly, patient encounters with health services result in a set of clinical codes that directly correlate to the care provided.

Top-level ICD codes represent the highest level of the hierarchy, with ICD-9/10 (ICD-10 being the later version) listing 19 and 21 chapters respectively. Clinically meaningful hierarchical subdivisions of each chapter provide further specialisation of a given condition.

Coding clinical text results in the assignment of a single primary diagnosis and further secondary diagnosis codes (World Health Organisation, 2011). The complexity of coding encounters largely stems from the substantial number of available codes. For example, ICD-10-CM is the US-specific extension to the standard ICD-10 and includes 72,000 codes. Although a significant portion of the hierarchy corresponds to rare conditions, 'common' conditions to code are still in the order of thousands.

Moreover, clinical text often contains specialist terminology, spelling mistakes, implicit mentions, abbreviations and bespoke grammatical rules. However, even qualified clinical coders are not permitted to infer codes that are not explicitly mentioned within the text. For example, a diagnostic test result that indicates a condition (with the condition not explicitly written), or a diagnosis that is written as 'questioned' or 'possible' cannot be coded.

Another factor contributing to the laborious nature of coding is the large amount of duplication present in EHRs, as a result of features such as copy & paste being made available to clinical staff. It has been reported that 20-78% of clinicians duplicate sections of records between notes (Bowman, 2013), subsequently producing an average data redundancy of 75% (Zhang et al., 2017).

2.2 MIMIC-III - a Clinical Coding Database

MIMIC-III (Johnson et al., 2016) is a de-identified database containing data from the intensive care unit of the Beth Israel Medical Deaconess Center, Boston, Massachusetts, USA, collected 2001-12. MIMIC-III is the world's largest resource of freely-accessible hospital data and contains demographics, laboratory test results, procedures, medications, caregiver notes, imaging reports, admission and discharge summaries, as well as mortality (both in and out of the hospital) data of 52,726 critical care patients. MIMIC provides an open-source platform for researchers to work on real patient data. At the time of writing, MIMIC-III has over 900 citations.

2.3 Automated Clinical Coding

Early ML work on automated clinical coding considered ensembles of simple text classifiers to predict codes from discharge summaries (Larkey and Croft, 1996). Rule-based models have also been formulated, by directly replicating coding manuals. A prominent example of rule-based models is the BioNLP 2007 shared task (Aronson et al., 2007), which supplied a gold standard labelled dataset of radiology reports. The dataset continues to be used to train and validate ML coding. For example, Kavuluru et al. (2015) used the dataset in addition to two US-based hospital EHRs. Although the two additional datasets used by Kavuluru et al. (2015) were not validated to a gold standard, they are reflective of the diversity found in clinical text. Their largest dataset contained 71,463 records, 60,238 distinct code combinations and had an average document length of 5303 words.

The majority of automated coding systems are trained and tested Using MIMIC-III. Perotte et al. (2014) trained hierarchical support vector machine models on the MIMIC-II EHR (Saeed et al., 2011), the earlier version of MIMIC. The models were trained using the full ICD-9-CM terminology, creating baseline results for subsequent models of 0.395 F1-micro score. Ayyar et al. (2016) used a long-short-term-memory (LSTM) neural network to predict ICD-9 codes in MIMIC-III. However, Ayyar et al. (2016) cannot be directly compared to former methods as the model only predicts the top nineteen level codes.

Methodological developments continued to use MIMIC-III with Tree-of-sequence LSTMs (Xie and Xing, 2018), hierarchical attention gated recurrent unit (HA-GRU) neural networks (Baumel et al., 2018) and convolutional neural networks with attention (CAML) (Mullenbach et al., 2018). The HA-GRU and CAML models were directly compared with (Perotte et al., 2014), achieving 0.405 and 0.539 F1-micro respectively. A recent empirical evaluation of ICD-9 coding methods predicted the top fifty ICD-9 codes from MIMIC-III, suggesting condensed memory networks as a superior network topology (Huang et al., 2018).

3 Semi-Supervised Extraction of Clinical Codes

In this section, we describe the data preprocessing, methodology and experimental design for evaluating the coding quality of MIMIC-III discharge summaries. We also describe the semi-supervised creation of a silver-standard dataset of clinical codes from unstructured EHR text based on MIMIC-III discharge summaries.

3.1 Data Preparation

Discharge summary reports are used to provide an overview for the given hospital episode. Automated coding systems often only use discharge reports as they contain the salient diagnostic text (Perotte et al., 2014; Baumel et al., 2018; Mullenbach et al., 2018) without over burdening the model. MIMIC-III discharge summaries are categorised distinctly from other clinical text. The text is often structured with section headings and content section delimiters such as line breaks. We identify Discharge Diagnosis (DD) sections in the majority of discharge summary reports 92% (n=48,898) using a simple rule based approach. These sections are lists of diagnoses assigned to the patient during admission. Xie and Xing (2018) previously used these sections to develop a matching algorithm from discharge diagnosis to ICD code descriptions with moderate success demonstrating state-of-the-art sensitivity (0.29) and specificity (0.33) scores. For the 8% (n=3,828) that are missing these sections we manually inspect a handful of examples and observe instances of patient death and administration errors. The SQL procedures used to extract the raw data from a locally built replica of the MIMIC-III database and the extraction logic for DDs are available open-source as part of this wider analysis[1].

Table 1 lists example extracted DDs. There is a large variation in structure, use of abbreviations and extensive use of clinical terms. Some DDs list the primary diagnosis alongside secondary diagnosis, whereas others simply list a set of conditions.

3.2 Semi-Supervised Named Entity Recognition and Linkage Tool

We use MedCAT (Kraljevic et al., 2019), a pre-trained named entity recognition and linking (NER+L) model, to identify and extract the corresponding ICD codes in a discharge summary note. MedCAT utilises a fast dictionary-based algorithm for direct text matches and a shallow neural network concept to learn fixed length distributed semantic vectors for ambiguous text spans. The method is conceptually similar to Word2Vec

---

[1]https://tinyurl.com/t7dxn3j

Extracted Discharge Diagnosis	Admission ID
CAD now s/p CABG HTN, DM, Osteoarthritis, Dyslipidemia	102894
Left convexity, tentorial, parafalcine Subdural hematoma	161919
Primary Diagnoses: 1. Acute ST segment Elevation Myocardial Infarction Secondary Diagnoses: 1. Hypertension 2. Hyperlipidemia	152382
Seizures.	132065

Table 1: Example discharge diagnosis subsections extracted from MIMIC-III discharge summaries

(Mikolov et al., 2013) in that word representations arc learnt by detecting correlations of context words, and learnt vectors exhibit the semantics of the underlying words. The tool can be trained in a unsupervised or a supervised manner. However, unlike Word2Vec that learns a single representation for each word, MedCAT enables the learning of 'concept' representations by accommodating synonymous terms, abbreviations or alternative spellings.

We use a MedCAT model pre-loaded with the Unified Medical Language System (Bodenreider, 2004) (UMLS). UMLS is a meta-thesaurus of medical ontologies that provides rich synonym lists that can be used for recognition and disambiguation of concepts. Mappings from UMLS to the ICD-9[2] taxonomy are then used to extract UMLS concept to ICD codes. Our large pre-trained MedCAT UMLS model contains ~1.6 million concepts. This model cannot be made publicly available due to constraints on the UMLS license, but can be trained in an an unsupervised method in ~1 week on MIMIC-III with standard CPU only hardware[3].

In an effort to keep our analysis tractable we limit our MedCAT model to only extract the 400 ICD-9 codes that occur most frequently in the dataset. This equates to 76% (n=48,2379) of total assigned codes (n=634,709). We exclude the other 6,441 codes that occur less frequently. Future work could consider including more of these codes.

<hr>

[2] https://bioportal.bioontology.org/ontologies/ICD9CM
[3] https://tinyurl.com/yadtnz3w

3.3 Code Prediction Datasets

We run our MedCAT model over each extracted DD subsection. The model assigns each token or sequence of tokens a UMLS code and therefore an associated ICD code. In our comparison of the MedCAT produced annotations with the MIMIC-III assigned codes we have 3 distinct datasets:

1. MedCAT does not identify a concept and the code has been assigned in MIMIC-III. Denoted **A_NP** for 'Assigned, Not Predicted'.

2. MedCAT identifies a concept and this matches with an assigned code in MIMIC-III. Denoted **P_A** for 'Predicted, Assigned'.

3. MedCAT identifies a concept and this does <u>not</u> match with an assigned code in MIMIC-III dataset. Denoted **P_NA** for 'Predicted, Not Assigned'.

We do not consider the case where both McdCAT and the existing MIMIC-III assigned codes have missed an assignable code as this would involve manual validation of all notes, and as previously discussed, is infeasible for a dataset of this size.

3.3.1 Producing the Silver Standard

Given the above initial datasets we produce our final silver-standard clinical coding dataset by:

1. Sampling from the missing predictions dataset (A_NP) to manually collect annotations where our out-of-the-box MedCAT model fails to recognise diagnoses.

2. Fine-tuning our MedCAT model with the collected annotations and re-running on the entire DD subsection dataset producing updated A_NP, P_A, P_NA datasets.

3. Sampling from P_NA and P_A and annotating predicted diagnoses to validate correctness of the MedCAT predicted codes.

4. Exclusion of any codes that fail manual validation step as they are not trustworthy predictions made by MedCAT.

We use the MedCATTrainer annotator (Searle et al., 2019) to both collect annotations (stage 1) and to validate predictions from MedCAT (stage 3). To collect annotations, we manually inspect 10 randomly sampled predictions for each of the 400 unique codes from A_NP and add further acronyms,

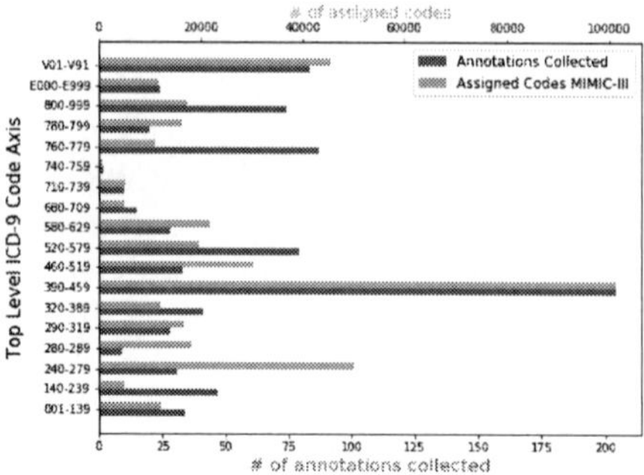

Figure 1: The distributions of manually annotated ICD-9 codes and the assigned codes in MIMIC-III grouped by top-level ICD-9 axis.

abbreviations, synonyms etc for diagnoses if they are present in the DD subsection to improve the underlying MedCAT model. To validate predictions from P_A and P_NA, we use the MedCAT-Trainer annotator to inspect 10 randomly sampled predictions for each of the 179 & 182 unique codes respectively found. We mark each prediction made by MedCAT as correct or incorrect and report results in Section 4.1.

4 Results

The following section presents the distribution of manually collected annotations from sampling A_NP, our validation of updated P_A and P_NA post MedCAT fine-tuning, and the final distribution of codes found in our produced silver standard dataset.

Adding annotations to selected text spans directly adds the spans to the MedCAT dictionary, thereby ensuring further text spans of the same content are annotated by the model - if the text span is unique. We collect 864 annotations after reviewing 4000 randomly sampled DD notes from the A_NP (Assigned, Not Predicted) dataset. 21.6% of DDs provide further annotations suggesting that the majority of missed codes lie outside the DD subsection, or are incorrectly assigned.

Figure 1 shows the distributions of manually collected code annotations and the current MIMIC-III set of clinical codes, grouped by their top-level axis as specified by ICD-9-CM hierarchy.

We collect proportionally consistent annotations for most groups, including the 390-459 chapter (Diseases Of The Circulatory System), which is the top occurring group in both scales. However, for groups such as 240-279 (endocrine, nutritional and metabolic diseases) and 460-519 (diseases of the respiratory system) we see proportionally fewer manually collected examples despite the high number of occurrence of codes assigned within MIMIC-III. We explain this by the DD subsection lacking appropriate detail to assign the specific code. For example codes under 250.* for diabetes mellitus and the various forms of complications are assigned frequently but often lack the appropriate level of detail specifying the type, control status and the manifestation of complication.

Using the manual amendments made on the 864 new annotations, we re-run the MedCAT model on the entire DD subsection dataset, producing updated P_NA, P_A and A_NP datasets. We acknowledge A_NP likely still includes cases of abbreviations, synonyms as we only subsampled 10 documents per code allowing for further improvements to the model.

The MedCAT fine-tuning process was run until convergence as measured by precision, recall and F1 achieving scores 0.90, 0.92 and 0.91 respectively on a held out a test-set with train/test splits 80/20. The fine-tuning code is made available[4]. Annotations are available upon request given the appropriate MIMIC-III licenses.

4.1 P_A & P_NA Validation

We use the MedCATTrainer interface to validate our MedCAT model predictions in the 'Predicted, Assigned' (P_A) and 'Predicted, Not Assigned' (P_NA) datasets. We sample (a maximum of) 10 unique predictions for each ICD-code resulting in 179 & 182 ICD-9 codes and 1588 & 1580 manually validated predictions from P_A and P_NA respectively. The validation of code assignment is performed by a medical informatics PhD student with no professional clinical coding experience and a qualified clinical coder, marking each term as correct or incorrect. We achieve good agreement with a Cohen's Kappa of 0.85 and 0.8 resulting in 95.51% and 87.91% marked correct for P_A and P_NA respectively. We exclude from further experiments all codes that fail this validation step as they are not trustworthy predictions made by MedCAT.

4.2 Aggregate Assigned Codes & Codes Silver Standard

We proportionally predict ~10% (n=42,837) of total assigned codes (n=432,770). We predict ~16%

---

[4]https://github.com/tomolopolis/MIMIC-III-Discharge-Diagnosis-Analysis/blob/master/Run_MedCAT.ipynb

of total assigned codes (n=258,953) if we only consider the 182 codes that resulted in at least one matched assignment to those present in the MIMIC-III assigned codes.

We label and gather our three datasets into a single table, with an extra column called 'validated', with values: 'yes' for codes that have matched with an assigned code (P_A), 'new_code' for newly discovered codes (P_NA), and 'no' for codes that we were not able to validate (A_NP). We have made this silver-standard dataset available alongside our analysis code[5].

4.3 Undercoding in MIMIC-III

This work aims to identify inconsistencies and variability in coding accuracy in the current MIMIC-III dataset. Ultimately to rigorously identify undercoding of clinical text full, double blind manual coding would be performed. However, as previously discussed, this is prohibitively expensive.

Comparing the codes predicted by MedCAT to the existing assigned codes enables the development of an understanding of specific groups of codes that exhibit possible undercoding. In this section we firstly show the effectiveness of our method in terms of DD subsection prediction coverage. We then present our predicted code distributions against the MIMIC-III assigned codes at the ICD code chapter level, highlighting the most prevalent missing codes and showing correlations between document length and prevalence.

4.3.1 Prediction Coverage

MedCAT provides predictions at the text span level, with only one concept prediction per span. We can therefore calculate the breadth of coverage of our predictions across all DD subsections. Figure 2 shows the proportion of DD subsection text that are included in code predictions. We note the 100% proportion (n=2105) is 75% larger than the next largest indicating that we are often utilising the entire DD subsection to suggest codes although the majority of the coverage distribution is around the 40-50% range.

We find a token length distribution of DD subsections with μ =14.54, σ =15.9, $Med = 10$ and $IQR = 14$ and a code extraction distribution with $\mu = 3.6$ and $\sigma = 3.1$, $Med = 3$ and $IQR = 4$ suggesting the DD subsections are complex and often list multiple conditions of which we identify, on average, 3 to 4 conditions.

[5]https://tinyurl.com/u8yae8n

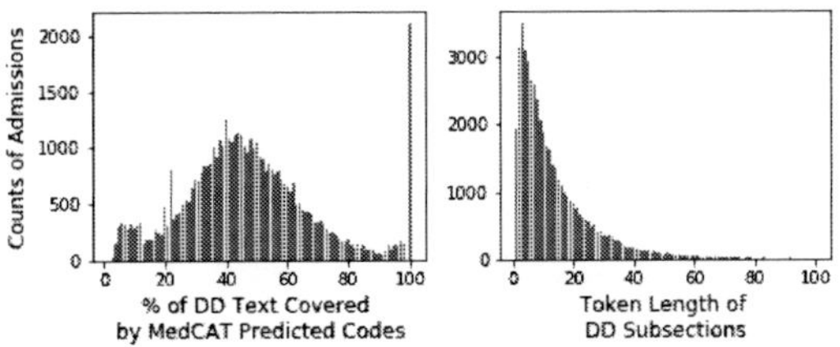

Figure 2: Left: Counts of admissions and the associated % of characters covered by MedCAT code predictions. Right:Distribution of DD token lengths

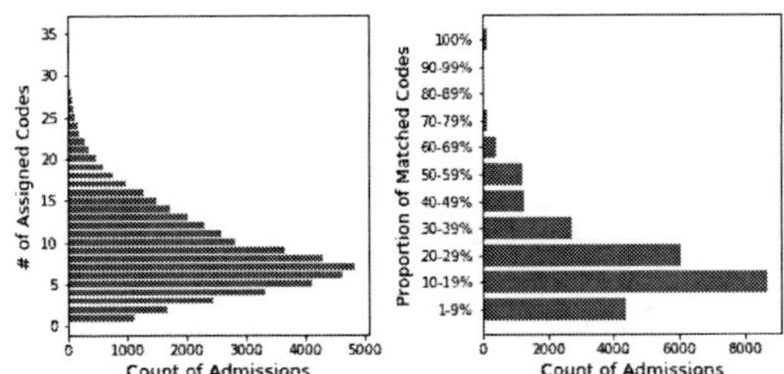

Figure 3: Proportions of matching predictions against total number of assigned codes per admission.

4.3.2 Predicted & Assigned

Figure 3 shows the distributions of the number of assigned codes and the proportion of matches grouped into buckets of 10% intervals. We see a high proportion of matches in assigned codes in the 1-40% range, indicating that although the DD subsection does contribute to the assigned ICD codes, many of the assigned codes are still missed. We exclude the admissions that had 0 matched codes and discuss this result further in Section 4.3.4.

If we order codes by the number of predicted and assigned we find the three highest occurring codes (4019, 41404, 4280) in MIMIC-III also rank highest in our predictions. However, we note that these three common codes only yield 25-39% of their total assigned occurrence, which could be explained by these chronic conditions not being listed in the DD subsection and referred elsewhere in the note. If we normalise predictions by their prevalence, we are most successful in matching specific conditions applicable to preterm newborns (7470, 7766), pulmonary embolism (41519) and liver cancer (1550), all of which we match between 69-55% but rank 114-305 in total prevalence. We suggest these diagnoses are either acute, or the primary cause of an ICU admission so will be specified in the DD subsection.

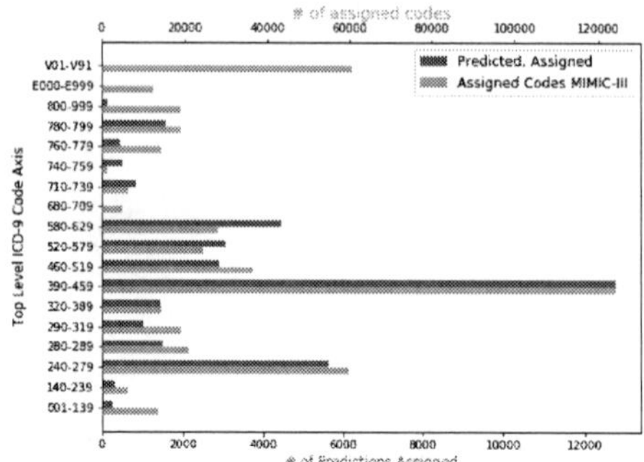

Figure 4: **Predicted, Assigned Codes** grouped by top-level code group vs total assigned codes

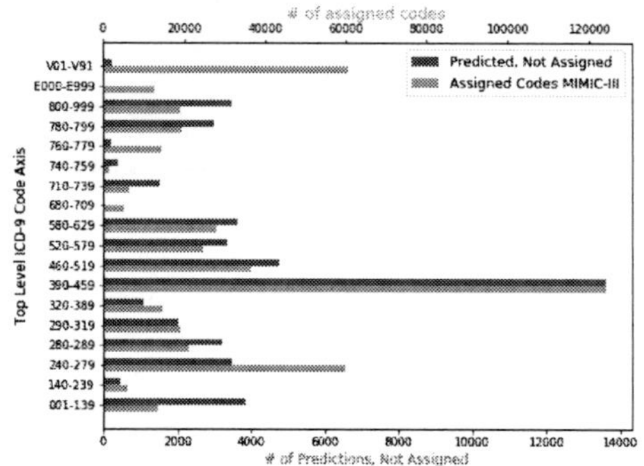

Figure 5: **Predicted, Not Assigned Codes** grouped by top-level code group vs total assigned codes

We also group the predicted codes into their respective top-level ICD-9 groups in Figure 4 and observe that predicted assigned codes display a similar distribution to total assigned codes. We quantify the difference in distributions via the Wasserstein metric or 'Earth Movers Distance'(Ramdas et al., 2015). This metric provides a single measure to compare the difference in our 3 datasets distributions when compared with the current assigned code distribution. We compute a small 2.7×10^{-3} distance between both distributions, suggesting our method proportionally identifies previously assigned codes from the DD subsection alone.

4.3.3 Predicted & Not Assigned

This dataset highlights codes that may have been missed from the current assigned codes.

Figure 5 shows that the distribution of predicted but not assigned codes is minimally different for most codes, supporting our belief that the MIMIC-III assigned codes are not wholly untrustworthy, but are likely under-coded in specific areas.

From this dataset we calculate how many examples of each code that has potentially been missed, or potentially under-coded. For the 10 most frequently assigned codes we see 0-35% missing occurrences. We also identify the most frequent code 4019 (Unspecified Essential Hypertension) has 16% or 3312 potentially missing occurrences.

To understand if DD subsection length impacts the occurrence of 'missed' codes we first calculate a Pearson-Correlation coefficient of 0.17 for DD subsection line length and counts of assigned codes over all admissions. This suggests a weak positive correlation between admission complexity and number of existing assigned codes.

In contrast we find a stronger positive correlation of 0.504 for predicted and not assigned codes and DD subsection line length. This implies that where an episode has a greater number of diagnoses or the complexity of an admission is greater, there is a likelihood to result in more codes being missed during routine collection.

We compute the Wasserstein metric between these two distributions at 1.6×10^{-2}. This demonstrates a degree of similarity between distributions albeit is 8x further from the Predicted and Assigned dataset distance presented in Section 4.3.2. We expect to see a larger distance here as we are detecting codes that are indicated in the text but have been missed during routine code assignment.

4.3.4 Assigned & Not Predicted

We observe that the distribution of assigned and not predicted codes largely mirrors the distribution of total codes assigned in MIMIC-III with a Wasserstein distance of 2.7×10^{-3} that is similar to the distance observed in in our Predicted and Assigned Section 4.3.2) dataset. This suggests that our method is proportionally consistent at not annotating codes that have likely been assigned from elsewhere in the admission, but may also be incorrectly assigned.

5 Discussion

On aggregate, the predicted codes by our MedCAT model suggest that the discharge diagnosis sections listed in 92% of all discharge notifications are not sufficient for full coding of an episode. Unsurprisingly, this confirms that clinicians infrequently document all codeable diagnoses within the discharge summary. Although, as previously stated, coders are not permitted to make clinical inferences. Therefore, to correctly assign a code, the diagnoses must be present within the documented patient episode within the structured or unstructured data.

However, the positive correlation between document length and number of predicted codes indicates that missed codes are more prevalent in highly complex cases with many diagnoses. From a coding workflow perspective, coders operate under strict time schedules and are required to code a minimum number of episodes each day. Therefore, it logically follows that the complexity of a case directly correlates to the number of codes missed during routine collection.

Looking at individual code groups we find 240-279 is not predicted proportionally with assigned codes both in P_A and P_NA. We explain this as follows. Firstly, DD subsections generally convey clinically important diagnoses for follow-up care. Certain codes such as (250.*) describe diabetes mellitus with specific complications, but the DD subsection will often only describe the diagnoses 'DMI' or 'DMII'. Secondly, ICU admissions are for individuals with severe illness and therefore are likely to have a high degree of co-morbidity. This is implied by the majority of patients (74%) are assigned between 4 and 16 codes.

We also observe E000-E999 and V01-V99 codes are disproportionately not predicted. However, this is expected given that both groups are supplementary codes that describe conditions or factors that contribute to an admission but would likely not be relevant for the DD subsection.

In contrast, we observe a disproportionately large number of predictions for 001-139 (Infectious and Parasitic Diseases). This is primarily driven by 0389 (Unspecified septicemia). A proportion of these predictions may be in error as the specific form of septicemia is likely described in more detail elsewhere in the note and therefore coded as the more specific form.

5.1 Method Reproducibility & Wider Utility

Inline with the suggestions of Johnson et al. (2017), the original authors of MIMIC-III, we have attempted to provide the research community all available materials to reproduce and build upon our experiments and method for the development of silver standard datasets. Specifically, we have made the following available as open-source: the SQL scripts to extract the raw data from a replica of the MIMIC-III database, the script required to parse DD subsections, an example script to build a pre-trained MedCAT model, the script required to run MedCAT on the DD subsections, load into the annotator and finally re-run MedCAT and perform experimental analysis alongside outputting the silver standard dataset[6].

Given these materials it is possible for researchers to replicate and build upon our method, or directly use the silver standard dataset in future work that investigates automated clinical coding using MIMIC-III. The silver standard dataset clearly marks if each assigned code has been validated or not, or if it is a new code according to our method.

6 Conclusions & Future Work

This work highlighted specific problems with using MIMIC-III as a dataset for training and testing an automated clinical coding system that would limit model performance within a real deployment.

We identified and deterministically extracted the discharge diagnosis (DD) subsections from discharge summaries. We subsequently trained an NER+L model (MedCAT) to extract ICD-9 codes from the DD subsections, comparing the results across the full set of assigned codes. We find our method covers 47% of all tokens, considering we only take 400 of the ~7k unique codes and perform minimal data cleaning of the DD subsection. We have shown in Section 4.3.2 and 4.3.3 that the MedCAT predicted codes are proportionally inline with assigned codes in MIMIC-III.

Interestingly, we found a 0.504 positive correlation between DD length and the number of codes predicted by MedCAT, but not assigned in MIMIC-III. This result can be understood by observing that the ICU admissions in MIMIC-III can be extremely complex, with up to 30 clinical codes assigned to a single episode. The DD subsections alone can contain up to 50 line items indicating highly complex cases where codes could easily be missed.

We found that the code group 390-459 (Diseases of the Circulatory System) is both the most assigned group and the group of codes where there are the most missing predictions from our model. Furthermore, codes such as Hypertension (4019), Sepsis and Septicemia (0389, 99591), Gastrointestinal hemorrhage (95789), Chronic Kidney disease (5859), anemia (2859) and Chronic obstructive asthma (49320) are all frequently assigned but

---

[6]https://github.com/tomolopolis/MIMIC-III-Discharge-Diagnosis-Analysis

are also the highest occurring conditions that appear in the DD diagnosis subsection but are not assigned in the MIMIC-III dataset. This suggests that MIMIC-III exhibits specific cases of undercoding, especially with codes that are frequently occurring in patients but are not likely to be the primary diagnosis for an admission to the ICU.

As we only use the DD section, there are many codes which likely appear elsewhere in the note that we cannot assign. Although 92% of discharge summaries contain DD subsections we only match $\sim 16\%$ of assigned codes. We suggest this is due to: our NER+L model lacking the ability to identify more synonyms and abbreviations for conditions, the DD subsections lacking enough detail to assign codes and in some occasions, little evidence to suggest a code assignment. Our textual span coverage, presented in Section 4.3.1 demonstrates that we often cover all available discharge diagnosis, although there is still room for improvement as the majority of the coverage distribution is around the 50% mark.

For future work we foresee applying the same method to either the entire discharge summary or more specific sections such as 'previous medical history' to surface chronic codeable diagnoses that could be validated against the current assigned code set. Researchers would however likely need to address false positive code predictions as clinical coding requires assigned codes to be from current conditions associated with an admission.

In conclusion, this work has found that frequently assigned codes in MIMIC-III display signs of undercoding up to 35% for some codes. With this finding we urge researchers to continue to develop automated clinical coding systems using MIMIC-III, but to also consider using our silver standard dataset or build on our method to further improve the dataset.

Acknowledgments

RD's work is supported by 1.National Institute for Health Research (NIHR) Biomedical Research Centre at South London and Maudsley NHS Foundation Trust and King's College London. 2. Health Data Research UK, which is funded by the UK Medical Research Council, Engineering and Physical Sciences Research Council, Economic and Social Research Council, Department of Health and Social Care (England), Chief Scientist Office of the Scottish Government Health and Social Care Directorates, Health and Social Care Research and Development Division (Welsh Government), Public Health Agency (Northern Ireland), British Heart Foundation and Wellcome Trust. 3. The National Institute for Health Research University College London Hospitals Biomedical Research Centre. This paper represents independent research part funded by the National Institute for Health Research (NIHR) Biomedical Research Centre at South London and Maudsley NHS Foundation Trust and King's College London. The views expressed are those of the author(s) and not necessarily those of the NHS, MRC, NIHR or the Department of Health and Social Care.

References

Alan R Aronson, Olivier Bodenreider, Dina Demner-Fushman, Kin Wah Fung, Vivian K Lee, James G Mork, Aurélie Névéol, Lee Peters, and Willie J Rogers. 2007. From indexing the biomedical literature to coding clinical text: Experience with MTI and machine learning approaches. In Proceedings of the Workshop on BioNLP 2007: Biological, Translational, and Clinical Language Processing, BioNLP '07, pages 105–112, Stroudsburg, PA, USA. Association for Computational Linguistics.

Sandeep Ayyar, O B Don, and W Iv. 2016. Tagging patient notes with icd-9 codes. In Proceedings of the 29th Conference on Neural Information Processing Systems.

Tal Baumel, Jumana Nassour-Kassis, Raphael Cohen, Michael Elhadad, and Nóemie Elhadad. 2018. Multi-Label classification of patient notes: Case study on ICD code assignment. In Workshops at the Thirty-Second AAAI Conference on Artificial Intelligence.

Olivier Bodenreider. 2004. The unified medical language system (UMLS): integrating biomedical terminology. Nucleic Acids Res., 32(Database issue):D267–70.

Sue Bowman. 2013. Impact of electronic health record systems on information integrity: quality and safety implications. Perspect. Health Inf. Manag., 10:1c.

E M Burns, E Rigby, R Mamidanna, A Bottle, P Aylin, P Ziprin, and O D Faiz. 2012. Systematic review of discharge coding accuracy.

CHKS Ltd. 2014. CHKS - insight for better healthcare: The quality of clinical coding in the NHS. https://bit.ly/34eU5g3. Accessed: 2019-5-10.

Matus Falis, Maciej Pajak, Aneta Lisowska, Patrick Schrempf, Lucas Deckers, Shadia Mikhael, Sotirios Tsaftaris, and Alison O'Neil. 2019. Ontological attention ensembles for capturing semantic concepts in ICD code prediction from clinical text.

Richárd Farkas and György Szarvas. 2008. Automatic construction of rule-based ICD-9-CM coding systems. BMC Bioinformatics, 9 Suppl 3:S10.

Toni Henderson, Jennie Shepheard, and Vijaya Sundararajan. 2006. Quality of diagnosis and procedure coding in ICD-10 administrative data. Med. Care, 44(11):1011–1019.

Jinmiao Huang, Cesar Osorio, and Luke Wicent Sy. 2018. An empirical evaluation of deep learning for ICD-9 code assignment using MIMIC-III clinical notes.

Alistair E W Johnson, Tom J Pollard, and Roger G Mark. 2017. Reproducibility in critical care: a mortality prediction case study. In Proceedings of the 2nd Machine Learning for Healthcare Conference, volume 68 of Proceedings of Machine Learning Research, pages 361–376, Boston, Massachusetts. PMLR.

Alistair E W Johnson, Tom J Pollard, Lu Shen, Li-Wei H Lehman, Mengling Feng, Mohammad Ghassemi, Benjamin Moody, Peter Szolovits, Leo Anthony Celi, and Roger G Mark. 2016. MIMIC-III, a freely accessible critical care database. Sci Data, 3:160035.

Ramakanth Kavuluru, Anthony Rios, and Yuan Lu. 2015. An empirical evaluation of supervised learning approaches in assigning diagnosis codes to electronic medical records. Artif. Intell. Med., 65(2):155–166.

Rae A Kokotailo and Michael D Hill. 2005. Coding of stroke and stroke risk factors using international classification of diseases, revisions 9 and 10. Stroke, 36(8):1776–1781.

Zeljko Kraljevic, Daniel Bean, Aurelie Mascio, Lukasz Roguski, Amos Folarin, Angus Roberts, Rebecca Bendayan, and Richard Dobson. 2019. MedCAT – medical concept annotation tool.

Leah S Larkey and W Bruce Croft. 1996. Combining classifiers in text categorization. In SIGIR, volume 96, pages 289–297.

Tomas Mikolov, Ilya Sutskever, Kai Chen, Greg S Corrado, and Jeff Dean. 2013. Distributed representations of words and phrases and their compositionality. In C J C Burges, L Bottou, M Welling, Z Ghahramani, and K Q Weinberger, editors, Advances in Neural Information Processing Systems 26, pages 3111–3119. Curran Associates, Inc.

James Mullenbach, Sarah Wiegreffe, Jon Duke, Jimeng Sun, and Jacob Eisenstein. 2018. Explainable prediction of medical codes from clinical text.

Adler Perotte, Rimma Pivovarov, Karthik Natarajan, Nicole Weiskopf, Frank Wood, and Noémie Elhadad. 2014. Diagnosis code assignment: models and evaluation metrics. J. Am. Med. Inform. Assoc., 21(2):231–237.

Aaditya Ramdas, Nicolas Garcia, and Marco Cuturi. 2015. On wasserstein two sample testing and related families of nonparametric tests.

Mohammed Saeed, Mauricio Villarroel, Andrew T Reisner, Gari Clifford, Li-Wei Lehman, George Moody, Thomas Heldt, Tin H Kyaw, Benjamin Moody, and Roger G Mark. 2011. Multiparameter intelligent monitoring in intensive care II: a public-access intensive care unit database. Crit. Care Med., 39(5):952–960.

Thomas Searle, Zeljko Kraljevic, Rebecca Bendayan, Daniel Bean, and Richard Dobson. 2019. MedCATTrainer: A biomedical free text annotation interface with active learning and research use case specific customisation. In Proceedings of the 2019 Conference on Empirical Methods in Natural Language Processing and the 9th International Joint Conference on Natural Language Processing (EMNLP-IJCNLP): System Demonstrations, pages 139–144, Stroudsburg, PA, USA. Association for Computational Linguistics.

World Health Organisation. 2011. International Statistical Classification of Diseases and Related Health Problems.

Pengtao Xie and Eric Xing. 2018. A neural architecture for automated ICD coding. In Proceedings of the 56th Annual Meeting of the Association for Computational Linguistics (Volume 1: Long Papers), pages 1066–1076.

Rui Zhang, Serguei V S Pakhomov, Elliot G Arsoniadis, Janet T Lee, Yan Wang, and Genevieve B Melton. 2017. Detecting clinically relevant new information in clinical notes across specialties and settings. BMC Med. Inform. Decis. Mak., 17(Suppl 2):68.

Comparative Analysis of Text Classification Approaches in Electronic Health Records

Aurelie Mascio[*] Zeljko Kraljevic[*]
Daniel Bean   Richard Dobson   Robert Stewart   Rebecca Bendayan   Angus Roberts

Department of Biostatistics
& Health Informatics
King's College London, UK

aurelie.mascio@kcl.ac.uk   zeljko.kraljevic@kcl.ac.uk

Abstract

Text classification tasks which aim at harvesting and/or organizing information from electronic health records are pivotal to support clinical and translational research. However these present specific challenges compared to other classification tasks, notably due to the particular nature of the medical lexicon and language used in clinical records.

Recent advances in embedding methods have shown promising results for several clinical tasks, yet there is no exhaustive comparison of such approaches with other commonly used word representations and classification models.

In this work, we analyse the impact of various word representations, text pre-processing and classification algorithms on the performance of four different text classification tasks. The results show that traditional approaches, when tailored to the specific language and structure of the text inherent to the classification task, can achieve or exceed the performance of more recent ones based on contextual embeddings such as BERT.

1   Introduction

Clinical text classification is an important task in natural language processing (NLP) (Yao et al., 2019), where it is critical to harvest data from electronic health records (EHRs) and facilitate its use for decision support and translational research. Thus, it is increasingly used to retrieve and organize information from the unstructured portions of EHRs (Mujtaba et al., 2019).

Examples include tasks such as: (1) detection of smoking status (Uzuner et al., 2008); (2) classification of medical concept mentions into family versus patient related (Dai, 2019); (3) obesity classification from free text (Uzuner, 2009); (4) identification of patients for clinical trials (Meystre et al., 2019).

Most of these tasks involve mapping mentions in narrative texts (e.g. "pneumonia") to their corresponding medical concepts (and concept ID) generally using the Unified Medical Language System (UMLS) (Bodenreider, 2004), and then training a classifier to identify these correctly (e.g. "pneumonia positive" versus "pneumonia negative") (Yao et al., 2019).

Text classification performed on medical records presents specific challenges compared to the general domain (such as newspaper texts), including dataset imbalance, misspellings, abbreviations or semantic ambiguity (Mujtaba et al., 2019).

Despite recent advances in NLP, including neural-network based word representations such as BERT (Devlin et al., 2019), few approaches have been extensively tested in the medical domain and rule-based algorithms remain prevalent (Koleck et al., 2019). Furthermore, there is no consensus on which word representation is best suited to specific downstream classification tasks (Si et al., 2019; Wang et al., 2018).

The purpose of this study is to analyse the impact of numerous word representation methods (bag-of-word versus traditional and contextual word embeddings) as well as classification approaches (deep learning versus traditional machine learning methods) on the performance of four different text classification tasks. To our knowledge this is the first paper to test a comprehensive range of word representation, text pre-processing and classification methods combinations on several medical text tasks.

---

[*] These two authors contributed equally.

Proceedings of the BioNLP 2020 workshop, pages 86–94
Online, July 9, 2020 ©2020 Association for Computational Linguistics

2   Materials & Methods

2.1   Datasets and text classification tasks

In order to conduct our analysis we derived text classification tasks from MIMIC-III (Multiparameter Intelligent Monitoring in Intensive Care) (Johnson et al., 2016), and the Shared Annotated Resources (ShARe)/CLEF dataset (Mowery et al., 2014). These datasets are commonly used for challenges in medical text mining and act as benchmarks for evaluating machine learning models (Purushotham et al., 2018).

MIMIC-III dataset   MIMIC-III (Johnson et al., 2016) is an openly available dataset developed by the MIT Lab for Computational Physiology. It comprises clinical notes, demographics, vital signs, laboratory tests and other data associated with 40,000 critical care patients.

We used MedCAT (Kraljevic et al., 2019) to prepare the dataset and annotate a sample of clinical notes from MIMIC-III with UMLS concepts (Bodenreider, 2004). We selected the concepts with the UMLS semantic type Disease or Syndrome (corresponding to T047), out of which we picked the 100 most frequent Concept Unique Identifier (CUIs, allowing to group mentions with the same meaning). For each concept we then randomly sampled 4 documents containing a mention of each concept, resulting in 400 documents with 2367 annotations in totals. The 100 most frequent concepts in these documents were manually annotated (and manually corrected in case of disagreement) for two text classification tasks:

- Status (affirmed/other, indicating if the disease is affirmed or negated/hypothetical);

- Temporality (current/other, indicating if the disease is current or past).

Such contextual properties are often critical in the medical domain in order to extract valuable information, as evidenced by the popularity of algorithms like ConText or NegEx (Harkema et al., 2009; Chapman et al., 2001).

Annotations were performed by two annotators, achieving an overall inter-annotator agreement above 90%. These annotations will be made publicly available.

ShARe/CLEF   (MIMIC-II)   dataset   The ShARe/CLEF annotated dataset proposed by Mowery et al. (2014) is based on 433 clinical records from the MIMIC-II database (Saeed et al., 2002). It was generated for community distribution as part of the Shared Annotated Resources (ShARe) project (Elhadad et al., 2013), and contains annotations including disorder mention spans, with several contextual attributes. For our analysis we derived two tasks from this dataset, focusing on two attributes, comprising 8075 annotations for each:

- Negation (yes/no, indicating if the disorder is negated or affirmed);

- Uncertainty (yes/no, indicating if the disorder is hypothetical or affirmed).

Text classification tasks   For both annotated datasets, we extracted from each document the portions of text containing a mention of the concepts of interest, keeping 15 words on each side of the mention (including line breaks). Each task is then made up of sequences comprising around 31 words, centered on the mention of interest, with its corresponding meta-annotation (status, temporality, negation, uncertainty), making up four text classification tasks, denoted:

- MIMIC | Status;

- MIMIC | Temporality;

- ShARe | Negation;

- ShARe | Uncertainty.

Table 1 summarizes the class distribution for each task.

Task	Class 1	Class 2	Total
MIMIC | Status (1: affirmed, 2: other)	1586 (67%)	781 (33%)	2367
MIMIC | Temporality (1: current, 2: other)	2026 (86%)	341 (14%)	2367
ShARe | Negation (1: yes, 2: no)	1470 (18%)	6605 (82%)	8075
ShARe | Uncertainty (1: yes, 2: no)	729 (9%)	7346 (91%)	8075

Table 1: Class distribution

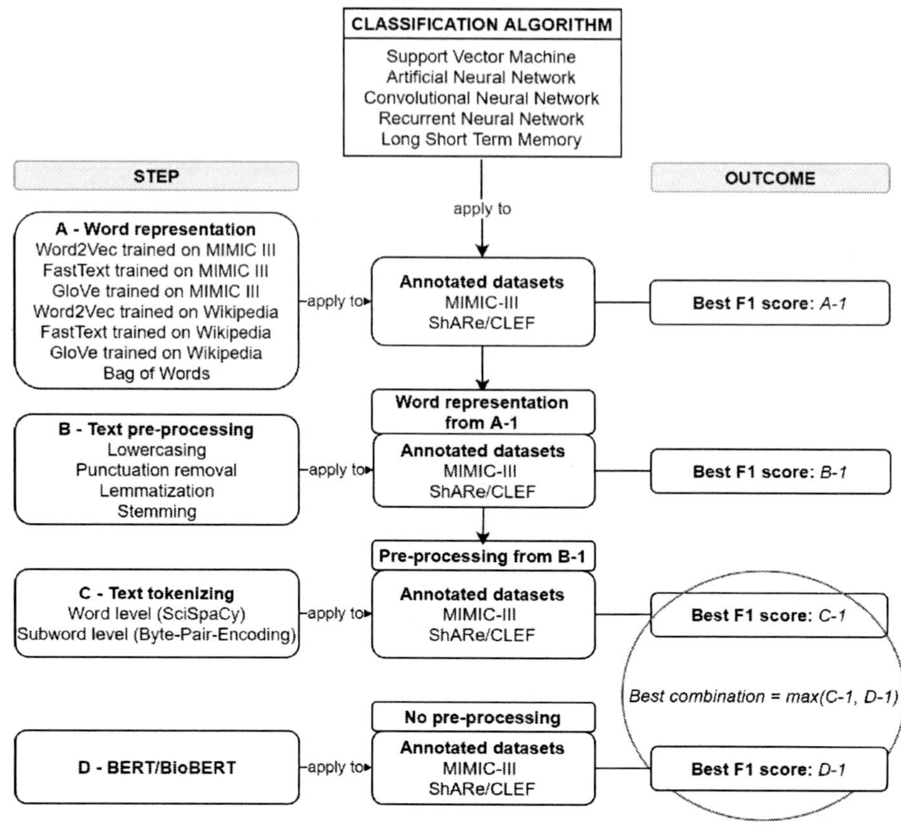

Figure 1: Main workflow

2.2 Evaluation steps and main workflow

We used the four different text classification tasks described in Section 2.1 in order to explore various combinations of word representation models (see Section 2.3), text pre-processing and tokenizing variations (Section 2.4) and classification algorithms (Section 2.5). In order to evaluate the different approaches we followed the steps detailed in Table 2 and Figure 1 for all four classification tasks.

For each step we measured the impact by evaluating the best possible combination, based on the average F1 score (weighted average score derived from 10-fold cross validation results).

2.3 Word representation models

Word embeddings as opposed to bag-of-words (BoW) present the advantage of capturing semantic and syntactic meaning by representing words as real valued vectors in a dimensional space (vectors that are close in that space will represent similar words). Contextual embeddings go one step further by capturing the context surrounding the word, whilst traditional embeddings assign a single representation to a given word.

For our analysis we considered four off-the-shelf embedding models, pre-trained on public domain data, and compared them to the same embedding models trained on biomedical corpora, as well as a BoW representation.

For the traditional embeddings we chose three

Step	Description	Outcome (best F1)
A	Run all bag-of-word and traditional embeddings + classification algorithms and select the best combination (using baseline methods for text pre-processing and tokenization)	A-1
B	Using A-1 as the new baseline model, test different pre-processing methods (lowercasing, punctuation removal, lemmatization, stemming)	B-1
C	Using B-1 as the new baseline model, compare various tokenizers (word and subword level)	C-1
D	Test contextual embedding approaches: BERT (base, uncased) and BioBERT	D-1

Table 2: Evaluation steps

commonly used algorithms, namely Word2Vec (Mikolov et al., 2013), GloVe (Pennington et al., 2014) and FastText (Bojanowski et al., 2017).

We used publicly available models pre-trained on Wikipedia and Google News for all three (Yamada et al., 2018).

To obtain medical specific models we trained all three on MIMIC-III clinical notes (covering 53,423 intensive care unit stays, including those used in the classification tasks) (Johnson et al., 2016). The following hyperparameters, aligned to off-the-shelf pre-trained models, were used: dimension of 300, window size of 10, minimum word count of 5, uncased, punctuation removed.

For the contextual embeddings we used BERT base (Devlin et al., 2019), and BioBERT (Lee et al., 2019) which are pre-trained respectively on general domain corpora and biomedical literature (PubMed abstracts and PMC articles).

Finally we used a BoW representation as a baseline approach.

2.4 Text pre-processing and tokenizers

In addition to pre-training several embedding models, we tested two different text tokenization methods, using the following types of tokenizers: (1) SciSpaCy (Neumann et al., 2019), a traditional tokenizer based on word detection; and (2) byte-pair-encoding (BPE) adapted to word segmentation that works on subword level (Gage, 1994; Sennrich et al., 2016).

For the word level tokenizer we chose SciSpaCy as it is specifically aimed at biomedical and scientific text processing. We further tested additional text pre-processing: lowercasing, punctuation removal, stopwords removal, stemming and lemmatization.

For the subword BPE tokenizer we used byte level byte-pair-encoding (BBPE) (Wang et al., 2019; Wolf et al., 2020). In this case the only pre-processing performed is lowercasing, whilst everything else including line breaks and spaces is left as is. This approach allows to limit the vocabulary size and is especially useful in the medical domain where a large number of words are very rare. We limited the number of words to 30522, a standard vocabulary size also used in BERT (Devlin et al., 2019).

2.5 Text classification algorithms

On all four classification tasks, we tested various machine learning algorithms which are widely used for clinical data mining tasks and achieve state-of-the-art performance (Yao et al., 2019), namely artificial neural network (ANN), convolutional neural network (CNN), recurrent neural network (RNN), bi-directional long short term memory (Bi-LSTM), and BERT (Devlin et al., 2019; Wolf et al., 2020). We compared these with a statistics-based approach as a baseline, using a Support Vector Machine (SVM) classifier, a popular method used for classification tasks (Cortes and Vapnik, 1995).

For Bi-LSTM and RNN, we tested both a standard approach and one that is configured to simulate attention on the medical entity of interest. This custom approach consisted in taking the representation of the network at the position of the entity of interest, which in most cases corresponds to the center for each sequence. We refer to this latter approach as custom Bi-LSTM and custom RNN.

For ANN and statistics-based models, which are limited by the size of the dataset and embeddings (300 dimensions x 31 words x 2300 or 8000 sequences), we chose to represent sequences by averaging the embeddings of the words composing each sequence. This representation method is commonly used and has proven efficient for various NLP applications (Kenter et al., 2016).

Furthermore, each of these models was tested using different sets of parameters (e.g. varying the support function, dropout, optimizer, as reported in Table 3), the ones producing the best performance were selected for further testing and are summarized in Table 3.

	SVM	ANN	CNN	RNN Bi-LSTM
Kernel or activation function	**Radial basis** Linear Poly Sigmoid	ReLU + sigmoid	ReLU (with max pooling)	N/A
Layers	N/A	2	3	2
Filters	N/A	N/A	128	N/A
Hidden units dimensions	N/A	100	N/A	300
Dropout	N/A	**0.5** 0	**0.5** 0	**0.5** 0
Optimizer	N/A	**Adam** Stochastic Gradient Descent	**Adam** Stochastic Gradient Descent	**Adam** Stochastic Gradient Descent
Learning rate	N/A	0.001	0.001	0.001
Epochs	tolerance: 0.001	5000	200	50

Table 3: Classifiers and corresponding parameters evaluated. Parameters highlighted in **bold** were the ones selected based on performance.

3   Results

3.1   Performance comparison for all embedding and algorithm combinations (steps A & D)

In this section we compare the performance of the different embeddings and classification approaches. We report the weighted average F1/precision/recall (weighted average value obtained from the 10-fold cross-validation results) for selected combinations on the four classification tasks in Tables 4 and 5 (full results in Appendix A.1).

For all word embedding methods tested (Word2Vec, GloVe, FastText), the ones trained on biomedical data show the best performance (see Table 4).

For classification algorithms, the best performance is obtained when using the custom Bi-LSTM model configured to target the biomedical concept of interest (see Table 5). Both contextual embeddings (BERT and BioBERT), whether trained on biomedical or general corpora, outperform any other combination of embedding/classification algorithm tested, and give results very close to the customized Bi-LSTM, as shown in Table 5.

This indicates that for tasks incorporating information about the position of the entity of interest in the text (e.g. ShARe which reports disorder mentions span offsets), the custom Bi-LSTM approach performs better than BioBERT, without necessitating any text pre-processing.

On the other hand, when looking at pure text classification, BioBERT shows better performance than a Bi-LSTM approach, and consequently may be preferred for tasks where the sequence of interest is not easily centered on a specific entity.

Finally, whilst the performance of BERT and BioBERT is relatively similar, BioBERT converges faster across all tasks tested.

3.2   Impact of text pre-processing (step B)

In addition to exploring various embeddings, we tested the impact of text pre-processing on classification task performance. In order to do so, we selected the best performing word embedding obtained in the previous step (Word2Vec trained on MIMIC-III, using SciSpacy tokenizer), and compared performances between all text cleaning variations (lowercasing, punctuation removal, stemming, lemmatization).

For each variant investigated, the same pre-processing settings were applied to prepare the annotated corpus as well as to the entire MIMIC-III dataset, which was then used to re-train Word2Vec. This ensured the same vocabulary was used across the embedding and sequences to classify for each experiment.

The results, summarized in Table 6, suggest that text pre-processing has a minor impact for all classification algorithms tested. Notably, stemming and lemmatization have a slightly negative impact on performance.

3.3   Impact of tokenizers (step C)

We tested the impact of tokenization on the performance of text classification tasks, focusing on SciSpacy and BBPE tokenizers, as they allow us to compare whole word versus subword unit methods. The results for the MIMIC | Status task (and using Word2Vec trained on MIMIC-III) are shown in Table 7, and indicate that the performances are roughly similar when using the BBPE tokenizer compared to SciSpacy.

Furthermore we compared both approaches in terms of speed and vocabulary size. Tokenizing text took on average 2.5 times longer with Scispacy (250 seconds to tokenize 100,000 medical notes for SciSpacy versus 99 seconds for BBPE, excluding model loading time). For the models trained on MIMIC-III corpus, Scispacy comprised 474,145 words, and BBPE 29,452 subword units.

3.4   Embeddings analysis: word similarities comparison

Finally, in order to analyse the differences between embeddings trained on general and medical corpora, we compared the semantic information captured by Word2Vec (using SciSpacy tokenizer and without any preliminary text pre-processing).

Table 8 explores word similarities by showing the top ten similar words for medical ("cancer") and non-medical ("concentration" and "attention") terms.

Notably, it highlights the numerous misspellings, abbreviations and domain-specific meanings contained in the medical lexicon, suggesting that general corpora such as Wikipedia may not be appropriate when working on data from medical records (and by implication, for other specific domains).

Model	Tokenizer	Embedding	F1-score (average from 10-fold cross validation)			
			MIMIC Status	MIMIC Temporality	ShARe Negation	ShARe Uncertainty
Bi-LSTM (custom)	SciSpacy	Wiki | Word2Vec	92.8%	97.3%	98.4%	96.7%
Bi-LSTM (custom)	SciSpacy	Wiki | GloVe	93.4%	97.2%	98.4%	97.2%
Bi-LSTM (custom)	SciSpacy	Wiki | FastText	93.6%	96.9%	98.6%	96.4%
Bi-LSTM (custom)	SciSpacy	MIMIC | Word2Vec	**94.5%**	**97.9%**	**98.7%**	**97.3%**
Bi-LSTM (custom)	SciSpacy	MIMIC | GloVe	93.9%	**97.9%**	**98.7%**	96.9%
Bi-LSTM (custom)	SciSpacy	MIMIC | FastText	93.7%	97.6%	98.5%	97.2%
BERT	WordPiece	BERTbase	91.5%	97.3%	98.2%	93.6%
BioBERT	WordPiece	BioBERT	93.4%	97.3%	98.5%	94.2%
SVM	SciSpacy	Wiki | Word2Vec	76.9%	94.8%	88.5%	85.9%
SVM	SciSpacy	Wiki | GloVe	78.6%	94.9%	88.8%	87.1%
SVM	SciSpacy	Wiki | FastText	78.1%	94.4%	88.7%	86.3%
SVM	SciSpacy	BoW	82.7%	96.0%	90.2%	91.7%
SVM	SciSpacy	MIMIC | Word2Vec	80.6%	95.1%	89.8%	90.2%
SVM	SciSpacy	MIMIC | GloVe	79.1%	94.1%	89.4%	87.6%
SVM	SciSpacy	MIMIC | FastText	79.6%	93.7%	88.9%	88.0%

Table 4: Comparison of embeddings (steps A & D)

Model	Tokenizer	Embedding	F1-score (average from 10-fold cross validation)			
			MIMIC Status	MIMIC Temporality	ShARe Negation	ShARe Uncertainty
Bi-LSTM	SciSpacy	MIMIC |Word2Vec	88.4%	97.1%	96.2%	94.1%
Bi-LSTM (custom)	SciSpacy	MIMIC |Word2Vec	**94.5%**	**97.9%**	**98.7%**	**97.3%**
BERT	WordPiece	BERTbase	91.5%	97.3%	98.2%	93.6%
BioBERT	WordPiece	BioBERT	93.4%	97.3%	98.5%	94.2%
ANN	SciSpacy	MIMIC |Word2Vec	80.9%	96.5%	88.6%	86.7%
CNN	SciSpacy	MIMIC |Word2Vec	84.6%	97.3%	92.0%	87.5%
RNN	SciSpacy	MIMIC |Word2Vec	77%	96.8%	94.0%	87.1%
RNN (custom)	SciSpacy	MIMIC |Word2Vec	89.5%	96.7%	97.9%	96.5%
SVM	SciSpacy	MIMIC |Word2Vec	80.6%	95.1%	89.8%	90.2%
ANN	SciSpacy	BoW	79.8%	94.8%	89.3%	89.3%
SVM	SciSpacy	BoW	82.7%	96%	90.2%	91.7%

Table 5: Comparison of classification algorithms (steps A & D)

Task	Embedding	Text pre-processing	F1-score (average from 10-fold cross validation)					
			SVM	ANN	RNN	RNN (custom)	CNN	Bi-LSTM (custom)
MIMIC | Status	MIMIC | Word2Vec	Lowercase (L)	**80.6%**	**80.9%**	77.0%	**89.5%**	84.6%	**94.5%**
MIMIC | Status	MIMIC | Word2Vec	L + punctuation removal (LP)	80.1%	80.0%	**80.2%**	86.1%	**84.7%**	94.4%
MIMIC | Status	MIMIC | Word2Vec	LP + lemmatizing	**80.6%**	79.6%	78.0%	86.3%	83.8%	94.1%
MIMIC | Status	MIMIC | Word2Vec	LP + stemming	80.4%	79.7%	79.4%	86.1%	84.1%	94.1%

Table 6: Comparison of text pre-processing methods (step B)

Task	Embedding	Tokenizer	F1-score (average from 10-fold cross validation)					
			SVM	ANN	RNN	RNN (custom)	CNN	Bi-LSTM (custom)
MIMIC | Status	MIMIC | Word2Vec	Sciscpacy	**80.6%**	**80.9%**	**77.0%**	**89.5%**	**84.6%**	94.5%
MIMIC | Status	MIMIC | Word2Vec	BBPE	78.8%	80.5%	76.5%	86.0%	84.3%	**94.7%**

Table 7: Comparison of tokenizing methods (step C)

Term: "cancer"		Term: "concentration"		Term: "attention"	
Word2Vec Medical	Word2Vec General	Word2Vec Medical	Word2Vec General	Word2Vec Medical	Word2Vec General
ca (0.78)	prostate (0.85)	hmf (0.51)	concentrations (0.71)	paid (0.43)	attentions (0.65)
carcinoma (0.78)	colorectal (0.82)	concentrations (0.49)	arbeitsdorf (0.67)	approximation (0.34)	notoriety (0.63)
cancer- (0.75)	melanoma (0.8)	formula (0.47)	vulkanwerft (0.65)	followup (0.33)	attracted (0.63)
caner (0.71)	pancreatic (0.8)	mct (0.47)	sophienwalde (0.64)	proximity (0.32)	criticism (0.63)
adenocarcinoma (0.71)	leukemia (0.79)	polycose (0.47)	lagerbordell (0.64)	short-term (0.31)	publicity (0.57)
ca- (0.64)	entity/breast_cancer (0.79)	virtue (0.45)	sterntal (0.64)	mangagement (0.31)	praise (0.57)
melanoma (0.64)	leukaemia (0.78)	corn (0.45)	dürrgoy (0.62)	atetntion (0.31)	aroused (0.56)
cancer;dehydration (0.63)	tumour (0.77)	dosage (0.44)	straflager (0.61)	attnetion (0.3)	acclaim (0.55)
cancer/sda (0.61)	cancers (0.76)	planimetry (0.44)	maidanek (0.61)	atention (0.3)	interest (0.55)
rcc (0.61)	ovarian (0.75)	equation (0.44)	szebnie (0.61)	non-rotated (0.3)	admiration (0.55)

Table 8: Comparison of word similarities between general and domain-specific embeddings

4 Discussion

This study compared the impact of various embedding and classification methods on four different text classification tasks. Notably we investigated the impact of pre-training embedding models on clinical corpora versus off-the-shelf models trained on general corpora.

The results suggest that using embeddings pre-trained for the specific task (clinical corpora in our case) leads to better performance with any classification algorithm tested. However, pre-training such embeddings is not necessarily feasible due to either data or technical constraints. In this case our results highlight that using off-the-shelf embeddings trained on large general corpora such as Wikipedia still produce acceptable performance. In particular BERTbase outperformed most algorithms tested, even when these were combined with clinical embeddings.
Additionally, BioBERT was not pre-trained on medical notes but on texts from a related domain (biomedical articles and abstracts as opposed to clinical records), and therefore excludes specificities inherent to the medical domain such as misspellings or technical jargon. Despite this, BioBERT's performance is only marginally below that of the best model (custom Bi-LSTM) combined with clinical embeddings.

The various experiments conducted on text pre-processing only lead to small variations in terms of performance, and even negatively impact the performance of several algorithms, for the text classification task and embedding model tested. Given the additional constraints required to perform this step (need to train embeddings on pre-processed texts and to clean input data) and the mixed results in performance, pre-processing does not appear to be essential.

Novel tokenization methods based on subword dictionaries, whilst not improving the performance, eliminate several shortcomings presented by SciSpacy and similar methods, notably its speed and vocabulary size.
In light of these limitations and the very small difference in performance for the task tested, BBPE appears to be a suitable alternative to traditional tokenizers and allows to reduce significantly computational costs.

Finally, custom Bi-LSTM outperforms BioBERT when it simulates attention on the entity of interest. However, this configuration requires information on the entity mention span, and then to center each document on this span. For some datasets, such information may either be readily available, or can be obtained by performing an additional named-entity extraction step. Unfortunately, many text classification tasks do not usually have this information, or may not rely on the specific entities/keywords required (e.g. sentiment analysis tasks). When Bi-LSTM is not customized, then both BERT models (trained on general and specific domains) produce the best performance, and consequently should be preferred for texts not easily allowing such customization.

5 Conclusion

In this article we have explored the performance of various word representation approaches (comparing bag-of-words to traditional and contextual embeddings trained on both specific and general corpora, combined with various text pre-processing and tokenizing methods) as well as classification algorithms on four different text classification tasks, all based on publicly available datasets.

A detailed performance comparison on these four tasks highlighted the efficacy of contextual embeddings when compared to traditional methods when no customization is possible, whether these embeddings are trained on specific or general corpora.
When combined with appropriate entity extraction tasks and specific domain embeddings, Bi-LSTM outperforms contextual embeddings. Across all classification algorithms, text pre-processing and tokenization approaches showed limited impact for the task and embedding tested, suggesting a rule of thumb to opt for the least time and resource intensive method.

Acknowledgments

We thank the anonymous reviewers for their valuable suggestions.
This work was supported by Health Data Research UK, an initiative funded by UK Research and Innovation, Department of Health and Social Care (England) and the devolved administrations, and leading medical research charities.

AM is funded by Takeda California, Inc.

RD, RS, AR are part-funded by the National Institute for Health Research (NIHR) Biomedical Research Centre at South London and Maudsley NHS Foundation Trust and King's College London.

RD is also supported by The National Institute for Health Research University College London Hospitals Biomedical Research Centre, and by the BigData@Heart Consortium, funded by the Innovative Medicines Initiative-2 Joint Undertaking under grant agreement No. 116074. This Joint Undertaking receives support from the European Union's Horizon 2020 research and innovation programme and EFPIA.

DB is funded by a UKRI Innovation Fellowship as part of Health Data Research UK MR/S00310X/1.

RB is funded in part by grant MR/R016372/1 for the King's College London MRC Skills Development Fellowship programme funded by the UK Medical Research Council (MRC) and by grant IS-BRC-1215-20018 for the National Institute for Health Research (NIHR) Biomedical Research Centre at South London and Maudsley NHS Foundation Trust and King's College London.

This paper represents independent research part funded by the National Institute for Health Research (NIHR) Biomedical Research Centre at South London and Maudsley NHS Foundation Trust and King's College London. The views expressed are those of the authors and not necessarily those of the NHS, the NIHR or the Department of Health and Social Care. The funders had no role in study design, data collection and analysis, decision to publish, or preparation of the manuscript.

References

Olivier Bodenreider. 2004. The Unified Medical Language System (UMLS): integrating biomedical terminology. *Nucleic Acids Research*, 32(suppl_1):D267–D270.

Piotr Bojanowski, Edouard Grave, Armand Joulin, and Tomas Mikolov. 2017. Enriching Word Vectors with Subword Information. *Transactions of the Association for Computational Linguistics*, 5:135–146.

Wendy W. Chapman, Will Bridewell, Paul Hanbury, Gregory F. Cooper, and Bruce G. Buchanan. 2001. A Simple Algorithm for Identifying Negated Findings and Diseases in Discharge Summaries. *Journal of Biomedical Informatics*, 34(5):301–310.

Corinna Cortes and Vladimir Vapnik. 1995. Support-vector networks. *Machine Learning*, 20(3):273–297.

Hong-Jie Dai. 2019. Family member information extraction via neural sequence labeling models with different tag schemes. *BMC Medical Informatics and Decision Making*, 19(10):257.

Jacob Devlin, Ming-Wei Chang, Kenton Lee, and Kristina Toutanova. 2019. BERT: Pre-training of Deep Bidirectional Transformers for Language Understanding. *arXiv:1810.04805 [cs]*. ArXiv: 1810.04805.

N Elhadad, W. W Chapman, T O'Gorman, M Palmer, and G Savova. 2013. The ShARe schema for the syntactic and semantic annotation of clinical texts.

Philip Gage. 1994. A new algorithm for data compression.

Henk Harkema, John N. Dowling, Tyler Thornblade, and Wendy W. Chapman. 2009. ConText: an algorithm for determining negation, experiencer, and temporal status from clinical reports. *Journal of Biomedical Informatics*, 42(5):839–851.

Alistair E. W. Johnson, Tom J. Pollard, Lu Shen, Liwei H. Lehman, Mengling Feng, Mohammad Ghassemi, Benjamin Moody, Peter Szolovits, Leo Anthony Celi, and Roger G. Mark. 2016. MIMIC-III, a freely accessible critical care database. *Scientific Data*, 3(1):1–9.

Tom Kenter, Alexey Borisov, and Maarten de Rijke. 2016. Siamese CBOW: Optimizing Word Embeddings for Sentence Representations. In *Proceedings of the 54th Annual Meeting of the Association for Computational Linguistics (Volume 1: Long Papers)*, pages 941–951, Berlin, Germany. Association for Computational Linguistics.

T.A. Koleck, C. Dreisbach, P.E. Bourne, and S. Bakken. 2019. Natural language processing of symptoms documented in free-text narratives of electronic health records: A systematic review. *Journal of the American Medical Informatics Association*, 26(4):364–379.

Zeljko Kraljevic, Daniel Bean, Aurelie Mascio, Lukasz Roguski, Amos Folarin, Angus Roberts, Rebecca Bendayan, and Richard Dobson. 2019. MedCAT – Medical Concept Annotation Tool. *arXiv:1912.10166 [cs, stat]*. ArXiv: 1912.10166.

Jinhyuk Lee, Wonjin Yoon, Sungdong Kim, Donghyeon Kim, Sunkyu Kim, Chan Ho So, and Jaewoo Kang. 2019. BioBERT: a pre-trained biomedical language representation model for biomedical text mining. *Bioinformatics*, page btz682. ArXiv: 1901.08746.

Stéphane M. Meystre, Paul M. Heider, Youngjun Kim, Daniel B. Aruch, and Carolyn D. Britten. 2019. Automatic trial eligibility surveillance based on unstructured clinical data. *International Journal of Medical Informatics*, 129:13–19.

Tomas Mikolov, Kai Chen, Greg Corrado, and Jeffrey Dean. 2013. Efficient Estimation of Word Representations in Vector Space. *arXiv:1301.3781 [cs].* ArXiv: 1301.3781.

Danielle L Mowery, Sumithra Velupillai, Brett R South, Lee Christensen, David Martinez, Liadh Kelly, Lorraine Goeuriot, Noemie Elhadad, Guergana Savova, and Wendy W Chapman. 2014. Task 2: ShARe/CLEF eHealth Evaluation Lab 2014. page 12.

Ghulam Mujtaba, Liyana Shuib, Norisma Idris, Wai Lam Hoo, Ram Gopal Raj, Kamran Khowaja, Khairunisa Shaikh, and Henry Friday Nweke. 2019. Clinical text classification research trends: Systematic literature review and open issues. *Expert Systems with Applications*, 116:494–520.

Mark Neumann, Daniel King, Iz Beltagy, and Waleed Ammar. 2019. ScispaCy: Fast and Robust Models for Biomedical Natural Language Processing. *Proceedings of the 18th BioNLP Workshop and Shared Task*, pages 319–327. ArXiv: 1902.07669.

Jeffrey Pennington, Richard Socher, and Christopher Manning. 2014. Glove: Global Vectors for Word Representation. In *Proceedings of the 2014 Conference on Empirical Methods in Natural Language Processing (EMNLP)*, pages 1532–1543, Doha, Qatar. Association for Computational Linguistics.

Sanjay Purushotham, Chuizheng Meng, Zhengping Che, and Yan Liu. 2018. Benchmarking deep learning models on large healthcare datasets. *Journal of Biomedical Informatics*, 83:112–134.

M. Saeed, C. Lieu, G. Raber, and R. G. Mark. 2002. MIMIC II: a massive temporal ICU patient database to support research in intelligent patient monitoring. *Computers in Cardiology*, 29:641–644.

Rico Sennrich, Barry Haddow, and Alexandra Birch. 2016. Neural Machine Translation of Rare Words with Subword Units. *arXiv:1508.07909 [cs].* ArXiv: 1508.07909.

Yuqi Si, Jingqi Wang, Hua Xu, and Kirk Roberts. 2019. Enhancing clinical concept extraction with contextual embeddings. *Journal of the American Medical Informatics Association*, 26(11):1297–1304.

Ozlem Uzuner, Ira Goldstein, Yuan Luo, and Isaac Kohane. 2008. Identifying patient smoking status from medical discharge records. *Journal of the American Medical Informatics Association: JAMIA*, 15(1):14–24.

Özlem Uzuner. 2009. Recognizing Obesity and Comorbidities in Sparse Data. *Journal of the American Medical Informatics Association : JAMIA*, 16(4):561–570.

Changhan Wang, Kyunghyun Cho, and Jiatao Gu. 2019. Neural Machine Translation with Byte-Level Subwords. *arXiv:1909.03341 [cs].* ArXiv: 1909.03341.

Y. Wang, L. Wang, M. Rastegar-Mojarad, S. Moon, F. Shen, N. Afzal, S. Liu, Y. Zeng, S. Mehrabi, S. Sohn, and H. Liu. 2018. Clinical information extraction applications: A literature review. *Journal of Biomedical Informatics*, 77:34–49.

Thomas Wolf, Lysandre Debut, Victor Sanh, Julien Chaumond, Clement Delangue, Anthony Moi, Pierric Cistac, Tim Rault, Rémi Louf, Morgan Funtowicz, and Jamie Brew. 2020. HuggingFace's Transformers: State-of-the-art Natural Language Processing. *arXiv:1910.03771 [cs].* ArXiv: 1910.03771.

Ikuya Yamada, Akari Asai, Hiroyuki Shindo, Hideaki Takeda, and Yoshiyasu Takefuji. 2018. Wikipedia2Vec: An Optimized Tool for Learning Embeddings of Words and Entities from Wikipedia. *arXiv:1812.06280 [cs].* ArXiv: 1812.06280.

Liang Yao, Chengsheng Mao, and Yuan Luo. 2019. Clinical text classification with rule-based features and knowledge-guided convolutional neural networks. *BMC Medical Informatics and Decision Making*, 19(3):71.

A   Appendices

A.1   Test set comparison of word representation, text pre-processing, tokenization and classification methods across tasks

Noise Pollution in Hospital Readmission Prediction:
Long Document Classification with Reinforcement Learning

Liyan Xu[†]   **Julien Hogan**[‡]   **Rachel Patzer**[‡]   **Jinho D. Choi**[†]

[†]Department of Computer Science, [‡]Department of Surgery
Emory University, Atlanta, US
{liyan.xu, julien.hogan, rpatzer, jinho.choi}@emory.edu

Abstract

This paper presents a reinforcement learning approach to extract noise in long clinical documents for the task of readmission prediction after kidney transplant. We face the challenges of developing robust models on a small dataset where each document may consist of over 10K tokens with full of noise including tabular text and task-irrelevant sentences. We first experiment four types of encoders to empirically decide the best document representation, and then apply reinforcement learning to remove noisy text from the long documents, which models the noise extraction process as a sequential decision problem. Our results show that the old bag-of-words encoder outperforms deep learning-based encoders on this task, and reinforcement learning is able to improve upon baseline while pruning out 25% text segments. Our analysis depicts that reinforcement learning is able to identify both typical noisy tokens and task-specific noisy text.

1 Introduction

Prediction of hospital readmission has always been recognized as an important topic in surgery. Previous studies have shown that the post-discharge readmission takes up tremendous social resources, while at least a half of the cases are preventable (Basu Roy et al., 2015; Jones et al., 2016). Clinical notes, as part of the patients' Electronic Health Records (EHRs), contain valuable information but are often too time-consuming for medical experts to manually evaluate. Thus, it is of significance to develop prediction models utilizing various sources of unstructured clinical documents.

The task addressed in this paper is to predict 30-day hospital readmission after kidney transplant, which we treat it as a long document classification problem without using specific domain knowledge. The data we use is the unstructured clinical documents of each patient up to the date of discharge.

In particular, we face three types of challenges in this task. First, the document size can be very long; documents associated with these patients can have tens of thousands of tokens. Second, the dataset is relatively small with fewer than 2,000 patients available, as kidney transplant is a non-trivial medical surgery. Third, the documents are noisy, and there are many target-irrelevant sentences and tabular data in various text forms (Section 2).

The lengthy documents together with the small dataset impose a great challenge on representation learning. In this work, we experiment four types of encoders: bag-of-words (BoW), averaged word embedding, and two deep learning-based encoders that are ClinicalBERT (Huang et al., 2019) and LSTM with weight-dropped regularization (Merity et al., 2018). To overcome the long sequence issue, documents are split into multiple segments for both ClinicalBERT and LSTM (Section 4).

After we observe the best performed encoders, we further propose to combine reinforcement learning (RL) to automatically extract out task-specific noisy text from the long documents, as we observe that many text segments do not contain predictive information such that removing these noise can potentially improve the performance. We model the noise extraction process as a sequential decision problem, which also aligns with the fact that clinical documents are received in time-sequential order. At each step, a policy network with strong entropy regularization (Mnih et al., 2016) decides whether to prune the current segment given the context, and the reward comes from a downstream classifier after all decisions have been made (Section 5).

Empirical results show that the best performed encoder is BoW, and deep learning approaches suffer from severe overfitting under huge feature space in contrast of the limited training data. RL is experimented on this BoW encoder, and able to improve upon baseline while pruning out around 25%

Proceedings of the BioNLP 2020 workshop, pages 95–104
Online, July 9, 2020 ©2020 Association for Computational Linguistics

Type	P	T	Description
CO	1,354	4,395.3	Report for every outpatient consultation before transplantation
DS	514	1,296.7	Summary at the time of discharge from every hospital admission happened before transplant
EC	1,110	1,073.6	Results of echocardiography
HP	1,422	3,025.1	Summary of the patient's medical history and clinical examination
OP	1,472	4,224.8	Report of surgical procedures
PG	1,415	13,723.4	Medical note during hospitalization summarizing the patient's medical status each day
SC	2,033	1,189.2	Report from the evaluation of each transplant candidate by the selection committee
SW	1,118	1,407.6	Report from encounters with social workers

Table 1: Statistics of our dataset with respect to different types of clinical notes. P: # of patients, T: avg. # of tokens, CO: Consultations, DS: Discharge Summary, EC: Echocardiography, HP: History and Physical, OP: Operative, PG: Progress, SC: Selection Conference, SW: Social Worker. The report for SC is written by the committee that consists of surgeons, nephrologists, transplant coordinators, social workers, etc. at the end of the transplant evaluation. All 8 types follow the approximately 3:7 positive-negative class distribution.

text segments (Section 6). Further analysis shows that RL is able to identify traditional noisy tokens with few document frequencies (DF), as well as task-irrelevant tokens with high DF but of little information (Section 7).

2 Data

This work is based on the Emory Kidney Transplant Dataset (EKTD) that contains structured chart data as well as unstructured clinical notes associated with 2,060 patients. The structured data comprises 80 features that are lab results before the discharge as well as the binary labels of whether each patient is readmitted within 30 days after kidney transplant or not where 30.7% patients are labeled as positive.

The unstructured data includes 8 types of notes such that all patients have zero to many documents for each note type. It is possible to develop a more accurate prediction model by co-training the structured and unstructured data; however, this work focuses on investigating the potentials of unstructured data only, which is more challenging.

2.1 Preprocessing

As the clinical notes are collected through various sources of EMRs, many noisy documents exist in EKTD such that 515 documents are HTML pages and 303 of them are duplicates. These documents are removed during preprocessing. Moreover, most documents contain not only written text but also tabular data, because some EMR systems can only export entire documents in the table format. While there are many tabular texts in the documents (e.g., lab results and prescription as in Table 2), it is impractical to write rules to filter them out, as the exported formats are not consistent across EMRs. Thus, any tokens containing digits or symbols, except for one-character tokens, are removed during

Lab Fishbone (BMP, CBC, CMP, Diff) and critical labs - Last 24 hours 03/08/2013 12:45 142(Na) 104(Cl) 70H(BUN) - 10.7L(Hgb) < 92(Glu) 6.5(WBC) 137L(Plt) 3.6(K) 26(CO2)

Table 2: An example of tabular text in EKTD.

preprocessing. Although numbers may provide useful features, most quantitative measurements are already included in the structured data so that those features can be better extracted from the structured data if necessary. The remaining tabular text contains headers and values that do not provide much helpful information and become another source of noise, which we handle by training a reinforcement learning model to identify them (Section 5).

Table 1 gives the statistics of each clinical note type after preprocessing. The average number of tokens is measured by counting tokens in all documents from the same note type of each patient. Given this preprocessed dataset, our task is to take all documents in each note type as a single input and predict whether or not the patient associated with those documents will be readmitted.

3 Related Work

Shin et al. (2019) presented ensemble models utilizing both the structured and the unstructured data in EKTD, where separate logistic regression (LR) models are trained on the structured data and each type of notes respectively, and the final prediction of each patient is obtained by averaging predictions from each models. Since some patients may lack documents from certain note types, prediction on these note types are simply ignored in the averaging process. For the unstructured notes, concatenation of Term Frequency-Inverse Document Frequency

(TF-IDF) and Latent Dirichlet Allocation (LDA) representation is fed into LR. However, we have found that the representation from LDA only contributes marginally, while LDA takes significantly more inferring time. Thus, we drop LDA and only use TF-IDF as our BoW encoder (Section 4.1).

Various deep learning models regarding text classification have been proposed in recent years. Pretrained language models like BERT have shown state-of-the-art performance on many NLP tasks (Devlin et al., 2019). ClinicalBERT is also introduced on the medical domain (Huang et al., 2019). However, deep learning approaches have two drawbacks on this particular dataset. First, deep learning requires large dataset to train, whereas most of our unstructured note types only have fewer than 2,000 samples. Second, these approaches are not designed for long documents, and difficult to keep long-term dependencies over thousands of tokens.

Reinforcement learning has been explored to combat data noise by previous work (Zhang et al., 2018; Qin et al., 2018) on the short text setting. A policy network makes decision left-to-right over tokens, and is jointly trained with another classifier. However, there is little investigation of using RL on the long text setting, as it still requires an effective encoder to give meaningful representation of long documents. Therefore, in our experiments, the first step is to select the best encoder, and then apply RL on the long document classification.

4 Document Representation

4.1 Bag-of-Words

For the baseline model, the bag-of-words representation with TF-IDF scores, excluding stopwords (Nothman et al., 2018), is fed into logistic regression (LR). The objective is to minimize the negative log likelihood of the gold label y_i:

$$-\frac{1}{m}\sum_{i=1}^{m}[y_i \log p(g_i) + (1-y_i) \log 1 - p(g_i)] \quad (1)$$

where g_i is the TF-IDF representation of D_i. In addition, we experiment two common techniques in the encoder to reduce feature space: token stemming, and document frequency cutoff.

4.2 Averaged Word Embedding

Word embeddings generated by fastText are used to establish another baseline, that utilizes subwords to better represent unseen terms (Bojanowski et al.,

2017). It is suitable for this task as unseen terms or misspellings frequently appear in these clinical notes. The averaged word embedding is used to represent the input document consisting of multiple notes, which gets fed into LR with the same training objective.

4.3 ClinicalBERT

Following Huang et al. (2019), the pretrained language BERT model (Devlin et al., 2019) is first tuned on the MIMIC-III clinical note corpus (Johnson et al., 2016), which has shown to provide better related word similarities in medical domains. Then, a dense layer is added on the CLS token of the last BERT layer. The entire parameters are fine-tuned to optimize the binary cross entropy loss, that is the same objective as Equation 1.

Since BERT has a limit on the input length, the input document of each patient is split into multiple subsequences. Each subsequence is within the BERT length limit, and serves as an independent sample with the same label of the patient. The training data is therefore noisily inflated. The final probability of readmission is computed as follows:

$$p(y_i = 1|g_i) = \frac{p_{\max}^{n_i} + p_{\text{mean}}^{n_i} n_i/c}{1 + n_i/c} \quad (2)$$

where g_i is the BERT representation of patient i, n_i is the corresponding number of subsequences, and c is a hyperparameter to control the influence of n_i. $p_{\max}^{n_i}$ and $p_{\text{mean}}^{n_i}$ are the max and mean probability across the subsequences, respectively.

The motivation behind balancing between the max and mean probability is that subsequences do not contain equal information. $p_{\max}^{n_i}$ represents the best potential, while longer text should give more importance to $p_{\text{mean}}^{n_i}$, because $p_{\max}^{n_i}$ is more easily affected by noise as the text length grows. Although Equation 2 seems intuitive, the use of pseudo labels on subsequences becomes another source of noise, especially when there are thousands of tokens; thus, the performance is uncertain. Section 6.2 provides detailed empirical analysis for this model.

4.4 Weight-dropped LSTM

We split documents of each patient into multiple short segments, and feed the segment representation to long short-term memory network (LSTM) at each time step:

$$h_j \leftarrow \text{LSTM}(s_j, h_{j-1}; \theta) \quad (3)$$

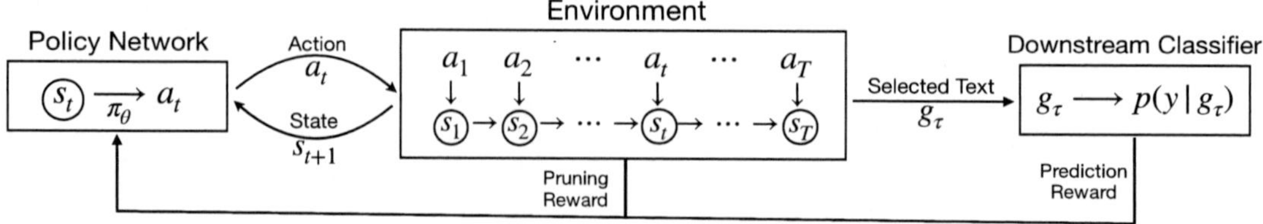

Figure 1: Overview of our reinforcement learning approach. Rewards are calculated and sent back to the policy network after all actions $a_{1:T}$ have been sampled for the given episode.

where h_j is the hidden state at time step j, s_j is the jth segment, and θ is the set of parameters. Although segmentation of documents is still necessary, no pseudo labels are needed. We get the segment representation by averaging its token embedding from the last layer of BERT. The final hidden state at each step j is the concatenated hidden states of a single-layer Bi-directional LSTM. After we get the hidden state for each segment, a max-pooling operation is performed on $h_{1:n}$ over the time dimension to obtain a fixed-length vector, similar to Kim (2014); Adhikari et al. (2019). A dense layer is immediately followed.

It is particularly important to strengthen regularization on this dataset with small sample size. Dropout (Srivastava et al., 2014) as a way of regularization has been shown effective in deep learning models, and Merity et al. (2018) has successfully applied dropout-like technique in LSTM: the use of DropConnect (Wan et al., 2013) is applied on the four hidden-to-hidden matrices, preventing overfitting from occurring on the recurrent weights.

5 Reinforcement Learning

Reinforcement learning is applied to the best performing encoder in Section 4 to prune noisy text, which can lead to comparable or even better performance, as many text segments in these clinical notes are found to be irrelevant to this task. Figure 1 describes the overview of our reinforcement learning approach. The pruning process is modeled as a sequential decision problem, for the fact that these notes are received in time-order. It consists of two separate components: a policy network, and a downstream classifier. To avoid having too many time steps, the policy is performed on the segment level instead of token level. For each patient, documents are split into short segments $g_{1:T} = \{g_1, g_2, \cdots, g_T\}$, and the policy network conducts a sequence of decisions $a_{1:T} = \{a_1, a_2, \cdots, a_T\}$ over segments. The downstream classifier is re-

sponsible for the reward, and the REINFORCE algorithm is used to train the policy (Williams, 1992).

State At each time step, the state s_t is the concatenation of two parts: the representation of previously selected text, and the current segment representation g_i. The previously selected text serves as the context and provides a prior importance. Both parts are represented by an effective encoder, e.g. the best performing encoder from Section 4.

Action The action space at each step is binary: {Keep, Prune}. If the action is Keep, the current segment is added to the selected text; otherwise, it is discarded. The final selected text for a patient is the concatenated segments selected by the policy.

Reward The reward comes at the end when all actions are sampled for the entire sequence. The final selected text is fed to the downstream classifier, and negative log-likelihood of the gold label is used as the reward R. In addition, we also include a reward term R_p to encourage pruning, as follows:

$$R_p = c \cdot \alpha \cdot [2\sigma(\frac{l}{\beta}) - 1] \qquad (4)$$

where c and β are hyperparameters to control the scale of R_p, l is the number of segments, α is the ratio of pruned segments $|\{a_k = \text{Prune}\}|/l$, σ is the sigmoid function. The value of the term $2\sigma(\frac{l}{\beta}) - 1$ falls into range $(0, 1)$. When l is small, it downgrades the encouragement of pruning; when l is large, it also gives an upper bound of R_p. Additionally, we apply exponential decay on the reward. The final reward is $d^l R + R_p$. d is the discount rate.

Policy Network The policy network maintains a stochastic policy $\pi(a_t|s_t; \theta)$:

$$\pi(a_t|s_t; \theta) = \sigma(W s_t + b) \qquad (5)$$

where θ is the set of policy parameters W and b, a_t and s_t are the action and state at the time step t respectively. During training, an action is sampled at

Encoder	CO	DS	EC	HP	OP	PG	SC	SW
Bag-of-Words (§4.1)	58.6	62.1	52.0	58.9	51.8	61.2	**59.3**	51.6
+ Cutoff	58.6	**62.3**	52.8	59.0	51.9	61.3	**59.3**	**51.9**
+ Stemming	**58.9**	61.8	**53.4**	**59.4**	51.9	**61.5**	**59.3**	51.6
Averaged Embedding (§4.2)	56.3	53.7	52.4	54.0	**53.4**	54.7	54.2	46.6
ClinicalBERT (§4.3)	51.9	53.3	-	52.7	-	-	52.3	-
Weight-dropped LSTM (§4.4)	53.7	55.8	-	54.2	-	-	54.5	-

Table 3: The Area Under the Curve (AUC) scores achieved by different encoders on the 5-fold cross-validation. See the caption in Table 1 for the descriptions of CO, DS, EC, HP, OP, PG, SC, and SW. For deep learning encoders, only four types are selected in experiments (Section 6.2).

each step with the probability from the policy. After the sampling is performed over the entire sequence, the delayed reward is computed. During evaluation, the action is picked by $\text{argmax}_a \pi(a|s_t; \theta)$.

The training is guided by the REINFORCE algorithm (Williams, 1992), which optimizes the policy to maximize the expected reward:

$$J(\theta) = \mathbb{E}_{a_{1:T} \sim \pi} R_{a_{1:T}} \tag{6}$$

and the gradient has the following form:

$$\nabla_\theta J(\theta) = \mathbb{E}_\tau \sum_{t=1}^{T} \nabla_\theta \log \pi(a_t|s_t; \theta) R_\tau \tag{7}$$

$$\approx \frac{1}{N} \sum_{i=1}^{N} \sum_{t=1}^{T} \nabla_\theta \log \pi(a_{it}|s_{it}; \theta) R_{\tau_i} \tag{8}$$

where τ represents the sampled trajectory $\{a_1, a_2, \cdots, a_T\}$, N is the number of sampled trajectories. R_{τ_i} here equals the delayed reward from the downstream classifier at the last step.

To encourage exploration and avoid local optima, we add the entropy regularization (Mnih et al., 2016) on the policy loss:

$$J_{reg}(\theta) = \frac{\lambda}{N} \sum_{i=1}^{N} \frac{1}{T_i} \nabla_\theta H(\pi(s_{it}; \theta)) \tag{9}$$

where H is the entropy, and λ is the regularization strength, T_i is the trajectory length.

Finally, the downstream classifier and policy network are warm-started by separate training, and then jointly trained together.

6 Experiments

Before experiments, we perform the preprocessing described in Section 2.1, and then randomly split patients in every note type by 5 folds to perform

cross-validation as suggested by Shin et al. (2019). To evaluate each fold F_i, 12.5% of the training set, that is the combined data of the other 4 folds, are held out as the development set and the best configuration from this development set is used to decode F_i. The same split is used across all experiments for fair comparison. Following Shin et al. (2019), the averaged Area Under the Curve (AUC) across these 5 folds is used as the evaluation metric.

6.1 Baseline

Bag-of-Words We first conduct experiments using the bag-of-words encoder (BoW; Section 4.1) to establish the baseline. Many experiments are performed on all note types using the vanilla TF-IDF, document frequency (DF) cutoff at 2 (removing all tokens whose DF ≤ 2), and token stemming. For every experiment, the class weight is assigned inversely proportional to class frequencies, and the inverse of regularization strength C is searched from $\{0.01, 0.1, 1, 10\}$, where the best results are achieved with $C = 1$ on the development set.

Table 3 describes the cross-validation results on every note type. The top AUC is 62.3%, which is within expectation given the difficulty of this task. Some note types are not as predictive as the others, such as Operative (OP) and Social Worker (SW), with the AUC under 52%. Most note types have the standard deviations in range 0.02 to 0.03.

In comparison to the previous work (Shin et al., 2019), we achieve 0.671 AUC combining both structured and unstructured data, despite without the use of LDA in our encoder.

Noise Observation The DF cutoff coupled with token stemming significantly reduce feature space for the BoW model. As shown in Table 4, the DF cutoff itself can achieve about 50% reduction of the feature space. Furthermore, applying the DF cutoff leads to slightly higher AUCs on most of the note types, despite almost a half of the tokens are removed from the vocabulary. This implies that there exists a large amount of noisy text that appears only in few documents, causing the models to be overfitted more easily. These results further verify our previous observation and strengthen the necessity to extract noise from these long documents using reinforcement learning (Section 6.3).

Averaged Word Embedding For the averaged word embedding encoder (AWE; Section 4.2), embeddings generated by FastText trained on the Common Crawl and the English Wikipedia with the 300 dimension is used.[1] AWE is outperformed by BoW on every note type except Operative (OP; Table 3). This empirical result implies that AWE over thousands of tokens is not so effective in generating the document representation so that the averaged embeddings are less discriminative than the sparse vectors generated by BoW for such long documents.

Type	Vanilla	+ Cutoff	+ Stemming
CO	28,213	15,022 (46.8)	12,243 (56.6)
DS	11,029	6,117 (44.5)	5,228 (52.6)
HP	20,245	11,276 (44.3)	9,329 (53.9)
SC	19,050	9,873 (48.2)	8,200 (57.0)

Table 4: The dimensions of the feature spaces used by each BoW model with respect to the four note types. The numbers in the parentheses indicate the percentage reduction from the vanilla model, respectively.

6.2 Deep Learning-based Encoders

For deep learning encoders, the four note types with good baseline performance ($\approx 60\%$ AUC) and reasonable sequence length (< 5000) are selected to use in the following experiments, which are Consultations (CO), Discharge Summary (DS), History and Physical (HP), and Selection Conference (SC) (see Tables 1 and 3).

Segmentation For both ClinicalBERT and the LSTM models, the input document is split into segments as described in Section 4.3. For LSTM, we set the maximum segment length to be 128 for CO and HP, 64 for DS and SC, to balance between segment length and sequence length. The segment length for ClinicalBERT is set to 318 (approaching 500 after BERT tokenization) to avoid noise brought by too many pseudo labels. More statistics about segmentation are summarized in Table 5.

For the ClinicalBERT, we use the PyTorch BERT implementation with the base configuration:[2] 768 embedding dimensions and 12 transformer layers, and we load the weights provided by Huang et al. (2019) whose language model has been finetuned on large-scale clinical notes.[3] We finetune the entire ClinicalBERT with batch size 4, learning rate 2×10^{-5}, and weight decay rate 0.01.

For the weight-dropped LSTM, we set the batch size to 64, the learning rate to 10^{-3}, the weight-drop rate to 0.5, and search the hidden state dimension from $\{128, 256, 512\}$ on the development set. Early stop is used for both approaches.

Type + Model	SEN	SEQ	INST
CO + BERT	318	14.8	11,376
CO + LSTM	128	36.8	948
DS + BERT	318	4.6	1,588
DS + LSTM	64	22.5	371
HP + BERT	318	10.1	8,364
HP + LSTM	128	27.3	987
SC + BERT	318	3.7	5,206
SC + LSTM	64	25.4	1,422

Table 5: SEN: maximum segment length (number of tokens) allowed by the corresponding model, SEQ: average sequence length (number of segments), INST: average number of samples in the training set.

Result Analysis Table 3 shows the final results achieved by the ClinicalBERT and LSTM models. The AUCs of both models experience a non-trivial drop from the baseline. After further investigation, the issue is that both models suffer from severe overfitting under the huge feature spaces, and struggle to learn generalized decision boundaries from this data. Figure 2 shows an example of the weak correlation between the training loss and the AUC scores on the development set.

As more steps are processed, the training loss gradually decreases to 0. However, the model has high variance and it does not necessarily give better performance on the development set as the training loss drops. This issue is more apparent with ClinicalBERT on CO because there are too many pseudo labels acting as noise, which makes it harder for the model to distinguish useful patterns from noise.

---

[1]https://fasttext.cc/docs/en/crawl-vectors.html

[2]https://github.com/huggingface/transformers
[3]https://github.com/kexinhuang12345/clinicalBERT

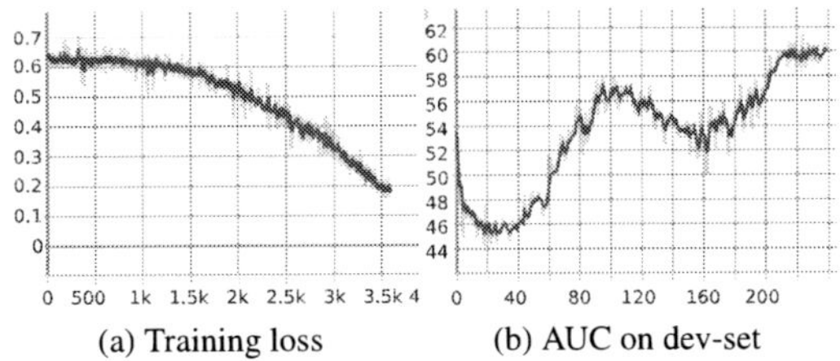

(a) Training loss   (b) AUC on dev-set

Figure 2: Training loss and AUC scores on the development set during the LSTM training on the CO type. The AUC scores depict high variance while showing weak correlation to the training loss.

6.3 Reinforcement Learning

According to Table 3, the BoW model achieves the best performance. Therefore, we decide to use TF-IDF to represent the long text of each patient, along with logistic regression as the classifier for reinforcement learning. Document segmentation is the same as LSTM (Table 5). During training, segments within each note are shuffled to reduce overfitting risks, and sequences with more than 36 segments are truncated.

The downstream classifier is warm-started by loading weights from the logistic regression model in the previous experiment. The policy network is then trained for 400 episodes while freezing the downstream classifier. After the warm start, both models are jointly trained. We set the number of sampling N as 10 episodes, learning rate 2×10^{-4}, and fix the scaling factor β in Equations 4 as 8, and discount rate as 0.95. Moreover, we search the reward coefficient c in $\{0.02, 0.1, 0.4\}$, and entropy coefficient λ in $\{2, 4, 6, 8\}$.

	CO	DS	HP	SC
Best	58.9	62.3	59.4	59.3
RL	59.8	62.4	60.6	60.2
Pruning	26%	5%	19%	23%

Table 6: The AUC scores and the pruning ratios of reinforcement learning (RL). Best: AUC scores from the best performing models in Table 3.

The AUC scores and the pruning ratios (the number of pruned segments divided by the sequence length) are shown in Table 6. Our reinforcement learning approach outperforms the best performing models in Table 3, achieving around 1% higher AUC scores on three note types, CO, HP, and SC, while pruning out up to 26% of the input documents.

Tuning Analysis We find that two hyperparameters are essential to the final success of reinforcement learning (RL). The first is the reward discount rate d. The scale of the policy gradient $\nabla_\theta J(\theta)$ depends on the sequence length T, while the delayed reward R_τ is always on the same scale regardless of T. Therefore, different sequence length across episodes causes turbulence on the policy gradient, leading to unstable training. It is important to apply reward decay to stabilize the scale of $\nabla_\theta J(\theta)$.

The second is the entropy regularization coefficient λ, which forces the model to add bias towards uncertainty. Without strong entropy regularization, the training is easy to fall into local optima in early stage, which is to keep all segments, as shown by Figure 3(a). $\lambda = 6$ gives the model descent incentive to explore aggressively, as shown by Figure 3(b), and finally leads to higher AUC.

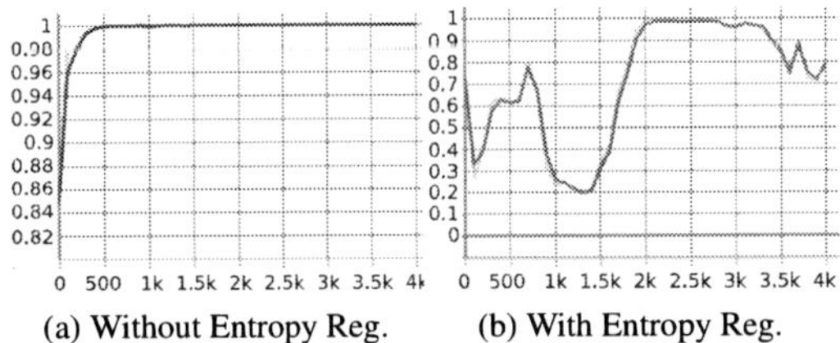

(a) Without Entropy Reg.   (b) With Entropy Reg.

Figure 3: Retaining ratios on the development set of SC while training the reinforcement learning model. Entropy regularization encourages more exploration.

7 Noise Analysis

To investigate the noise extracted by RL, we analyze the pruned segments on the validation sets of the Consultations type (CO), and compare the results with other basic noise removal techniques.

Qualitative Analysis Table 7 demonstrates the potential of the learned policy to automatically identify noisy text from the long documents. The original notes of shown examples are tabular text with headers and values, mostly lab results and medical prescription. After the data cleaning step, the text becomes broken and does not make much sense for humans to evaluate. The learned policy can identify noisy segments by looking at the presence of headers such as "lab fishbone", "lab report", and certain medical terms that frequently appear in tabular reports such as "chloride", "creatinine", "hemoglobin", "methylprednisolone", etc. We find that many pruned segments have strong indicators

lab fishbone (bmp , cbc , **cmp** , **diff**) and critical labs - last hours (not an official **lab report** . please see flowsheet (or printed official **lab** reports) for official **lab results** .) (na) (cl) h (**bun**) - (hgb) (glu) (wbc) (plt) ( ) h (**cr**) (hct) na = not applicable a = abnormal (ftn) = **footnote** .
laboratory studies : sodium , **potassium** , **chloride** , . , **bun** , **creatinine** , **glucose** . **total** bilirubin 1 , phos of , calcium , **ast** 9 , **alt** , alk phos . parathyroid hormone level . white blood cell count , **hemoglobin** , hematocrit , **platelets** . inr , ptt , and pt .
methylprednisolone ivpb : mg , **ivpb** , give in surgery , routine , / , infuse over : minute . **mycophenolate mofetil** : mg = 4 cap , po , capsule , once , now , / , stop date / , ml . documented medications documented accupril : mg , po , **qday** , 0 refill , substitution allowed .

Table 7: Examples of pruned segments by the learned policy. Tokens that have feature importance lower than -0.001 (towards *Prune* action) are marked bold.

the **social worker** met with this pleasant year old **caucasian male** on this date for kidney transplant **evaluation** . the patient was **alert** , oriented and easily **engaged** in conversation with the **social worker** today . he resides in atlanta with his spouse of years , who he describes as very **supportive** .
he **reports** occasional alcohol **drinks** per month but **denies** any illicit drug **use** . he has a **grade** education . he has been **married** for years . he is working full - time while on **peritoneal dialysis** as a business asset manager . he has **medicare** and an aarp **prescriptions** supplement . family history : mother **deceased** at age with complications of **obesity** , **high blood pressure** and **heart** disease .

Table 8: Examples of kept segments by the learned policy. Tokens that have feature importance greater than 0.0005 (towards *Keep* action) are marked bold.

of headers and specific medical terms, which appear mostly in tabular text rather than written notes.

Table 8 shows examples that are kept by the policy. Tokens that contribute towards *Keep* action are words related with human and social life, such as "social worker", "engaged", "drinks", "married", "medicare", and terms related with health conditions, such as "obesity", "heart", "high blood pressure". These terms indeed appear mostly in written text rather than tabular data.

In addition, we also notice that the policy is able to remove certain duplicate segments. Medical professionals sometimes repeat certain description from previous notes to a new document, causing duplicate content. The policy learns to make use of the already selected context, and assigns negative coefficients to certain tokens. Duplicate segments are only selected once if the segment contains many tokens that have opposite feature importance in the context and segment vectors.

Quantitative Analysis We examine tokens that are pruned by RL and compare with document frequency (DF) cutoff. We select 3000 unique tokens in the vocabulary that have the top negative feature importance (towards *Prune* action) in the segment vector of CO. Figure 4 shows the DF distribution of these tokens.

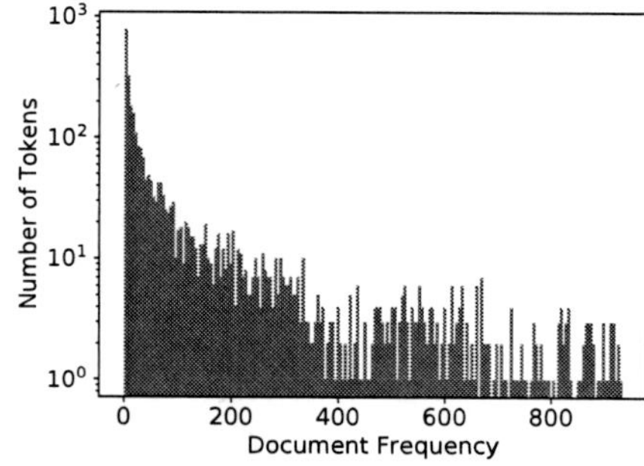

Figure 4: Log scale distribution on document frequency of tokens with top negative feature importance.

We observe that the majority of those tokens have small DF values. It shows that the learned policy is able to identify certain tokens with small DF values as noise, which aligns with DF cutoff. Moreover, the distribution also shows a non-trivial amount of tokens with large DF values, demonstrating that RL can also identify task-specific noisy tokens that commonly appear in documents, which in this case are certain tokens in noisy tabular text.

Either RL or DF cutoff achieves higher AUC while reducing input features, proving that given the small sample size, the extracted text is more likely to cause overfit than being generalizable pattern, which also verifies our initial hypothesis.

8 Conclusion

In this paper, we address the task of 30-day readmission prediction after kidney transplant, and propose to improve the performance by applying reinforcement learning with noise extraction capability. To overcome the challenge of long document representation with a small dataset, four different encoders are experimented. Empirical results show that bag-of-words is the most suitable encoder, surpassing overfitted deep learning models, and reinforcement learning is able to improve the performance, while being able to identify both traditional noisy tokens that appear in few documents, and task-specific noisy text that commonly appear.

Acknowledgments

We gratefully acknowledge the support of the National Institutes of Health grant R01MD011682, *Reducing Disparities among Kidney Transplant Recipients*. Any opinions, findings, and conclusions or recommendations expressed in this material are those of the authors and do not necessarily reflect the views of the National Institutes of Health.

References

Ashutosh Adhikari, Achyudh Ram, Raphael Tang, and Jimmy Lin. 2019. Rethinking complex neural network architectures for document classification. In *Proceedings of the 2019 Conference of the North American Chapter of the Association for Computational Linguistics: Human Language Technologies, Volume 1 (Long and Short Papers)*, pages 4046–4051, Minneapolis, Minnesota. Association for Computational Linguistics.

Senjuti Basu Roy, Ankur Teredesai, Kiyana Zolfaghar, Rui Liu, David Hazel, Stacey Newman, and Albert Marinez. 2015. Dynamic hierarchical classification for patient risk-of-readmission. In *Proceedings of the 21th ACM SIGKDD International Conference on Knowledge Discovery and Data Mining*, KDD '15, page 1691–1700, New York, NY, USA. Association for Computing Machinery.

Piotr Bojanowski, Edouard Grave, Armand Joulin, and Tomas Mikolov. 2017. Enriching word vectors with subword information. *Transactions of the Association for Computational Linguistics*, 5:135–146.

Jacob Devlin, Ming-Wei Chang, Kenton Lee, and Kristina Toutanova. 2019. BERT: Pre-training of deep bidirectional transformers for language understanding. In *Proceedings of the 2019 Conference of the North American Chapter of the Association for Computational Linguistics: Human Language Technologies, Volume 1 (Long and Short Papers)*, pages 4171–4186, Minneapolis, Minnesota. Association for Computational Linguistics.

Kexin Huang, Jaan Altosaar, and Rajesh Ranganath. 2019. Clinicalbert: Modeling clinical notes and predicting hospital readmission. *CoRR*, abs/1904.05342.

Alistair EW Johnson, Tom J Pollard, Lu Shen, H Lehman Li-wei, Mengling Feng, Mohammad Ghassemi, Benjamin Moody, Peter Szolovits, Leo Anthony Celi, and Roger G Mark. 2016. Mimic-iii, a freely accessible critical care database. *Scientific data*, 3:160035.

Caroline Jones, Robert Hollis, Tyler Wahl, Brad Oriel, Kamal Itani, Melanie Morris, and Mary Hawn. 2016. Transitional care interventions and hospital readmissions in surgical populations: A systematic review. *The American Journal of Surgery*, 212.

Yoon Kim. 2014. Convolutional neural networks for sentence classification. In *Proceedings of the 2014 Conference on Empirical Methods in Natural Language Processing (EMNLP)*, pages 1746–1751, Doha, Qatar. Association for Computational Linguistics.

Stephen Merity, Nitish Shirish Keskar, and Richard Socher. 2018. Regularizing and optimizing LSTM language models. In *International Conference on Learning Representations*.

Volodymyr Mnih, Adria Puigdomenech Badia, Mehdi Mirza, Alex Graves, Timothy Lillicrap, Tim Harley, David Silver, and Koray Kavukcuoglu. 2016. Asynchronous methods for deep reinforcement learning. In *Proceedings of The 33rd International Conference on Machine Learning*, volume 48 of *Proceedings of Machine Learning Research*, pages 1928–1937, New York, New York, USA. PMLR.

Joel Nothman, Hanmin Qin, and Roman Yurchak. 2018. Stop word lists in free open-source software packages. In *Proceedings of Workshop for NLP Open Source Software (NLP-OSS)*, pages 7–12, Melbourne, Australia. Association for Computational Linguistics.

Pengda Qin, Weiran Xu, and William Yang Wang. 2018. Robust distant supervision relation extraction via deep reinforcement learning. In *Proceedings of the 56th Annual Meeting of the Association for Computational Linguistics (Volume 1: Long Papers)*, pages 2137–2147, Melbourne, Australia. Association for Computational Linguistics.

Bonggun Shin, Julien Hogan, Andrew B. Adams, Raymond J. Lynch, and Jinho D. Choi. 2019. Multimodal ensemble approach to incorporate various types of clinical notes for predicting readmission. In *2019 IEEE EMBS International Conference on Biomedical & Health Informatics (BHI)*.

Nitish Srivastava, Geoffrey Hinton, Alex Krizhevsky, Ilya Sutskever, and Ruslan Salakhutdinov. 2014. Dropout: A simple way to prevent neural networks from overfitting. *Journal of Machine Learning Research*, 15(56):1929–1958.

Li Wan, Matthew Zeiler, Sixin Zhang, Yann LeCun, and Rob Fergus. 2013. Regularization of neural networks using dropconnect. In *Proceedings of the 30th International Conference on International Conference on Machine Learning - Volume 28*, ICML'13, page III–1058–III–1066. JMLR.org.

Ronald J. Williams. 1992. Simple statistical gradient-following algorithms for connectionist reinforcement learning. *Mach. Learn.*, 8(3–4):229–256.

Tianyang Zhang, Minlie Huang, and Li Zhao. 2018. Learning structured representation for text classification via reinforcement learning. In *AAAI Conference on Artificial Intelligence*.

Evaluating the Utility of Model Configurations and Data Augmentation on Clinical Semantic Textual Similarity

Yuxia Wang   **Fei Liu**   **Karin Verspoor**   **Timothy Baldwin**

School of Computing and Information Systems

The University of Melbourne

Victoria, Australia

`{yuxiaw, fliu3}@student.unimelb.edu.au`

`karin.verspoor@unimelb.edu.au  tb@ldwin.net`

Abstract

In this paper, we apply pre-trained language models to the Semantic Textual Similarity (STS) task, with a specific focus on the clinical domain. In low-resource setting of clinical STS, these large models tend to be impractical and prone to overfitting. Building on BERT, we study the impact of a number of model design choices, namely different fine-tuning and pooling strategies. We observe that the impact of domain-specific fine-tuning on clinical STS is much less than that in the general domain, likely due to the concept richness of the domain. Based on this, we propose two data augmentation techniques. Experimental results on N2C2-STS[1] demonstrate substantial improvements, validating the utility of the proposed methods.

1 Introduction

Semantic Textual Similarity (STS) is a language understanding task, involving assessing the degree of semantic equivalence between two pieces of text based on a graded numerical score (Corley and Mihalcea, 2005). It has application in tasks such as information retrieval (Hliaoutakis et al., 2006), question answering (Hoogeveen et al., 2018), and summarization (AL-Khassawneh et al., 2016). In this paper, we focus on STS in the clinical domain, in the context of a recent task within the framework of N2C2 (the National NLP Clinical Challenges)[1], which makes use of the extended MedSTS data set (Wang et al., 2018), referring to N2C2-STS, with limited annotated sentences pairs (1.6K) that are rich in domain terms.

Neural STS models typically consist of encoders to generate text representations, and a regression layer to measure the similarity score (He et al., 2015; Mueller and Thyagarajan, 2016; He and Lin,

---

[1] `https://portal.dbmi.hms.harvard.edu/` `projects/n2c2-2019-t1/`

2016; Reimers and Gurevych, 2019). These architectures require a large amount of training data, an unrealistic requirement in low resource settings.

Recently, pre-trained language models (LMs) such as GPT-2 (Radford et al., 2018) and BERT (Devlin et al., 2019) have been shown to benefit from pre-training over large corpora followed by fine tuning over specific tasks. However, for small-scale datasets, only limited fine-tuning can be done. For example, GPT-2 achieved strong results across four large natural language inference (NLI) datasets, but was less successful over the small-scale RTE corpus (Bentivogli et al., 2009), performing below a multi-task biLSTM model. Similarly, while the large-scale pre-training of BERT has led to impressive improvements on a range of tasks, only very modest improvements have been achieved on STS tasks such as STS-B (Cer et al., 2017) and MRPC (Dolan and Brockett, 2005) (with 5.7k and 3.6k training instances, resp.). Compared to general-domain STS benchmarks, labeled clinical STS data is more scarce, which tends to cause overfitting during fine-tuning. Moreover, further model scaling is a challenge due to GPU/TPU memory limitations and longer training time (Lan et al., 2019). This motivates us to search for model configurations which strike a balance between model flexibility and overfitting.

In this paper, we study the impact of a number of model design choices. First, following Reimers and Gurevych (2019), we study the impact of various pooling methods on STS, and find that convolution filters coupled with max and mean pooling outperform a number of alternative approaches. This can largely be attributed to their improved model expressiveness and ability to capture local interactions (Yu et al., 2019). Next, we consider different parameter fine-tuning strategies, with varying degrees of flexibility, ranging from keeping all parameters frozen during training to allowing all pa-

Proceedings of the BioNLP 2020 workshop, pages 105–111

Online, July 9, 2020 ©2020 Association for Computational Linguistics

rameters to be updated. This allows us to identify the optimal model flexibility without over-tuning, thereby further improving model performance.

Finally, inspired by recent studies, including sentence ordering prediction (Lan et al., 2019) and data-augmented question answering (Yu et al., 2019), we focus on data augmentation methods to expand the modest amount of training data. We first consider segment reordering (SR), in permuting segments that are delimited by commas or semicolons. Our second method increases linguistic diversity with back translation (BT). Extensive experiments on N2C2-STS reveal the effectiveness of data augmentation on clinical STS, particularly when combined with the best parameter fine-tuning and pooling strategies identified in Section 3, achieving an absolute gain in performance.

2   Related Work

2.1   Model Configurations

In pre-training, a spectrum of design choices have been proposed to optimize models, such as the pre-training objective, training corpus, and hyperparameter selection. Specific examples of objective functions include masked language modeling in BERT, permutation language modeling in XLNet (Yang et al., 2019), and sentence order prediction (SOP) in ALBERT (Lan et al., 2019). Additionally, RoBERTa (Liu et al., 2019) explored benefits from a larger mini-batch size, a dynamic masking strategy, and increasing the size of the training corpus (16G to 160G). However, all these efforts are targeted at improving downstream tasks indirectly by optimizing the capability and generalizability of LMs, while adapting a single fully-connected layer to capture task features.

Sentence-BERT (Reimers and Gurevych, 2019) makes use of task-specific structures to optimize STS, concentrating on computational and time efficiency, and is evaluated on relatively larger datasets in the general domain. For evaluating the impact of number of layers transferred to the supervised target task from the pre-trained language model, GPT-2 has been analyzed on two datasets. However, they are both large: MultiNLI (Williams et al., 2018) with >390k instances, and RACE (Lai et al., 2017) with >97k instances. These tasks also both involve reasoning-related classification, as opposed to the nuanced regression task of STS.

2.2   Data Augmentation

Synonym replacement is one of the most commonly used data augmentation methods to simulate linguistic diversity, but it introduces ambiguity if accurate context-dependent disambiguation is not performed. Moreover, random selection and replacement of a single word used in general texts is not plausible for term-rich clinical text, resulting in too much semantic divergence (e.g *patient* to *affected role* and *discharge to home* to *spark to home*). By contrast, replacing a complete mention of the concept can increase error propagation due to the prerequisite concept extraction and normalization.

Random insertion, deletion, and swapping of words have been demonstrated to be effective on five text classification tasks (Wei and Zou, 2019). But those experiments targeted topic prediction, in contrast to semantic reasoning such as STS and MultiNLI. Intuitively, they do not change the overall topic of a text, but can skew the meaning of a sentence, undermining the STS task. Swapping an entire semantic segment may mitigate the risk of introducing label noise to the STS task.

Compared to semantic and syntactic distortion potentially caused by aforementioned methods, back translation (BT) (Sennrich et al., 2016) — translating to a target language then back to the original language — presents fluent augmented data and reliable improvements for tasks demanding for adequate semantic understanding, such as low-resource machine translation (Xia et al., 2019) and question answering (Yu et al., 2019). This motivates our application of BT on low-resource clinical STS, to bridge linguistic variation between two sentences. This work represents the first exploration of applying BT for STS.

3   STS Model Configurations

In this section, we study the impact of a number of model design choices on BERT for STS, using a 12-layer base model initialized with pretrained weights.

3.1   Hierarchical Convolution (HConv)

The resource-poor and concept-rich nature of clinical STS makes it difficult to train a large model end-to-end on sentence pairs. To address this, most recent studies have made use of pre-trained language models, such as BERT. The most straightforward way to use BERT is the feature-based approach, where the output of the last transformer block is taken as input to the task-specific classifier. Many have proposed the use of a dummy CLS token to generate the feature vector, where CLS is a special symbol added in front of every sequence during pre-training, with its final hidden state always used as the aggregate sequence representation for classification tasks, referring to CLS pooling. Other types of pooling, such as mean and max pooling, are investigated by Reimers and Gurevych (2019).

However, this results in inferior performance as shown in the first row of Table 1.[2] As a consequence, the best strategy for extracting feature vectors to represent a sentence remains an open question.

In this work, we first experiment with the feature-based approach, coupled with convolutional filters. This is inspired by the use of convolutional filters in QANet (Yu et al., 2019) to capture local interactions. The difference lies in where convolutional filters are applied. With QANet, multiple conv filters are incorporated into each transformer encoder block to process the input from the previous layer. In contrast, HConv-BERT is largely based on BERT, with the addition of a single task-specific classifier placed on top of BERT consisting of conv filters organised in a hierarchical fashion. This results in a much simplified model, making HConv-BERT less prone to overfitting.

Specifically, we run a collection of convolutional filters with a kernel of size $k \in [2, 4]$, each with $J = 768$ output channels (indexed by $j \in [1, J]$), over the temporal axis (indexed by $i \in [1, T]$):

$$c_{i,k_j} = \mathbf{w}_{k_j} * \mathbf{x}_{i:i+k-1} + b_{k_j} \qquad (1)$$

$$\mathbf{c}_{i,k} = [c_{i,k_1}; \ldots; c_{i,k_J}] \qquad (2)$$

where $\mathbf{x}_{i:i+k-1}$ is the output BERT features for the token span i to $i + k - 1$, $*$ is the convolution operation, $\mathbf{w}_{k_j}$ and b_{k_j} are the convolution filter and bias term for the j-th kernel of size k, and $[\mathbf{a}; \mathbf{b}]$ denotes the concatenation of $\mathbf{a}$ and $\mathbf{b}$.

To capture interactions between distant elements, we feed the output $\mathbf{c}_{i,k}$ into another convolution layer of kernel size 2 with $M = 128$ output channels (indexed by $m \in [1, M]$):

$$c^k_{i,m} = \mathbf{w}_m * \mathbf{c}_{i:i+1,k} + b_m \qquad (3)$$

$$\mathbf{c}^k_i = \left[c^k_{i,1}; \ldots; c^k_{i,M} \right] \qquad (4)$$

---

[2]Due to space constrains, we limit our comparison to the CLS pooling strategy, based on the observation of little improvements when using other types of pooling (mean, max) and concatenation, or sequence processing recurrent units.

Model	SICK-R	STS-B	N2C2-STS
Feature-based:			
CLS-BERT	53.6/52.1	49.3/67.9	14.6/28.4
HConv-BERT	80.1/73.6	83.0/83.2	79.4/74.4
Fine-tuning:			
CLS-BERT	88.6/82.9	90.0/89.6	86.7/81.9
HConv-BERT	88.7/83.5	90.1/89.6	87.7/80.7

Table 1: Pearson and Spearman correlation (r/ρ) between the predicted score and the gold labels for three STS datasets using the feature-based approach (upper half) and fine-tuning (bottom half) with CLS-BERT and HConv-BERT. Performance is reported by convention as $r/\rho \times 100$.

where $\mathbf{c}_{i:i+1,k}$ is the output of the first convolutional layer over the span i to $i + 1$ as defined in Equation (2), and $\mathbf{w}_m$ and b_m are the filter and bias term for the second convolutional layer with, a kernel size of 2 and output dimension of $M = 128$.

Lastly, we extract feature vectors by max and mean pooling over the temporal axis and then concatenation:

$$\mathbf{v}^k_{\max} = \max\left(\mathbf{c}^k_i\right) \qquad \mathbf{v}^k_{\text{mean}} = \text{avg}\left(\mathbf{c}^k_i\right) \qquad (5)$$

$$\mathbf{v} = \left[\mathbf{v}^2_{\max}; \mathbf{v}^3_{\max}; \mathbf{v}^4_{\max}; \mathbf{v}^2_{\text{avg}}; \mathbf{v}^3_{\text{avg}}; \mathbf{v}^4_{\text{avg}} \right]. \qquad (6)$$

The upper half of Table 1 shows that the proposed hierarchical convolutional (HConv) architecture provides substantial performance gains.

3.2 Model Flexibility

We also evaluate the utility of this mechanism in the fine-tuning setting with varying modelling flexibility. Concretely, we progressively increase the number of trainable parameters by transformer blocks. That is, for the base BERT model with 12 layers, we allow errors to be back-propagated through the last l layers while keeping the rest $(12 - l)$ fixed.

The results on STS-B and N2C2-STS are shown in Figure 1. We observe performance crossover of HConv and CLS-pooling on both datasets as the number of trainable transformer layers increases. While HConv reaches peak performance before the crossover, CLS-pooling often requires more blocks to be trainable to achieve comparable accuracy, rendering the model much slower. Notably, the proposed mechanism peaks with much fewer trainable blocks on N2C2-STS than STS-B. We speculate that this is due to the size difference between the two datasets. To verify this hypothesis, we further look into the relationship between the number of

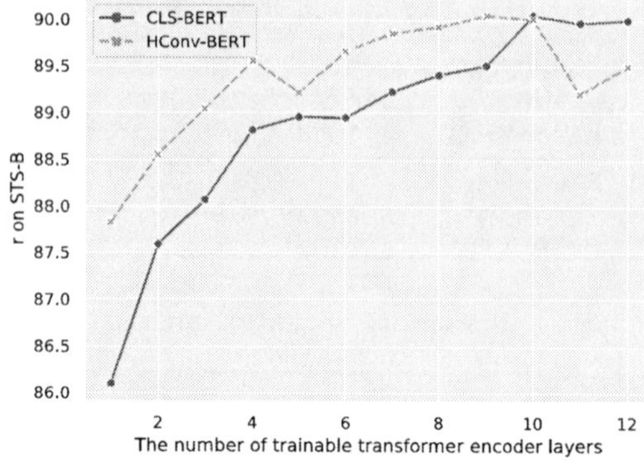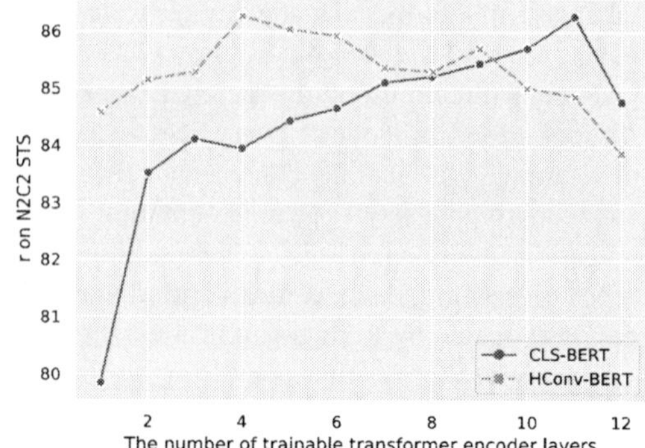

Figure 1: Evaluation of CLS-BERT and HConv-BERT over datasets from the general (STS-B) and clinical (N2C2) domains. r refers to Pearson correlation. N2C2-STS is split into 1233 and 409 instances for training and dev.

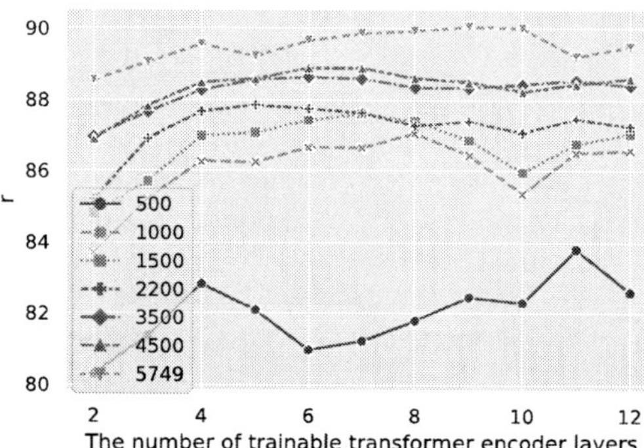

Figure 2: Impact of number of trainable transformer blocks based on HConv-BERT over different data size, randomly sampled from STS-B, ranging from 500 to full set $(5, 749)$.

trainable transformer blocks and training data size. In Figure 2, we observe performance degradation as the size of training data shrinks, with the models trained on the full set achieving far superior Pearson correlation to those trained on the smaller subsets. Zooming into the curve representing each subset, we find that peak performance is attained at different points depending on data size: with the smallest dataset (500 instances), the number of parameter updates is also limited. Only updating the top few layers of transformer blocks is simply not enough to make the model fully adapt to the task. It is therefore beneficial to allow the model access to more trainable layers (e.g., 11) to improve performance.

Based on this, we set the number of trainable blocks to 6 for SICK-R (consisting of $4, 500$ training instances), as presented in the bottom half of

Table 1, with HConv outperforming CLS-pooling.

4 Data Augmentation

The accuracy of an STS model unsurprisingly depends on the amount of labeled data. This is reflected in Figure 2, where models trained with more data outperform those with fewer training instances. In this section, we propose two data augmentation methods, namely segment reordering (SR) and back translation (BT), to address the data sparsity issue in clinical STS.

Segment reordering. Clinical texts often consist of text segments describing multiple events and patient symptoms. Each segment is often an independent semantic unit, separated by commas or semicolons. Inspired by the random word swapping of Wei and Zou (2019), we exploit this property and propose a heuristic, named segment reordering (SR), to generate permutations of the original sequence based on these segments. While we expect this to introduce some noise to the training data, our hypothesis is that the increase in training data size will outweigh this. For instance, consider the text *new confusion or inability to stay alert and awake; feeling like you are going to pass out*. Flipping the order of the two segments *new confusion or inability to stay alert and awake* and *feeling like you are going to pass out* will not hinder the overall understanding of the text. More formally, for a given pair of sentences S_1 and S_2, each consisting of a sequence of segments $S_1 = \{s_{11}, \ldots, s_{1m}\}$ and $S_2 = \{s_{21}, \ldots, s_{2n}\}$, we generate a new pair by randomly permuting the segment order, effectively doubling the size of the training corpus.

Back translation. Inspired by the work of Yu et al. (2019), we make use of machine translation tools to perform back translation (BT). Here, we choose Chinese as the pivot language as it is linguistically distant to English and supported by mature commercial translation solutions. That is, we first translate from English to Chinese and then back to English. We use Google Translate to translate each sentence in a sentence pair from English to Chinese, and Baidu Translation[3] to translate back to English. For example, for the original sentence *negative for cough and stridor*, the backtranslated result is *bad for coughing and wheezing*. We apply this to each sentence pair, doubling the amount of training data.

5 Experiments

5.1 Experimental Setup

We evaluate the effectiveness of SR and BT on N2C2-STS with four baseline models: $BERT_{base}$ (Devlin et al., 2019) and $BERT_{clinical}$ (Alsentzer et al., 2019), both using CLS-pooling and consisting of 12 layers; $ConvBERT_{base}$, based on $BERT_{base}$ with hierarchical convolution and fine-tuning over the last 4 layers (consistent with our findings of the best model configuration in Section 3); and $ConvBERT_{STS-B}$, where we take $ConvBERT_{base}$ and fine-tune first over STS-B, before N2C2-STS.

We split the training partition of N2C2-STS into $1,233$ (train) and 409 (dev) instances, and report results on the test set (412 instances).

5.2 Results

Experimental results are presented in Table 2. We see clear benefits of the two proposed data augmentation methods, consistently boosting performance across all categories, with BT providing larger gains than SR. This is likely caused by the rather naïve implementation of SR, resulting in unnatural segment sequences. A possible fix to this is to further filter out such irregular statements with a language model pre-trained on clinical corpora. We leave this for future work.

It is impressive that the best-performing configuration $ConvBERT_{STS-B}$ + BT is capable of achieving comparable results with the state-of-the-art IBM-N2C2, an approach heavily reliant on external, domain-specific resources, and an ensemble of multiple pre-trained language models.

[3] https://fanyi.baidu.com/

Model	r	ρ
IBM-N2C2	90.1	—
$BERT_{base}$	86.7	81.9
+ SR	87.1	80.8
+ BT	87.2	81.7
$BERT_{clinical}$	86.1	81.4
+ SR	87.4	82.7
+ BT	88.6	82.4
$Conv1dBERT_{base}$	87.7	80.7
+ SR	88.0	81.4
+ BT	88.1	82.2
$Conv1dBERT_{STS-B}$	87.9	82.5
+ SR	88.6	**83.1**
+ BT	**89.4**	83.0

Table 2: Pearson r and Spearman ρ on N2C2-STS for models with and without segment reordering ("SR") and back translation ("BT").

We additionally conduct a cross-domain experiment on BIOSSES (Soğancıoğlu et al., 2017), a biomedical literature STS dataset comprising 100 sentence pairs derived from the Text Analysis Conference Biomedical Summarization task with scores ranging from 0 (complete unrelatedness) to 4 (exact equivalence). Specifically, baseline model Pooling $BERT_{base}$ and proposed $ConvBERT_{STS-B}$ + BT are both fine-tuned on N2C2-STS, and then applied with no further training to BIOSSES. Despite the increase in task difficulty, the proposed method demonstrates strong generalisability, outperforming the baseline by an absolute gain of 2.4 and 3.9 to 85.42/82.83 (r/ρ).

6 Conclusions

In this paper, we have presented an empirical study of the impact of a number of model design choices on a BERT-based approach to clinical STS. We have demonstrated that the proposed hierarchical convolution mechanism outperforms a number of alternative conventional pooling methods. Also, we have investigated parameter fine-tuning strategies with varying degrees of flexibility, and identified the optimal number of trainable transformer blocks, thereby preventing over-tuning. Lastly, we have verified the utility of two data augmentation methods on clinical STS. It may be interesting to see the impact of leveraging target languages other than Chinese in BT, which we leave for future work.

References

Yazan Alaya AL-Khassawneh, Naomie Salim, and Adekunle Isiaka Obasae. 2016. Sentence similarity techniques for automatic text summarization. *Journal of Soft Computing and Decision Support Systems*, 3(3):35–41.

Emily Alsentzer, John Murphy, William Boag, Wei-Hung Weng, Di Jindi, Tristan Naumann, and Matthew McDermott. 2019. Publicly available clinical BERT embeddings. In *Proceedings of the 2nd Clinical Natural Language Processing Workshop*, pages 72–78, Minneapolis, Minnesota, USA.

Luisa Bentivogli, Peter Clark, Ido Dagan, and Danilo Giampiccolo. 2009. The fifth pascal recognizing textual entailment challenge. In *TAC*.

Daniel Cer, Mona Diab, Eneko Agirre, Iñigo Lopez-Gazpio, and Lucia Specia. 2017. SemEval-2017 task 1: Semantic textual similarity multilingual and crosslingual focused evaluation. In *Proceedings of the 11th International Workshop on Semantic Evaluation (SemEval-2017)*, pages 1–14, Vancouver, Canada.

Courtney Corley and Rada Mihalcea. 2005. Measuring the semantic similarity of texts. In *Proceedings of the ACL workshop on empirical modeling of semantic equivalence and entailment*, pages 13–18. Association for Computational Linguistics.

Jacob Devlin, Ming-Wei Chang, Kenton Lee, and Kristina Toutanova. 2019. BERT: Pre-training of deep bidirectional transformers for language understanding. In *Proceedings of the 2019 Conference of the North American Chapter of the Association for Computational Linguistics: Human Language Technologies, Volume 1 (Long and Short Papers)*, pages 4171–4186, Minneapolis, Minnesota.

William B Dolan and Chris Brockett. 2005. Automatically constructing a corpus of sentential paraphrases. In *Proceedings of the Third International Workshop on Paraphrasing (IWP2005)*.

Hua He, Kevin Gimpel, and Jimmy Lin. 2015. Multi-perspective sentence similarity modeling with convolutional neural networks. In *Proceedings of the 2015 Conference on Empirical Methods in Natural Language Processing*, pages 1576–1586, Lisbon, Portugal.

Hua He and Jimmy Lin. 2016. Pairwise word interaction modeling with deep neural networks for semantic similarity measurement. In *Proceedings of the 2016 Conference of the North American Chapter of the Association for Computational Linguistics: Human Language Technologies*, pages 937–948, San Diego, California.

Angelos Hliaoutakis, Giannis Varelas, Epimenidis Voutsakis, Euripides GM Petrakis, and Evangelos Milios. 2006. Information retrieval by semantic similarity. *International Journal on Semantic Web and Information Systems (IJSWIS)*, 2(3):55–73.

Doris Hoogeveen, Andrew Bennett, Yitong Li, Karin M Verspoor, and Timothy Baldwin. 2018. Detecting misflagged duplicate questions in community question-answering archives. In *Twelfth International AAAI Conference on Web and Social Media*.

Guokun Lai, Qizhe Xie, Hanxiao Liu, Yiming Yang, and Eduard Hovy. 2017. RACE: Large-scale ReAding comprehension dataset from examinations. In *Proceedings of the 2017 Conference on Empirical Methods in Natural Language Processing*, pages 785–794, Copenhagen, Denmark. Association for Computational Linguistics.

Zhenzhong Lan, Mingda Chen, Sebastian Goodman, Kevin Gimpel, Piyush Sharma, and Radu Soricut. 2019. Albert: A lite BERT for self-supervised learning of language representations. *arXiv preprint arXiv:1909.11942*.

Yinhan Liu, Myle Ott, Naman Goyal, Jingfei Du, Mandar Joshi, Danqi Chen, Omer Levy, Mike Lewis, Luke Zettlemoyer, and Veselin Stoyanov. 2019. Roberta: A robustly optimized bert pretraining approach. *arXiv preprint arXiv:1907.11692*.

Jonas Mueller and Aditya Thyagarajan. 2016. Siamese recurrent architectures for learning sentence similarity. In *Thirtieth AAAI Conference on Artificial Intelligence*.

Alec Radford, Karthik Narasimhan, Tim Salimans, and Ilya Sutskever. 2018. Improving language understanding by generative pre-training. *URL https://s3-us-west-2.amazonaws.com/openai-assets/researchcovers/languageunsupervised/language understanding paper.pdf*.

Nils Reimers and Iryna Gurevych. 2019. Sentence-BERT: Sentence embeddings using Siamese BERT-networks. In *Proceedings of the 2019 Conference on Empirical Methods in Natural Language Processing and the 9th International Joint Conference on Natural Language Processing (EMNLP-IJCNLP)*, pages 3980–3990, Hong Kong, China.

Rico Sennrich, Barry Haddow, and Alexandra Birch. 2016. Improving neural machine translation models with monolingual data. In *Proceedings of the 54th Annual Meeting of the Association for Computational Linguistics (Volume 1: Long Papers)*, pages 86–96, Berlin, Germany.

Gizem Soğancıoğlu, Hakime Öztürk, and Arzucan Özgür. 2017. BIOSSES: a semantic sentence similarity estimation system for the biomedical domain. *Bioinformatics*, 33(14):i49–i58.

Yanshan Wang, Naveed Afzal, Sunyang Fu, Liwei Wang, Feichen Shen, Majid Rastegar-Mojarad, and Hongfang Liu. 2018. MedSTS: a resource for clinical semantic textual similarity. *Language Resources and Evaluation*, pages 1–16.

Jason Wei and Kai Zou. 2019. EDA: Easy data augmentation techniques for boosting performance on text classification tasks. In *Proceedings of the 2019 Conference on Empirical Methods in Natural Language Processing and the 9th International Joint Conference on Natural Language Processing (EMNLP-IJCNLP)*, pages 6381–6387, Hong Kong, China.

Adina Williams, Nikita Nangia, and Samuel Bowman. 2018. A broad-coverage challenge corpus for sentence understanding through inference. In *Proceedings of the 2018 Conference of the North American Chapter of the Association for Computational Linguistics: Human Language Technologies, Volume 1 (Long Papers)*, pages 1112–1122, New Orleans, Louisiana. Association for Computational Linguistics.

Mengzhou Xia, Xiang Kong, Antonios Anastasopoulos, and Graham Neubig. 2019. Generalized data augmentation for low-resource translation. In *Proceedings of the 57th Annual Meeting of the Association for Computational Linguistics*, pages 5786–5796, Florence, Italy. Association for Computational Linguistics.

Zhilin Yang, Zihang Dai, Yiming Yang, Jaime Carbonell, Russ R Salakhutdinov, and Quoc V Le. 2019. Xlnet: Generalized autoregressive pretraining for language understanding. In *Advances in neural information processing systems*, pages 5754–5764.

Adams Wei Yu, David Dohan, Minh-Thang Luong, Rui Zhao, Kai Chen, Mohammad Norouzi, and Quoc V Le. 2019. QANet: Combining local convolution with global self-attention for reading comprehension. In *The Sixth International Conference on Learning Representations (ICLR)*.

Entity-Enriched Neural Models for Clinical Question Answering

Bhanu Pratap Singh Rawat[1,*], **Wei-Hung Weng**[2,3,#], **So Yeon Min**[2,3,§],
Preethi Raghavan[3,4,†], **Peter Szolovits**[2,3,‡]
[1]UMass-Amherst, [2]MIT CSAIL, [3]MIT-IBM Watson AI Lab, [4]IBM Research, Cambridge
[*]`brawat@umass.edu`, [#]`ckbjimmy@mit.edu`, [§]`symin95@mit.edu`
[†]`praghav@us.ibm.com`, [‡]`psz@mit.edu`

Abstract

We explore state-of-the-art neural models for question answering on electronic medical records and improve their ability to generalize better on previously unseen (paraphrased) questions at test time. We enable this by learning to predict logical forms as an auxiliary task along with the main task of answer span detection. The predicted logical forms also serve as a rationale for the answer. Further, we also incorporate medical entity information in these models via the ERNIE (Zhang et al., 2019a) architecture. We train our models on the large-scale emrQA dataset and observe that our multi-task entity-enriched models generalize to paraphrased questions $\sim$ 5% better than the baseline BERT model.

1 Introduction

The field of question answering (QA) has seen significant progress with several resources, models and benchmark datasets. Pre-trained neural language encoders like BERT (Devlin et al., 2019) and its variants (Seo et al., 2016; Zhang et al., 2019b) have achieved near-human or even better performance on popular open-domain QA tasks such as SQuAD 2.0 (Rajpurkar et al., 2016). While there has been some progress in biomedical QA on medical literature (Šuster and Daelemans, 2018; Tsatsaronis et al., 2012), existing models have not been similarly adapted to clinical domain on electronic medical records (EMRs).

Community-shared large-scale datasets like emrQA (Pampari et al., 2018) allow us to apply state-of-the-art models, establish benchmarks, innovate and adapt them to clinical domain-specific needs. emrQA enables question answering from electronic medical records (EMRs) where a question is asked by a physician against a patient's medical record

---

[*]The author did this work while interning at MIT-IBM Watson AI Lab.

Context: The patient had an elective termination of her pregnancy on [DATE]. The work-up for the extent of the patient's disease included mri scan of the cervical and thoracic spine which revealed multiple metastatic lesions in the vertebral bodies; A T 3 lesion extending from the body to the right neural for amina with foraminal obstruc tion. An abdominal and pelvic ct scan with iv contrast revealed bilateral pulmo nary nodules and bilateral pleural effusions, extensive liver metastases, narrowing of the intra hepatic ivc and distention of the azygous system suggestive of ivc obstructi on by liver metastases.
Question: How was the patient's extensive liver metastases diagnosed?
Paraphrase: What diagnosis was used for the patient's extensive liver metastases?
Logical Form: {LabEvent (x) [date=x, result=x] OR ProcedureEvent (x) [date=x, result=x] OR VitalEvent (x) [date=x, result=x]} reveals ConditionEvent (|problem|)
Answer: An abdominal and pelvic ct scan with iv contrast

Figure 1: A synthetic example of a clinical context, question, its logical form and the expected answer.

(clinical notes). Thus, we adapt these models for EMR QA while focusing on model generalization via the following. (1) learning to predict the logical form (a structured semantic representation that captures the answering needs corresponding to a natural language question) along with the answer and (2) incorporating medical entity embeddings into models for EMR QA. We now examine the motivation behind these.

A physician interacting with a QA system on EMRs may ask the same question in several different ways; a physician may frame a question as: *"Is the patient allergic to penicillin?"* whereas the other could frame it as *"Does penicillin cause any allergic reactions to the patient?"*. Since paraphrasing is a common form of generalization in natural language processing (NLP) (Bhagat et al., 2009), a QA model should be able to generalize well to such paraphrased question variants that may not be seen during training (and avoid simply memorizing the questions). However, current state-of-the-art models do not consider the use of meta-information such as the semantic parse or logical form of the questions in unstructured QA. In order to give the model the ability to understand the semantic information about answering needs of a question, we frame our problem in a multitask learning setting where the primary task is extractive QA and the

Proceedings of the BioNLP 2020 workshop, pages 112–122
Online, July 9, 2020 ©2020 Association for Computational Linguistics

auxiliary task is the logical form prediction of the question.

Fine-tuning on medical copora (MIMIC-III, PubMed (Johnson et al., 2016; Lee et al., 2020)) helps models like BERT align their representations according to medical vocabulary (since they are previously trained on open-domain corpora such as WikiText (Zhu et al., 2015)). However, another challenge for developing EMR QA models is that different physicians can use different medical terminology to express the same entity; e.g., "heart attack" vs. "myocardial infarction". Mapping these phrases to the same UMLS semantic type[1] as *disease or syndrome (dsyn)* provides common information between such medical terminologies. Incorporating such entity information about tokens in the context and question can further improve the performance of QA models for the clinical domain.

Our contributions are as follows:

1. We establish state-of-the-art benchmarks for EMR QA on a large clinical question answering dataset, emrQA (Pampari et al., 2018)

2. We demonstrate that incorporating an auxiliary task of predicting the logical form of a question helps the proposed models generalize well over unseen paraphrases, improving the overall performance on emrQA by $\sim 5\%$ over BERT (Devlin et al., 2019) and by $\sim 3.5\%$ over clinicalBERT (Alsentzer et al., 2019). We support this hypothesis by running our proposed model over both emrQA and another clinical QA dataset, MADE (Jagannatha et al., 2019).

3. The predicted logical form for unseen paraphrases helps in understanding the model better and provides a rationale (explanation) for why the answer was predicted for the provided question. This information is critical in *clinical domain* as it provides an accompanying answer justification for clinicians.

4. We incorporate medical entity information by including entity embeddings via the ERNIE (Zhang et al., 2019a) architecture (Zhang et al., 2019a) and observe that the model accuracy and ability to generalize goes up by $\sim 3\%$ over BERT_{base}(Devlin et al., 2019).

---

[1] https://metamap.nlm.nih.gov/SemanticTypesAndGroups.shtml

2   Problem Formulation

We formulate the EMR QA problem as a reading comprehension task. Given a natural language *question* (asked by a physician) and a *context*, where the context is a set of contiguous sentences from a patient's EMR (unstructured clinical notes), the task is to predict the answer span from the given context. Along with the (question, context, answer) triplet, also available as input are *clinical entities* extracted from the question and context. Also available as input is the, *logical form* (LF) that is a structured representation that captures answering needs of the question through entities, attributes and relations required to be in the answer (Pampari et al., 2018). A question may have multiple paraphrases where all paraphrases map to the same LF (and the same answer, fig. 1).

3   Methodology

In this section, we briefly describe BERT (Devlin et al., 2019), ERNIE (Zhang et al., 2019a) and our proposed model.

3.1   Bidirectional Encoder Representations from Transformers (BERT)

BERT (Devlin et al., 2019) uses multi-layer bidirectional Transformer (Vaswani et al., 2017) networks to encode contextualised language representations. BERT representations are learned from two tasks: masked language modeling (Taylor, 1953) and next sentence prediction task. We chose BERT model as pre-trained BERT models can be fine-tuned with just one additional inference layer and it achieved state-of-the-art results for a wide range of tasks such as question answering, such as SQuAD (Rajpurkar et al., 2016, 2018), and multiple language inference tasks, such as MultiNLI (Williams et al., 2017). *clinicalBERT* (Alsentzer et al., 2019) yielded superior performance on clinical-related NLP tasks such as i2b2 named entity recognition (NER) challenges (Uzuner et al., 2011). It was created by further fine-tuning of BERT_{base} with biomedical and clinical corpus (MIMIC-III) (Johnson et al., 2016).

3.2   Enhanced Language Representation with Informative Entities (ERNIE)

We adopt the ERNIE framework (Zhang et al., 2019a) to integrate the entity-level clinical concept information into the BERT architecture, which has not yet been explored in the previous works.

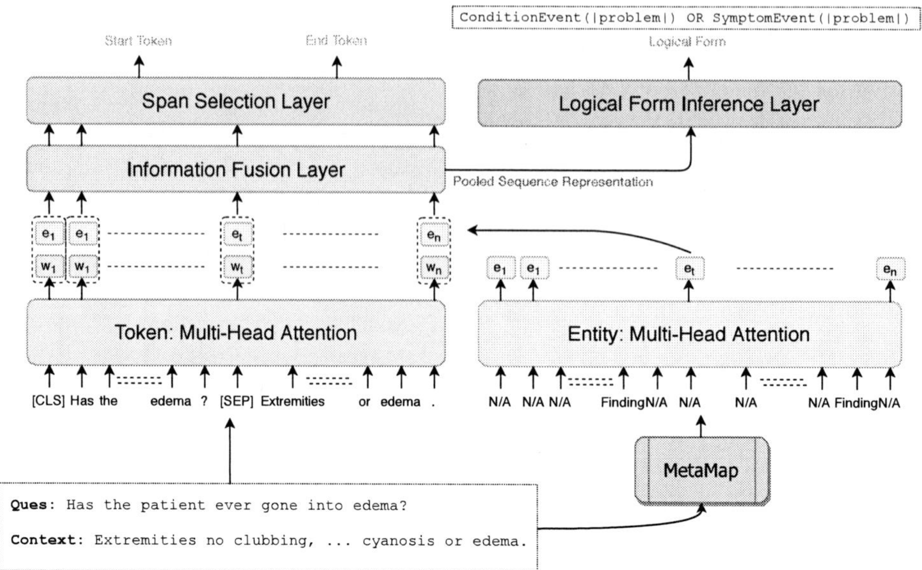

Figure 2: The network architecture of our multi-task learning question answering model (*M-cERNIE*). The question and context are provided to a multi-head attention model (*orange*) and are also passed through MetaMap to extract clinical entities which are passed through a separate multi-head attention (*yellow*). The token and entity representations are then passed through an information fusion layer (*blue*) to extract entity-enriched token representations which are then used for answer span prediction. The pooled sequence representation from the information fusion layer is passed through logical form inference layer to predict the logical form.

ERNIE has shown significant improvement in different entity typing and relation classification tasks, as it utilises the extra entity information which is provided from knowledge graphs. ERNIE uses BERT for extracting contextualized token embeddings and a multi-head attention model to generate entity embeddings. These two set of embeddings are aligned and provided as an input to an information fusion layer which provides entity-enriched token embeddings. For a token (w_j) and its aligned entity ($e_k = f(w_j)$), the information fusion process is as follows:

$$h_j = \sigma(W_t^{(i)} w_j^{(i)} + W_e^{(i)} e_k^{(i)} + b^{(i)}) \quad (1)$$

Here h_j represents the entity enriched token embedding, σ is the non-linear activation function, W_t refers to an affine layer for token embeddings and W_e refers to an affine layer for entity embeddings. For the tokens without corresponding entities, the information fusion process becomes:

$$h_j = \sigma(W_t^{(i)} w_j^{(i)} + b^{(i)}) \quad (2)$$

Initially, each entity embedding is assigned randomly and is fine-tuned along with token embeddings throughout the training procedure. The

ERNIE architecture would be applicable to the model even if the logical forms are not available.

3.3 Multi-task Learning for Extractive QA

In order to improve the ability of a QA model to generalize better over paraphrases, it helps to provide the model information about the logical form that links these paraphrases. Since the answer to all the paraphrased questions is the same (and hence, logical form is the same), we constructed a multi-task learning framework to incorporate the logical form information into the model. Thus, along with predicting the answer span, we added an auxiliary task to also predict the corresponding logical form of the question. Multi-task learning provides an inductive bias to enhance the primary task's performance via auxiliary tasks (Weng et al., 2019). In our setting, the primary task is span detection of the answer and the auxiliary task is logical form prediction for both emrQA and MADE (both datasets are explained in detail in § 4). The final loss for our model is defined as:

$$\mathcal{L}_{model} = \omega \mathcal{L}_{lf} + (1 - \omega) \mathcal{L}_{span}, \quad (3)$$

where ω is the weightage given to the loss of auxillary task ($\mathcal{L}_{lf}$), logical form prediction. $\mathcal{L}_{span}$

114

is loss for answer span prediction and $\mathcal{L}_{model}$ is the final loss for our proposed model. The multi-task learning model can work with both BERT and ERNIE as the base model. Figure 2 depicts the proposed multi-task model to predict both the answer and logical form given a question and ERNIE architecture that is used to learn entity-enriched token embeddings.

4   Datasets

We used *emrQA*[2] and *MADE*[3] datasets for our experiments. We provide a brief summary of each dataset and the methodology followed to split these datasets into train and test sets.

emrQA   The emrQA corpus (Pampari et al., 2018) is the only community-shared clinical QA dataset that consists of questions, posed by physicians against electronic medical records (EMRs) of a patient, along with their answers. The dataset was developed by leveraging existing annotations available for other clinical natural language processing (NLP) tasks (i2b2 challenge datasets (Uzuner et al., 2011)). It is a credible resource for clinical QA as logical forms that are generated by a physician help slot fill question templates and extract corresponding answers from annotated notes. Multiple question templates can be mapped to the same logical form (LF), as shown in Table 1, and are referred to as *paraphrases* of each other.

LF: MedicationEvent (|medication|) [dosage=x]
How much |medication| does the patient take per day?   What is her current dose of |medication|?   What is the current dose of the patient's |medication|?   What is the current dose of |medication|?   What is the dosage of |medication|?   What was the dosage prescribed of |medication|?

Table 1: A logical form (LF) and its respective question templates (paraphrases).

The *emrQA* corpus has over $1M+$ question, logical form, and answer/evidence triplets, an example of a context, question, its logical form and a paraphrase is shown in Fig 1. The evidences are the sentences from the clinical note that are relevant to a particular question. There are total 30 logical forms in the emrQA dataset [4].

MADE   *MADE* 1.0 (Jagannatha et al., 2019) dataset was hosted as an adverse drug reactions (ADRs) and medication extraction challenge from EMRs. This dataset was converted into a QA dataset by following the same procedure as enumerated in the literature of *emrQA* (Pampari et al., 2018). *MADE* QA dataset is smaller than *emrQA*, as *emrQA* consists of multiple datasets taken from i2b2 (Uzuner et al., 2011) whereas *MADE* only has specific relations and entity mentions to that of ADRs and medications. This resulted in a clinical QA dataset which has different properties as compared to *emrQA*. *MADE* also has lesser number of logical forms (8 LFs) as compared to *emrQA* because of fewer entities and relations. The 8 LFs for MADE are provided in Appendix B.

4.1   Train/test splits

The emrQA dataset is generated using a semi-automated process that normalizes real physician questions to create question templates, associates expert annotated logical forms with each template and slot fills them using annotations for various NLP tasks from i2b2 challenge datasets (for e.g., fig. 1). emrQA is rich in paraphrases as physicians often tend to express the same information need in different ways. As shown in Table. 1, all paraphrases of a question map to the same logical form. Thus, if a model has observed some of the paraphrases it should be able to generalize to the others effectively with the help of their shared logical form "MedicationEvent (\|medication\|) [dosage=x]". In order to simulate this, and test the true capability of the model to generalize to unseen paraphrased questions, we create a splitting scheme and refer to it as *paraphrase-level* split.

Paraphrase-level split

The basic idea is that some of question templates would be observed by the model during training and remaining would be used during validation and testing. The steps taken for creating this split are enumerated below:

1. First, the clinical notes are separated into train, val and test sets. Then the question, logical form and context triplets are generated for each set resulting in the full dataset. Here the context is the set of contiguous sentences from the EMR.

2. Then for each logical form (LF), 70% of its

[2]https://github.com/panushri25/emrQA
[3]https://bio-nlp.org/index.php/projects/39-nlp-challenges
[4]https://github.com/panushri25/emrQA/blob/master/templates/templates-all.csv

corresponding question templates are chosen for train dataset and the rest are kept for validation and test dataset. Considering the LF shown in Table 1, four of the question templates (QT_{tr}) would be assigned for training and two ($QT_{v/t}$) of them would be assigned for validation/testing. So any sample in training dataset whose question is generated from the question template set $Q_{v/t}$ would be discarded. Similarly, any sample with a question generated from the question template set Q_{tr} would be discarded.

3. To compare the generalizability performance of our model, we keep the training dataset with both set of question templates ($QT_{tr} + QT_{v/t}$) as well. Essentially, a baseline model which has observed all the question templates ($QT_{tr} + QT_{v/t}$) should be able to perform better on the $QT_{v/t}$ set as compared to a model which has only observed QT_{tr} set. This comparison would help us in measuring the improvement in performance with the help of logical forms even when a set of question templates are not observed by the model.

The dataset statistics for both *emrQA* and *MADE* are shown in Table 2. The training set with both question template sets ($QT_{tr} + QT_{v/t}$) is shown with '(r)' appended as suffix, as it is essentially a *random* split, whereas the training set with the question template (QT_{tr}) is appended with '(pl)' for *paraphrase-level* split.

Datasets	Split	Train	Val.	Test
emrQA	# Notes	433	44	47
	# Samples (pl)	133,589	21,666	19,401
	# Samples (r)	198,118	21,666	19,401
MADE	# Notes	788	88	213
	# Samples (pl)	73,224	4,806	9,235
	# Samples (r)	113,975	4,806	9,235

Table 2: Train, validation and test data splits.

5 Experiments

In this section, we briefly discuss the experimental settings, clinical entity extraction method, implementation details of our proposed model and evaluation metrics for our experiments.

5.1 Experimental Setting

As a reading comprehension style task, the model has to identify the span of the answer given the question-context pair. For both *emrQA* and *MADE* dataset, the span is marked as the answer to the question and the sentence is marked as the evidence. Hence, we perform extractive question answering at two levels: *sentence* and *paragraph*.

Sentence setting: For this setting, the evidence sentence which contains the answer span is provided as the context to the question and the model has to predict the span of the answer, given the question.

Paragraph setting: Clinical notes are noisy and often contain incomplete sentences, lists and embedded tables making it difficult to segment paragraphs in notes. Hence, we decided to define the context as evidence sentence and $15 - 20$ sentences around it. We randomly chose the length of the paragraph (l_{para}) and another number less than the length of the paragraph ($l_{pre} < l_{para}$). We chose l_{pre} contiguous sentences which exist prior to the evidence sentence in the EMR and ($l_{para} - l_{pre}$) sentences after the evidence sentence. We adopted this strategy because the model could have benefited from the information that the evidence sentence is exactly in the middle of a fixed length paragraph. The model has to predict the span of the answer from the l_{para} sentences long paragraph (context) given the question.

The datasets are appended by '-p' and '-s' for *paragraph* and *sentence* settings respectively. The *sentence* setting is a relatively easier setting, for the model, compared to the *paragraph* setting because the scope of the answer is narrowed down to lesser number of tokens and there is less noise. For both settings, as also mentioned in § 4, we kept the train set where all the question templates (paraphrases) are observed by the model during training and that is referred with '(r)' prefix, suggesting 'random' selection and no filtering based on question templates (paraphrases). All these dataset abbreviations are shown in the first column of Table 3.

5.2 Extracting Entity Information

`MetaMap` (Aronson, 2001) uses a knowledge-intensive approach to discover different clinical concepts referred to in the text according to unified medical language system (UMLS) (Bodenreider, 2004). The clinical ontologies, such as SNOMED (Spackman et al., 1997) and RxNorm (Liu et al., 2005), embedded in `MetaMap` are quite useful in extracting ~ 127 entities across diagnosis, medication, procedure and sign/symptoms. We shortlisted these entities (semantic types) by mapping them to the entities which were used for creating logical forms of the questions as these are the main entities for which the question has been posed. The selected entities are: acab, aggp, anab, anst, bpoc, cgab, clnd, diap, emod, evnt, fndg, inpo, lbpr, lbtr, phob, qnco, sbst, sosy and topp. Their descriptions are provided in Appendix C.

These filtered entities (Table 7), extracted from MetaMap, are provided to ERNIE. A separate embedding space is defined for the entity embeddings which are passed through a multi-head attention layer (Vaswani et al., 2017) before interacting with token embeddings in the information fusion layer. The entity-enriched token embeddings are then used to predict the span of the answer from the context. We fine-tuned these entity embeddings along with the token embeddings, as opposed to using learned entities and not fine-tuning during downstream tasks (Zhang et al., 2019a). The architecture is illustrated in Fig 2.

5.3 Implementation Details

The BERT model was released with pre-trained weights as $BERT_{base}$ and $BERT_{large}$. $BERT_{base}$ has lesser number of parameters but achieved state-of-the-art results on a number of open-domain NLP tasks. We performed our experiments with $BERT_{base}$ and hence, from here onwards we refer to $BERT_{base}$ as *BERT*. A fine-tuned version of $BERT_{base}$ on clinical notes was released as clinicalBERT (*cBERT*) (Alsentzer et al., 2019). We use *cBERT* as the multi-head attention model for getting the token representations in ERNIE. We refer to this version of ERNIE, with entities from MetaMap, as *cERNIE* for clinical ERNIE. Our final multi-task learning model, incorporated with an auxillary task of predicting logical forms, is referred to as *M-cERNIE* for multi-task clinical ERNIE. The code for all the models is provided at
`https://github.com/emrQA/bionlp_acl20`.

Evaluation Metrics For our extractive question answering task, we utilised exact match and F1-score for evaluation as per earlier literature (Rajpurkar et al., 2016).

6 Results and Discussion

In this section, we compare the results of all the models that we introduced in § 3. With the help of different experiments, we try to analyse whether the induced entity and logical form information

Dataset	Model	F1-score	Exact Match
emrQA-s (pl)	BERT	72.13	65.81
	cBERT	74.75 (+2.62)	67.25 (+1.44)
	cERNIE	77.39 (+5.26)	70.17 (+4.36)
	M-cERNIE	**79.87 (+7.74)**	**71.86 (+6.05)**
emrQA-s (r)	cBERT	**82.34**	**74.58**
emrQA-p (pl)	BERT	64.19	56.30
	cBERT	65.45 (+1.26)	57.58 (+1.28)
	cERNIE	66.15 (+1.96)	59.80 (+3.5)
	M-cERNIE	**67.21 (+3.02)**	**61.22 (+4.92)**
emrQA-p (r)	cBERT	**72.51**	**65.14**
MADE-s (pl)	BERT	68.45	60.73
	cBERT	70.19 (+1.74)	62.00 (+1.27)
	cERNIE	71.51 (+3.06)	65.31 (+4.58)
	M-cERNIE	**73.83 (+5.38)**	**67.53 (+6.8)**
MADE-s (r)	cBERT	**73.70**	**65.54**
MADE-p (pl)	BERT	63.39	57.49
	cBERT	64.97 (+1.58)	58.94 (+1.45)
	cERNIE	**65.71 (+2.32)**	**60.55 (+3.06)**
	M-cERNIE	64.58 (+1.19)	59.39 (+1.9)
MADE-p (r)	cBERT	**66.89**	**61.27**

Table 3: F1-score and exact match values for Models on *emrQA* and *MADE*. The '-s' suffix refers to the *sentence* setting and '-p' refers to the *paragraph* setting for the context provided in our reading comprehension style QA task. The '(pl)' refers to the *paraphrase-level* and '(r)' refers to the *random* split as explained in § 4. *BERT* refers to $BERT_{base}$, *cBERT* refers to *clinicalBERT*, *cERNIE* refers to *clinicalERNIE* and *M-cERNIE* refers to the multi-task learning *clinicalERNIE* model.

help the model in achieving better performance or not. We also analyse the logical form predictions to understand whether it provides a rationale for the answer predicted by our proposed model. The compiled results for all the models are shown in Table 3. The hyper-parameter values for the best performing models are provided in Appendix A.

Does clinical entity information improve models' performance? Across all settings, the F1-score of *cERNIE* improves by $\sim 2-5\%$ over *BERT* and $\sim 0.75 - 3\%$ over *cBERT*. The exact match performance improved by $\sim 3 - 4.5$ over *BERT* and $1.5 - 3.25\%$ over *cBERT*. Also, as expected, the performance in *sentence setting (-s)* improved relatively more than it did in *paragraph-setting*. The entity-enriched tokens help in identifying the tokens which are required by the question. For example, in Fig. 3, the token 'infiltrative' in the question as well as the context get highlighted with the help of the identified entity 'topp' (therapeutic or preventive procedure) and then relevant tokens in the context, chest x ray, get highlighted with the relevant entity 'diap' (diagnostic procedure). This information aids the model in narrowing down its focus to highlighted diagnostic procedures in the context for answer extraction.

Figure 3: An example of a question, context, their extracted entities and expected answer.

Does logical form information help the model generalize better?   In order to answer this question, we compared the performance of our *M-cERNIE* model to *cERNIE* model and observed an improvement of $1.1 - 2.5\%$ in F1-score and an improvement of $1.4 - 1.8\%$ in exact match performance. Here as well, the performance improvement is more for *sentence setting (-s)* as compared to the *paragraph setting (-p)*. This helps the model in understanding the information need expressed in the question and helps in narrowing down its focus to certain tokens as the candidate answer. As seen in example 3, the logical form helps in understanding that the 'dose' of 'medication' needs to be extracted from the context where 'dose' was already highlighted with the help of the entity embedding of 'qnco'.

Overall, the performance of our proposed model improves the F1-score by $1.2 - 7.7\%$ and exact-match by $3.1 - 6.8\%$ over *BERT* model. Thus, embedding clinical entity information with the help of further fine-tuning, entity-enriching and logical form prediction help the model in performing better over the unseen paraphrases by a significant margin. For *emrQA*, the performance of *M-cERNIE* is still below the upper bound performance of the *cBERT* model which is achieved when all the question templates are observed (*emrQA-s/p (r)*) by the model but for *MADE*, in *sentence setting (-s)*, the performance of *M-cERNIE* is even better than the upper bound model performance. For *MADE-p* dataset the performance dropped a little when the LF prediction information is added to the model which might be because *MADE-p* only has 8 logical forms (Appendix B) in total, resulting in low variety between the questions. Thus, the auxiliary task did not add much value to the learning of the base model (*cERNIE*) at paragraph level.

Does the model provide a supporting rationale via logical form (LF) prediction?   We analyzed the performance of *M-cERNIE* on *MADE-s* and *emrQA-s* datasets for logical form prediction, as we saw most improvement in *sentence setting (-s)*. We calculated macro-weighted precision, recall and F1-score for logical form classification. The model achieved a F1-score of $\sim 0.45 - 0.59$ for both datasets, as shown in Table 4, *exact* match setting. We analysed the confusion matrix of predicted LF and observed that the model mainly gets confused between the logical forms which convey similar semantic information as shown in Fig. 4.

Figure 4: Two similar questions with different logical forms (LFs) but overlapping answer conditions.

As we can see in Fig. 4 that both logical forms refer to quite similar information, hence, we decided to obtain performance metrics (precision, recall and F1-score) in *relaxed* setting. We designed this *relaxed* setting to create a more realistic setting, where the *tokens* of predicted and actual logical forms are matched rather than the whole logical form. An example of logical form tokenization is shown in Fig. 5.

Figure 5: Tokenized logical form (LF).

The model achieves a F1-score of 0.92 for *emrQA-s* and 0.84 for *MADE-s* in relaxed setting (Table 4). This suggests that the model can efficiently identify important semantic information from the question, which is critical for efficient QA. During inference, the *M-cERNIE* models yield a rationale regarding a new test question (unseen paraphrase) by predicting the logical form of the question as an auxiliary task. For ex, the LF in Fig. 1 provides a rationale that any *lab* or *procedure* event related to the *condition* event needs to be extracted from the EMR for diagnosis.

Setting	Dataset	Precision	Recall	F1-score
Exact	emrQA	0.65	0.61	0.59
	MADE	0.47	0.52	0.45
Relaxed	emrQA	0.93	0.91	0.92
	MADE	0.83	0.85	0.84

Table 4: Precision, Recall and F1-score for logical form prediction.

Can logical form information be induced in multi-class QA tasks as well? To answer this question, we performed another experiment where the model has to classify the evidence sentences from the non-evidence sentences making it a two-class classification task. The model would be provided a tuple of question and a sentence and it has to predict whether the sentence is evidence or not? The final loss of the model ($\mathcal{L}_{model}$) changes to:

$$\mathcal{L}_{model} = \omega\mathcal{L}_{lf} + (1 - \omega)\mathcal{L}_{evidence} \quad (4)$$

where ω is the weightage given to the loss of auxiliary task ($\mathcal{L}_{lf}$), logical form prediction. $\mathcal{L}_{evidence}$ is loss for evidence classification and $\mathcal{L}_{model}$ is the final loss for our proposed model. We conducted our experiments on *emrQA* dataset as evidence sentences were provided in it. In the multi-class setting, the `[CLS]` token representation would be used for evidence classification as well as logical form prediction.

Dataset	Model	Precision	Recall	F1-score
emrQA	cBERT	0.67	0.99	0.76
	cERNIE	0.69	0.98	0.78 (+0.02)
	M-cERNIE	0.73	0.99	**0.82 (+0.06)**

Table 5: Macro-weighted precision, recall and F1-score of Proposed Models on Test Dataset (Multi-choice QA). For the model names, c: clinical; M: multitask.

The multi-task entity enriched model (*M-cERNIE*) achieved an absolute improvement of 6% over *cBERT* and 4% over *cERNIE*. This suggests that the inductive bias introduced via LF prediction does help in improving the overall performance of the model for multi-class QA as well.

7 Related Work

In the general domain, BERT-based models are on the top of different leader boards across various tasks, including QA tasks (Rajpurkar et al., 2018, 2016). The authors of (Nogueira and Cho, 2019) applied BERT to the MS-MARCO passage retrieval QA task and observed improvement over state of the art results. (Nogueira et al., 2019) further extended the work by combining BERT with re-ranking of predictions for queries that will be issued for each document. However, BERT-based models have not been adapted to answering physician questions on EMRs.

In case of domain-specific QA, logical forms or semantic parse are typically used to integrate the domain knowledge associated with a KB-based (knowledge base) structured QA datasets, where a model is learnt for mapping a natural language question to a LF. GeoQuery (Zelle and Mooney, 1996), and ATIS (Dahl et al., 1994), are the oldest known manually generated question-LF annotations on closed-domain databases. QALD (Lopez et al., 2013), FREE 917 (Cai and Yates, 2013), SIMPLEQuestions (Bordes et al., 2015) contain hundreds of hand-crafted questions and their corresponding database queries. Prior work has also used LFs as a way to generate questions via crowdsourcing (Wang et al., 2015). WEBQuestions (Berant et al., 2013) contains thousands of questions from Google search where the LFs are learned as latent representations in helping answer questions from Freebase. Prior work has not investigated the utility of logical forms in unstructured QA, especially as a means to generalize the QA model across different paraphrases of a question.

There have been efforts on using multi-task learning for efficient question answering, such as the authors of (McCann et al., 2018) tried to learn multiple tasks together resulting in an overall boost in the performance of the model on SQuAD (Rajpurkar et al., 2016). Similarly, the authors of (Lu et al., 2019) also utilised the information across different tasks which lie at the intersection of vision and natural language processing to improve the performance of their model across all tasks. The authors of (Rawat et al., 2019) utilised weak supervision to the model while predicting the answer but not much work has been done to incorporate the logical form of the question for unstructured question answering in a multi-task setting. Hence, we decided to explore this direction and incorporate the structured semantic information of the questions for extractive question answering.

8 Conclusion

The proposed entity-enriched QA models trained with an auxiliary task improve over the state-of-the-art models by about $3 - 6\%$ across the large-scale clinical QA dataset, emrQA (Pampari et al., 2018) (as well as MADE (Jagannatha et al., 2019)). We also show that multitask learning for logical forms along with the answer results in better generalizing over unseen paraphrases for EMR QA. The predicted logical forms also serve as an accompanying justification to the answer and help in adding credibility to the predicted answer for the physician.

Acknowledgement

This work is supported by MIT-IBM Watson AI Lab, Cambridge, MA USA.

References

Emily Alsentzer, John R Murphy, Willie Boag, Wei-Hung Weng, Di Jin, Tristan Naumann, and Matthew McDermott. 2019. Publicly available clinical bert embeddings. *NAACL-HLT Clinical NLP Workshop.*

Alan R Aronson. 2001. Effective mapping of biomedical text to the umls metathesaurus: the metamap program. In *AMIA*, page 17.

Jonathan Berant, Andrew Chou, Roy Frostig, and Percy Liang. 2013. Semantic parsing on freebase from question-answer pairs. In *Proceedings of the 2013 Conference on Empirical Methods in Natural Language Processing*, pages 1533–1544.

Rahul Bhagat, Eduard Hovy, and Siddharth Patwardhan. 2009. Acquiring paraphrases from text corpora. In *Proceedings of the fifth international conference on Knowledge capture*, pages 161–168.

Olivier Bodenreider. 2004. The unified medical language system (umls): integrating biomedical terminology. *Nucleic acids research*, 32(suppl_1):D267–D270.

Antoine Bordes, Nicolas Usunier, Sumit Chopra, and Jason Weston. 2015. Large-scale simple question answering with memory networks. *arXiv preprint arXiv:1506.02075.*

Qingqing Cai and Alexander Yates. 2013. Large-scale semantic parsing via schema matching and lexicon extension. In *ACL*, volume 1, pages 423–433.

Deborah A Dahl, Madeleine Bates, Michael Brown, William Fisher, Kate Hunicke-Smith, David Pallett, Christine Pao, Alexander Rudnicky, and Elizabeth Shriberg. 1994. Expanding the scope of the atis task: The atis-3 corpus. In *HLT*, pages 43–48.

Jacob Devlin, Ming-Wei Chang, Kenton Lee, and Kristina Toutanova. 2019. BERT: Pre-training of deep bidirectional transformers for language understanding. *NAACL-HLT.*

Abhyuday Jagannatha, Feifan Liu, Weisong Liu, and Hong Yu. 2019. Overview of the first natural language processing challenge for extracting medication, indication, and adverse drug events from electronic health record notes (made 1.0). *Drug safety*, 42(1):99–111.

Alistair EW Johnson, Tom J Pollard, Lu Shen, H Lehman Li-wei, Mengling Feng, Mohammad Ghassemi, Benjamin Moody, Peter Szolovits, Leo Anthony Celi, and Roger G Mark. 2016. Mimic-iii, a freely accessible critical care database. *Scientific data*, 3:160035.

Jinhyuk Lee, Wonjin Yoon, Sungdong Kim, Donghyeon Kim, Sunkyu Kim, Chan Ho So, and Jaewoo Kang. 2020. Biobert: a pre-trained biomedical language representation model for biomedical text mining. *Bioinformatics*, 36(4):1234–1240.

Simon Liu, Wei Ma, Robin Moore, Vikraman Ganesan, and Stuart Nelson. 2005. Rxnorm: prescription for electronic drug information exchange. *IT professional*, 7(5):17–23.

Vanessa Lopez, Christina Unger, Philipp Cimiano, and Enrico Motta. 2013. Evaluating question answering over linked data. *Web Semantics: Science, Services and Agents on the World Wide Web*, 21:3–13.

Jiasen Lu, Vedanuj Goswami, Marcus Rohrbach, Devi Parikh, and Stefan Lee. 2019. 12-in-1: Multi-task vision and language representation learning. *arXiv preprint arXiv:1912.02315.*

Bryan McCann, Nitish Shirish Keskar, Caiming Xiong, and Richard Socher. 2018. The natural language decathlon: Multitask learning as question answering. *arXiv preprint arXiv:1806.08730.*

Rodrigo Nogueira and Kyunghyun Cho. 2019. Passage re-ranking with bert. *arXiv preprint arXiv:1901.04085.*

Rodrigo Nogueira, Wei Yang, Jimmy Lin, and Kyunghyun Cho. 2019. Document expansion by query prediction. *arXiv preprint arXiv:1904.08375.*

Anusri Pampari, Preethi Raghavan, Jennifer Liang, and Jian Peng. 2018. emrqa: A large corpus for question answering on electronic medical records. *EMNLP.*

Pranav Rajpurkar, Robin Jia, and Percy Liang. 2018. Know what you don't know: Unanswerable questions for squad. *arXiv preprint arXiv:1806.03822.*

Pranav Rajpurkar, Jian Zhang, Konstantin Lopyrev, and Percy Liang. 2016. SQuAD: 100,000+ questions for machine comprehension of text. *arXiv preprint arXiv:1606.05250.*

Bhanu Pratap Singh Rawat, Fei Li, and Hong Yu. 2019. Naranjo question answering using end-to-end multitask learning model. In *Proceedings of the 25th ACM SIGKDD International Conference on Knowledge Discovery & Data Mining*, pages 2547–2555. ACM.

Minjoon Seo, Aniruddha Kembhavi, Ali Farhadi, and Hannaneh Hajishirzi. 2016. Bidirectional attention flow for machine comprehension. *arXiv preprint arXiv:1611.01603.*

Kent A Spackman, Keith E Campbell, and Roger A Côté. 1997. Snomed rt: a reference terminology for health care. In *Proceedings of the AMIA annual fall symposium*, page 640. American Medical Informatics Association.

Simon Šuster and Walter Daelemans. 2018. Clicr: a dataset of clinical case reports for machine reading comprehension. *arXiv preprint arXiv:1803.09720*.

Wilson L Taylor. 1953. "cloze procedure": A new tool for measuring readability. *Journalism Bulletin*, 30(4):415–433.

George Tsatsaronis, Michael Schroeder, Georgios Paliouras, Yannis Almirantis, Ion Androutsopoulos, Eric Gaussier, Patrick Gallinari, Thierry Artieres, Michael R Alvers, Matthias Zschunke, et al. 2012. Bioasq: A challenge on large-scale biomedical semantic indexing and question answering. In *2012 AAAI Fall Symposium Series*.

Özlem Uzuner, Brett R South, Shuying Shen, and Scott L DuVall. 2011. 2010 i2b2/va challenge on concepts, assertions, and relations in clinical text. *Journal of the American Medical Informatics Association*, 18(5):552–556.

Ashish Vaswani, Noam Shazeer, Niki Parmar, Jakob Uszkoreit, Llion Jones, Aidan N Gomez, Łukasz Kaiser, and Illia Polosukhin. 2017. Attention is all you need. In *Advances in neural information processing systems*, pages 5998–6008.

Yushi Wang, Jonathan Berant, and Percy Liang. 2015. Building a semantic parser overnight. In *ACL-IJCNLP*, volume 1, pages 1332–1342.

Wei-Hung Weng, Yuannan Cai, Angela Lin, Fraser Tan, and Po-Hsuan Cameron Chen. 2019. Multimodal multitask representation learning for pathology biobank metadata prediction. *arXiv preprint arXiv:1909.07846*.

Adina Williams, Nikita Nangia, and Samuel R Bowman. 2017. A broad-coverage challenge corpus for sentence understanding through inference. *arXiv preprint arXiv:1704.05426*.

John M Zelle and Raymond J Mooney. 1996. Learning to parse database queries using inductive logic programming. In *Proceedings of the national conference on artificial intelligence*, pages 1050–1055.

Zhengyan Zhang, Xu Han, Zhiyuan Liu, Xin Jiang, Maosong Sun, and Qun Liu. 2019a. Ernie: Enhanced language representation with informative entities. *arXiv preprint arXiv:1905.07129*.

Zhuosheng Zhang, Yuwei Wu, Junru Zhou, Sufeng Duan, and Hai Zhao. 2019b. Sg-net: Syntax-guided machine reading comprehension. *arXiv preprint arXiv:1908.05147*.

Yukun Zhu, Ryan Kiros, Rich Zemel, Ruslan Salakhutdinov, Raquel Urtasun, Antonio Torralba, and Sanja Fidler. 2015. Aligning books and movies: Towards story-like visual explanations by watching movies and reading books. In *Proceedings of the IEEE international conference on computer vision*, pages 19–27.

A   Model Hyper-parameters

Most of the hyper-parameters across our models remained same: learning rate: $2e-5$, weight decay: $1e-5$, warm-up proportion: 10% and hidden dropout probability: 0.1. The parameters that varied across models for different datasets are enumerated in the Table 6. The hyper-parametsrs provided in Table 6 are for all models in a particular dataset. This also suggests that even after adding an auxiliary task, the proposed model doesn't need a lot of hyper-parameter tuning.

Dataset	Entity Embedding Dim	Auxiliary Task Wt.
emrQA-rel	100	0.3
BoolQ	90	0.3
emrQA	100	0.3
MADE	80	0.2

Table 6: Hyper-parameter values across different datasets.

B   Logical forms (LFs) for MADE dataset

1. MedicationEvent (|medication|) [sig=x]
2. MedicationEvent (|medication|) causes ConditionEvent (x) OR SymptomEvent (x)
3. MedicationEvent (|medication|) given ConditionEvent (x) OR SymptomEvent (x)
4. [ProcedureEvent (|treatment|) given/conducted ConditionEvent (x) OR SymptomEvent (x)] OR [MedicationEvent (|treatment|) given ConditionEvent (x) OR SymptomEvent (x)]
5. MedicationEvent (x) CheckIfNull ([enddate]) OR MedicationEvent (x) [enddate>currentDate] OR ProcedureEvent (x) [date=x] given ConditionEvent (|problem|) OR SymptomEvent (|problem|)
6. MedicationEvent (x) CheckIfNull ([enddate]) OR MedicationEvent (x) [enddate>currentDate] given ConditionEvent (|problem|) OR SymptomEvent (|problem|)
7. MedicationEvent (|treatment|) OR ProcedureEvent (|treatment|) given ConditionEvent (x) OR SymptomEvent (x)
8. MedicationEvent (|treatment|) OR ProcedureEvent (|treatment|) improves/worsens/causes ConditionEvent (x) OR SymptomEvent (x)

C   Selected entities from MetaMap

The list of selected semantic types in the form of entities and their brief descriptors are provided in Table 7.

Semantic Type	Description
acab	Acquired Abnormality
aggp	Age Group
anab	Anatomical Abnormality
anst	Anatomical Structure
bpoc	Body Part, Organ, or Organ Component
cgab	Congenital Abnormality
clnd	Clinical Drug
diap	Diagnostic Procedure
emod	Experimental Model of Disease
evnt	Event
fndg	Finding
inpo	Injury or Poisoning
lbpr	Laboratory Procedure
lbtr	Laboratory or Test Result
phob	Physical Object
qnco	Quantitative Concept
sbst	Substance
sosy	Sign or Symptom
topp	Therapeutic or Preventive Procedure

Table 7: Selected semantic types as per MetaMap and their brief descriptions.

Evidence Inference 2.0: More Data, Better Models

Jay DeYoung[*Ψ], Eric Lehman[*Ψ], Ben Nye[Ψ], Iain J. Marshall[Φ], and Byron C. Wallace[Ψ]

[*]Equal contribution
[Ψ]Khoury College of Computer Sciences, Northeastern University
[Φ]Kings College London
{deyoung.j,lehman.e,nye.b,b.wallace}@northeastern.edu, mail@ijmarshall.com

Abstract

How do we most effectively treat a disease or condition? Ideally, we could consult a database of evidence gleaned from clinical trials to answer such questions. Unfortunately, no such database exists; clinical trial results are instead disseminated primarily via lengthy natural language articles. Perusing all such articles would be prohibitively time-consuming for healthcare practitioners; they instead tend to depend on manually compiled *systematic reviews* of medical literature to inform care.

NLP may speed this process up, and eventually facilitate immediate consult of published evidence. The *Evidence Inference* dataset (Lehman et al., 2019) was recently released to facilitate research toward this end. This task entails inferring the comparative performance of two treatments, with respect to a given outcome, from a particular article (describing a clinical trial) and identifying supporting evidence. For instance: Does this article report that *chemotherapy* performed better than *surgery* for *five-year survival rates* of operable cancers? In this paper, we collect additional annotations to expand the Evidence Inference dataset by 25%, provide stronger baseline models, systematically inspect the errors that these make, and probe dataset quality. We also release an *abstract only* (as opposed to full-texts) version of the task for rapid model prototyping. The updated corpus, documentation, and code for new baselines and evaluations are available at `http://evidence-inference.ebm-nlp.com/`.

1 Introduction

As reports of clinical trials continue to amass at rapid pace, staying on top of all current literature to inform evidence-based practice is next to impossible. As of 2010, about seventy clinical trial reports were published daily, on average (Bastian et al., 2010). This has risen to over one hundred thirty trials per day.[1] Motivated by the rapid growth in clinical trial publications, there now exist a plethora of tools to partially automate the systematic review task (Marshall and Wallace, 2019). However, efforts at fully integrating the PICO framework into this process have been limited (Eriksen and Frandsen, 2018). What if we could build a database of **P**articipants,[2] **I**nterventions, **C**omparisons, and **O**utcomes studied in these trials, and the findings reported concerning these? If done accurately, this would provide direct access to which treatments the evidence supports. In the near-term, such technologies may mitigate the tedious work necessary for manual synthesis.

Recent efforts in this direction include the EBM-NLP project (Nye et al., 2018), and Evidence Inference (Lehman et al., 2019), both of which comprise annotations collected on reports of Randomized Control Trials (RCTs) from PubMed.[3] Here we build upon the latter, which tasks systems with inferring findings in full-text reports of RCTs with respect to particular interventions and outcomes, and extracting evidence snippets supporting these.

We expand the Evidence Inference dataset and evaluate transformer-based models (Vaswani et al., 2017; Devlin et al., 2018) on the task. Concretely, **our contributions** are:

- We describe the collection of an additional 2,503 unique 'prompts' (see Section 2) with matched full-text articles; this is a 25% expansion of the original evidence inference dataset that we will release. We additionally have collected an *abstract-only* subset of data intended to facilitate rapid iterative design of models,

---

[1]See `https://ijmarshall.github.io/sote/`.
[2]We omit Participants in this work as we focus on the document level task of inferring study result directionality, and the Participants are inherent to the study, i.e., studies do not typically consider multiple patient populations.
[3]`https://pubmed.ncbi.nlm.nih.gov/`

Proceedings of the BioNLP 2020 workshop, pages 123–132
Online, July 9, 2020 ©2020 Association for Computational Linguistics

as working over full-texts can be prohibitively time-consuming.

- We introduce and evaluate new models, achieving SOTA performance for this task.

- We ablate components of these models and characterize the types of errors that they tend to still make, pointing to potential directions for further improving models.

2 Annotation

In the *Evidence Inference* task (Lehman et al., 2019), a model is provided with a full-text article describing a randomized controlled trial (RCT) and a 'prompt' that specifies an *Intervention* (e.g., aspirin), a *Comparator* (e.g., placebo), and an *Outcome* (e.g., duration of headache). We refer to these as ICO prompts. The task then is to infer whether a given article reports that the Intervention resulted in a *significant increase*, *significant decrease*, or produced *no significant difference* in the Outcome, as compared to the Comparator.

Our annotation process largely follows that outlined in Lehman et al. (2019); we summarize this briefly here. Data collection comprises three steps: (1) prompt generation; (2) prompt and article annotation; and (3) verification. All steps are performed by Medical Doctors (MDs) hired through Upwork.[4] Annotators were divided into mutually exclusive groups performing these tasks, described below.

Combining this new data with the dataset introduced in Lehman et al. (2019) yields in total 12,616 unique prompts stemming from 3,346 unique articles, increasing the original dataset by 25%.[5] To acquire the new annotations, we hired 11 doctors: 1 for prompt generation, 6 for prompt annotation, and 4 for verification.

2.1 Prompt Generation

In this collection phase, a single doctor is asked to read an article and identify triplets of interventions, comparators, and outcomes; we refer to these as ICO prompts. Each doctor is assigned a unique article, so as to not overlap with one another. Doctors were asked to find a maximum of 5 prompts per article as a practical trade-off between the expense of exhaustive annotation and acquiring annotations

over a variety of articles. This resulted in our collecting 3.77 prompts per article, on average. We asked doctors to derive at least 1 prompt from the body (rather than the abstract) of the article. A large difficulty of the task stems from the wide variety of treatments and outcomes used in the trials: 35.8% of interventions, 24.0% of comparators, and 81.6% of outcomes are unique to one another.

In addition to these ICO prompts, doctors were asked to report the relationship between the intervention and comparator with respect to the outcome, and cite what span from the article supports their reasoning. We find that 48.4% of the collected prompts can be answered using only the abstract. However, 63.0% of the evidence spans supporting judgments (provided by both the prompt generator and prompt annotator), are from outside of the abstract. Additionally, 13.6% of evidence spans cover more than one sentence in length.

2.2 Prompt Annotation

Following the guidelines presented in Lehman et al. (2019), each prompt was assigned to a single doctor. They were asked to report the difference between the specified intervention and comparator, with respect to the given outcome. In particular, options for this relationship were: "increase", "decrease", "no difference" or "invalid prompt." Annotators were also asked to mark a span of text supporting their answers: a rationale. However, unlike Lehman et al. (2019), here, annotators were not restricted via the annotation platform to only look at the abstract at first. They were free to search the article as necessary.

Because trials tend to investigate multiple interventions and measure more than one outcome, articles will usually correspond to multiple — potentially many — valid ICO prompts (with correspondingly different findings). In the data we collected, 62.9% of articles comprise at least two ICO prompts with different associated labels (for the same article).

2.3 Verification

Given both the answers and rationales of the prompt generator and prompt annotator, a third doctor — the verifier — was asked to determine the validity of both of the previous stages.[6] We estimate the accuracy of each task with respect to these verification labels. For prompt generation, answers

---

[4] http://upwork.com.

[5] We use the first release of the data by Lehman et al., which included 10,137 prompts. A subsequent release contained 10,113 prompts, as the authors removed prompts where the answer and rationale were produced by different doctors.

[6] The verifier can also discard low-quality or incorrect prompts.

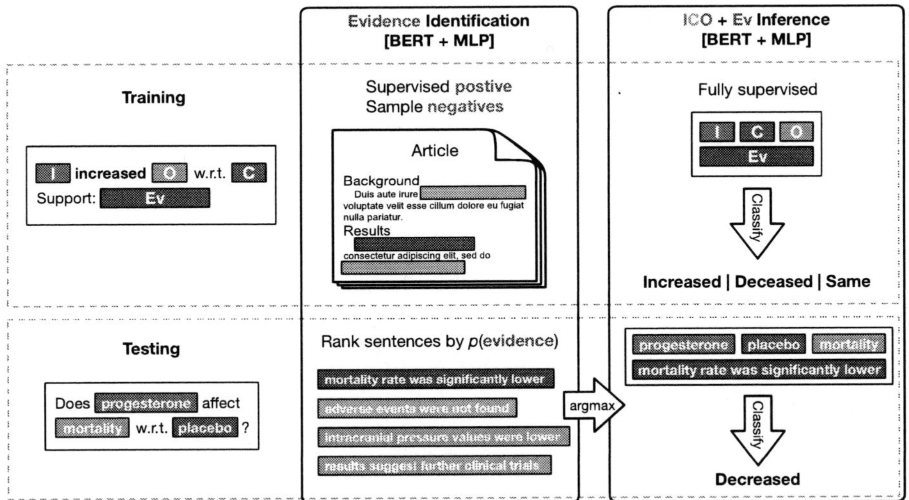

Figure 1: BERT to BERT pipeline. Evidence identification and classification stages are trained separately. The identifier is trained via negative samples against the positive instances, the classifier via only those same positive evidence spans. Decoding assigns a score to every sentence in the document, and the sentence with the highest evidence score is passed to the classifier.

were 94.0% accurate, and rationales were 96.1% accurate. For prompt annotation, the answers were 90.0% accurate, and accuracy of the rationales was 88.8%. The drop in accuracy between prompt generation answers and prompt annotation answers is likely due to confusion with respect to the scope of the intervention, comparator, and outcome.

We additionally calculated agreement statistics amongst the doctors across all stages, yielding a Krippendorf's α of $\alpha = 0.854$. In contrast, the agreement between prompt generator and annotator (excluding verifier) had a $\alpha = 0.784$.

2.4 Abstract Only Subset

We subset the articles and their content, yielding 9,680 of 24,686 annotations, or approximately 40%. This leaves 6375 prompts, 50.5% of the total.

3 Models

We consider a simple BERT-based (Devlin et al., 2018) pipeline comprising two independent models, as depicted in Figure 1. The first *identifies* evidence bearing sentences within an article for a given ICO. The second model then *classifies* the reported findings for an ICO prompt using the evidence extracted by this first model. These models place a dense layer on top of representations yielded from (Gururangan et al., 2020), [7] a variant of RoBERTa (Liu et al., 2019) pre-trained over

scientific corpora,[8] followed by a Softmax.

Specifically, we first perform sentence segmentation over full-text articles using `ScispaCy` (Neumann et al., 2019). We use this segmentation to recover evidence bearing sentences. We train an evidence *identifier* by learning to discriminate between evidence bearing sentences and randomly sampled non-evidence sentences.[9] We then train an evidence *classifier* over the evidence bearing sentences to characterize the trial's finding as reporting that the Intervention *significantly decreased, did not significantly change,* or *significantly increased* the Outcome compared to the Comparator in an ICO. When making a prediction for an (ICO, document) pair we use the highest scoring evidence sentence from the identifier, feeding this to the evidence classifier for a final result. Note that the evidence classifier is conditioned on the ICO frame; we prepend the ICO embedding (from Biomed RoBERTa) to the embedding of the identified evidence snippet. Reassuringly, removing this signal degrades performance (Table 1).

For all models we fine-tuned the underlying BERT parameters. We trained all models using the Adam optimizer (Kingma and Ba, 2014) with a BERT learning rate 2e-5. We train these models for 10 epochs, keeping the best performing version on a nested held-out set with respect to

---

[7]An earlier version of this work used SciBERT (Beltagy et al., 2019); we preserve these results in Appendix C.

[8]We use the `[CLS]` representations.

[9]We train this via negative sampling because the vast majority of sentences are not evidence-bearing.

macro-averaged f-scores. When training the evidence identifier, we experiment with different numbers of random samples per positive instance. We used `Scikit-Learn` (Pedregosa et al., 2011) for evaluation and diagnostics, and implemented all models in `PyTorch` (Paszke et al., 2019). We additionally reproduce the end-to-end system from Lehman et al. (2019): a gated recurrent unit (Cho et al., 2014) to encode the document, attention (Bahdanau et al., 2015) conditioned on the ICO, with the resultant vector (plus the ICO) fed into an MLP for a final significance decision.

4 Experiments and Results

Our main results are reported in Table 1. We make a few key observations. First, the gains over the prior state-of-the-art model — which was not BERT based — are substantial: 20+ absolute points in F-score, even beyond what one might expect to see shifting to large pre-trained models.[10] Second, conditioning on the ICO prompt is key; failing to do so results in substantial performance drops. Finally, we seem to have reached a plateau in terms of the performance of the BERT pipeline model; adding the newly collected training data does not budge performance (evaluated on the augmented test set). This suggests that to realize stronger performance here, we perhaps need a less naive architecture that better models the domain. We next probe specific aspects of our design and training decisions.

Impact of Negative Sampling As negative sampling is a crucial part of the pipeline, we vary the number of samples and evaluate performance. We provide detailed results in Appendix A, but to summarize briefly: we find that two to four negative samples (per positive) performs the best for the end-to-end task, with little change in both AUROC and accuracy of the best fit evidence sentence. This is likely because the model needs only to maximize discriminative capability, rather than calibration.

Distribution Shift In addition to comparable Krippendorf-α values computed above, we measure the impact of the new data on pipeline performance. We compare performance of the pipeline with all data "Biomed RoBERTa (BR) Pipeline" vs. just the old data "Biomed RoBERTA (BR) BERT Pipeline 1.0" in Table 1. As performance stays relatively constant, we believe the new data

Model	Cond?	P	R	F
BR Pipeline	✓	.784	.777	.780
BR Pipeline	✗	.513	.510	.510
BR Pipeline abs.	✓	.776	.777	.776
Baseline	✓	.526	.516	.514
Diagnostics:				
BR Pipeline 1.0	✓	.762	.764	.763
Baseline 1.0	✓	.531	.519	.520
BR ICO Only		.522	.515	.511
BR Oracle Spans	✓	.851	.853	.851
BR Oracle Sentence	✓	.845	.843	.843
BR Oracle Spans	✗	.806	.812	.808
BR Oracle Sentence	✗	.802	.795	.797
BR Oracle Spans abs.	✓	.830	.823	.824
Baseline Oracle 1.0	✓	.740	.739	.739
Baseline Oracle	✓	.760	.761	.759

Table 1: **Classification Scores**. BR Pipeline: Biomed RoBERTa BERT Pipeline. *abs*: Abstracts only. *Baseline*: model from Lehman et al. (2019). **Diagnostic models**: *Baseline* scores Lehman et al. (2019), BR Pipeline when trained using the Evidence Inference 1.0 data, BR classifier when presented with only the ICO element, an entire human selected evidence span, or a human selected evidence sentence. Full document BR models are trained with four negative samples; abstracts are trained with sixteen; Baseline oracle span results from Lehman et al. (2019). In all cases: *'Cond?'* indicates whether or not the model had access to the ICO elements; P/R/F scores are macro-averaged.

to be well-aligned with the existing release. This also suggests that the performance of the current simple pipeline model may have plateaued; better performance perhaps requires inductive biases via domain knowledge or improved strategies for evidence identification.

Oracle Evidence We report two types of Oracle evidence experiments - one using ground truth evidence spans "Oracle *spans*", the other using *sentences* for classification. In the former experiment, we choose an arbitrary evidence span[11] for each prompt for decoding. For the latter, we arbitrarily choose a sentence contained within a span. Both experiments are trained to use a matching classifier. We find that using a span versus a sentence causes a marginal change in score. Both diagnostics provide an upper bound on this model type, improve over the original Oracle baseline by approximately 10 points. Using Oracle evidence as opposed to a trained evidence identifier leaves an end-to-end performance gap of approximately 0.08 F1 score.

---

[10]To verify the impact of architecture changes, we experiment with randomly initialized and fine-tuned BERTs. We find that these perform worse than the original models in all instances and elide more detailed results.

[11]Evidence classification operates on a single sentence, but an annotator's selection is *span* based. Furthermore, the prompt annotation stage may produce different evidence spans than prompt generation.

Ev. Cls	ID Acc.	Predicted Class		
		Sig $\ominus$	Sig $\sim$	Sig $\oplus$
Sig $\ominus$	.667	.684	.153	.163
Sig $\sim$	.674	.060	.840	.099
Sig $\oplus$	.652	.085	.107	.808

Table 2: Breakdown of the conditioned Biomed RoBERTa pipeline model mistakes and performance by evidence class. ID Acc. is the "identification accuracy", or percentage of . To the right is a confusion matrix for end-to-end predictions. 'Sig $\ominus$' indicates significantly decreased, 'Sig $\sim$' indicates no significant difference, 'Sig $\oplus$' indicates significantly increased.

Conditioning As the pipeline can optionally condition on the ICO, we ablate over both the ICO and the actual document text. We find that using the ICO alone performs about as effectively as an unconditioned end-to-end pipeline, 0.51 F1 score (Table 1). However, when fed Oracle sentences, the unconditioned pipeline performance jumps to 0.80 F1. As shown in Table 3 (Appendix A), this large decrease in score can be attributed to the model losing the ability to identify the correct evidence sentence.

Mistake Breakdown We further perform an analysis of model mistakes in Table 2. We find that the BERT-to-BERT model is somewhat better at identifying *significantly decreased* spans than it is at identifying spans for the *significantly increased* or *no significant difference evidence* classes. Spans for the *no significant difference* tend to be classified correctly, and spans for the *significantly increased* category tend to be confused in a similar pattern to the *significantly decreased class*. End-to-end mistakes are relatively balanced between all possible confusion classes.

Abstract Only Results We report a full suite of experiments over the abstracts-only subset in Appendix B. We find that the pipeline models perform similarly on the abstract-only subset; differing in score by less than .01F1. Somewhat surprisingly, we find that the abstracts oracle model falls behind the full document oracle model, perhaps due to a difference in language reporting general results vs. more detailed conclusions.

5 Conclusions and Future Work

We have introduced an expanded version of the Evidence Inference dataset. We have proposed and evaluated BERT-based models for the evidence inference task (which entails identifying snippets of evidence for particular ICO prompts in long documents and then classifying the reported finding on the basis of these), achieving state of the art results on this task.

With this expanded dataset, we hope to support further development of NLP for assisting Evidence Based Medicine. Our results demonstrate promise for the task of automatically inferring results from Randomized Control Trials, but still leave room for improvement. In our future work, we intend to jointly automate the identification of ICO triplets and inference concerning these. We are also keen to investigate whether pre-training on related scientific 'fact verification' tasks might improve performance (Wadden et al., 2020).

Acknowledgments

We thank the anonymous BioNLP reviewers.

This work was supported by the National Science Foundation, CAREER award 1750978.

References

Dzmitry Bahdanau, Kyunghyun Cho, and Yoshua Bengio. 2015. Neural machine translation by jointly learning to align and translate. In *3rd International Conference on Learning Representations, ICLR 2015, San Diego, CA, USA, May 7-9, 2015, Conference Track Proceedings*.

Hilda Bastian, Paul Glasziou, and Iain Chalmers. 2010. Seventy-five trials and eleven systematic reviews a day: how will we ever keep up? *PLoS Med*, 7(9):e1000326.

Iz Beltagy, Arman Cohan, and Kyle Lo. 2019. Scibert: Pretrained contextualized embeddings for scientific text. *arXiv preprint arXiv:1903.10676*.

Kyunghyun Cho, Bart van Merriënboer, Caglar Gulcehre, Dzmitry Bahdanau, Fethi Bougares, Holger Schwenk, and Yoshua Bengio. 2014. Learning phrase representations using RNN encoder–decoder for statistical machine translation. In *Proceedings of the 2014 Conference on Empirical Methods in Natural Language Processing (EMNLP)*, pages 1724–1734, Doha, Qatar. Association for Computational Linguistics.

Jacob Devlin, Ming-Wei Chang, Kenton Lee, and Kristina Toutanova. 2018. Bert: Pre-training of deep bidirectional transformers for language understanding. *arXiv preprint arXiv:1810.04805*.

Mette Brandt Eriksen and Tove Faber Frandsen. 2018. The impact of patient, intervention, comparison, outcome (PICO) as a search strategy tool on literature search quality: a systematic review. *Journal of the Medical Library Association*, 106(4).

Suchin Gururangan, Ana Marasovic, Swabha Swayamdipta, Kyle Lo, Iz Beltagy, Doug Downey, and Noah A. Smith. 2020. Don't stop pretraining: Adapt language models to domains and tasks.

Diederik P. Kingma and Jimmy Ba. 2014. Adam: A method for stochastic optimization. *CoRR*, abs/1412.6980.

Eric Lehman, Jay DeYoung, Regina Barzilay, and Byron C. Wallace. 2019. Inferring which medical treatments work from reports of clinical trials. In *Proceedings of the 2019 Conference of the North American Chapter of the Association for Computational Linguistics: Human Language Technologies, Volume 1 (Long and Short Papers)*, pages 3705–3717, Minneapolis, Minnesota. Association for Computational Linguistics.

Yinhan Liu, Myle Ott, Naman Goyal, Jingfei Du, Mandar Joshi, Danqi Chen, Omer Levy, Mike Lewis, Luke Zettlemoyer, and Veselin Stoyanov. 2019. Roberta: A robustly optimized bert pretraining approach.

Iain J. Marshall and Byron C. Wallace. 2019. Toward systematic review automation: A practical guide to using machine learning tools in research synthesis. *Systematic Reviews*, 8(1):163.

Mark Neumann, Daniel King, Iz Beltagy, and Waleed Ammar. 2019. Scispacy: Fast and robust models for biomedical natural language processing.

Benjamin Nye, Junyi Jessy Li, Roma Patel, Yinfei Yang, Iain Marshall, Ani Nenkova, and Byron Wallace. 2018. A corpus with multi-level annotations of patients, interventions and outcomes to support language processing for medical literature. In *Proceedings of the 56th Annual Meeting of the Association for Computational Linguistics (Volume 1: Long Papers)*, pages 197–207, Melbourne, Australia. Association for Computational Linguistics.

Adam Paszke, Sam Gross, Francisco Massa, Adam Lerer, James Bradbury, Gregory Chanan, Trevor Killeen, Zeming Lin, Natalia Gimelshein, Luca Antiga, et al. 2019. Pytorch: An imperative style, high-performance deep learning library. In *Advances in Neural Information Processing Systems*, pages 8024–8035.

F. Pedregosa, G. Varoquaux, A. Gramfort, V. Michel, B. Thirion, O. Grisel, M. Blondel, P. Prettenhofer, R. Weiss, V. Dubourg, J. Vanderplas, A. Passos, D. Cournapeau, M. Brucher, M. Perrot, and E. Duchesnay. 2011. Scikit-learn: Machine learning in Python. *Journal of Machine Learning Research*, 12:2825–2830.

Ashish Vaswani, Noam Shazeer, Niki Parmar, Jakob Uszkoreit, Llion Jones, Aidan N Gomez, Łukasz Kaiser, and Illia Polosukhin. 2017. Attention is all you need. In *Advances in neural information processing systems*, pages 5998–6008.

David Wadden, Kyle Lo, Lucy Lu Wang, Shanchuan Lin, Madeleine van Zuylen, Arman Cohan, and Hannaneh Hajishirzi. 2020. Fact or fiction: Verifying scientific claim. In *Association for Computational Linguistics (ACL)*.

Appendix

A Negative Sampling Results

We report negative sampling results for Biomed RoBERTa pipelines in Table 3 and Figure 2.

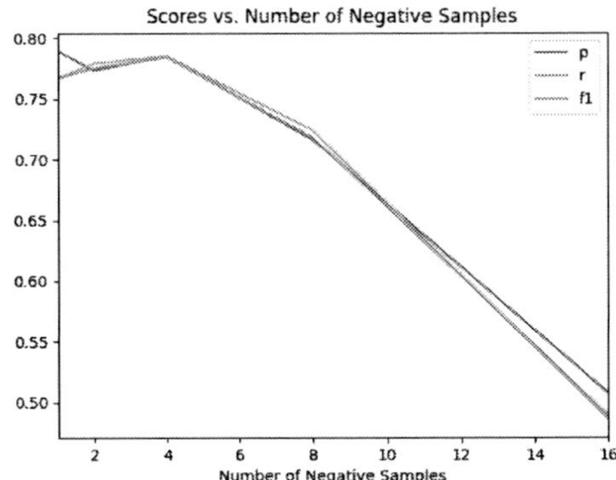

Figure 2: End to end pipeline scores for different negative sampling strategies with Biomed RoBERTa.

Neg, samples	Cond?	AUROC	Top1 Acc
1	✓	0.973	0.682
2	✓	0.972	0.700
4	✓	0.972	0.671
8	✓	0.961	0.492
16	✓	0.590	0.027
1	✗	0.915	0.236
2	✗	0.921	0.226
4	✗	0.925	0.251
8	✗	0.899	0.165
16	✗	0.508	0.015

Table 3: Evidence Inference v2.0 evidence identification validation scores varying across negative sampling strategies using Biomed RoBERTa in the pipeline.

B Abstract Only Results

We repeat the experiments described in Section 4. Our primary findings are that the abstract-only task is easier and sixteen negative samples perform better than four. Otherwise results follow a similar trend to the full-document task. We document these in Table 4, 5, 6 and Figure 3.

C SciBERT Results

We report original SciBERT results in Tables 7, 8, 9 and Figures 4, 5. Table 7 contains the Biomed RoBERTa numbers for comparison. Note that original SciBERT experiments use the evidence inference v1.0 dataset as v2.0 collection was incomplete

Model	Cond?	P	R	F
BR Pipeline	✓	.776	.777	.776
BR Pipeline	✗	.513	.510	.510
Diagnostics:				
ICO Only		.545	.543	.537
Oracle Spans	✓	.830	.823	.824
Oracle Sentence	✓	.845	.843	.843
Oracle Spans	✗	.814	.809	.809
Oracle Sentence	✗	.802	.795	.797

Table 4: **Classification Scores**. Biomed RoBERTa Abstract only version of Table 1. All evidence identification models trained with sixteen negative samples.

Neg. Samples	Cond?	AUROC	Top1 Acc
1	✓	0.983	0.647
2	✓	0.982	0.664
4	✓	0.981	0.680
8	✓	0.978	0.656
16	✓	0.980	0.673
1	✗	0.944	0.351
2	✗	0.953	0.373
4	✗	0.947	0.334
8	✗	0.938	0.273
16	✗	0.947	0.308

Table 5: Abstract only (v2.0) evidence identification validation scores varying across negative sampling strategies using Biomed RoBERTa.

at the time experiment configurations were determined. Biomed RoBERTa experiments use the v2.0 set for calibration. We find that Biomed RoBERTa generally performs better, with a notable exception in performance on abstracts-only Oracle span classification.

C.1 Negative Sampling Results

We report SciBERT negative sampling results in Table 9 and Figure 4.

C.2 Abstract Only Results

We repeat the experiments described in Section 4 and report results in Tables 10, 11, 12 and Figure 5. Our primary findings are that the abstract-only task is easier and eight negative samples perform better than four. Otherwise results follow a similar trend to the full-document task.

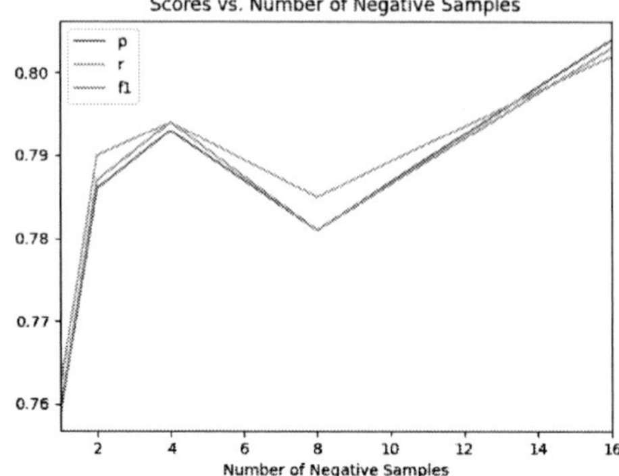

Figure 3: End to end pipeline scores on the abstract-only subset for different negative sampling strategies with Biomed RoBERTa.

Ev. Cls	ID Acc.	Conf. Cls		
		Sig $\ominus$	Sig $\sim$	Sig $\oplus$
Sig $\ominus$	.728	.761	.067	.172
Sig $\sim$	.691	.130	.802	.068
Sig $\oplus$	.573	.123	.109	.768

Table 6: Breakdown of the abstract-only conditioned Biomed RoBERTa pipeline model mistakes and performance by evidence class. ID Acc. is breakdown by final evidence truth. To the right is a confusion matrix for end-to-end predictions.

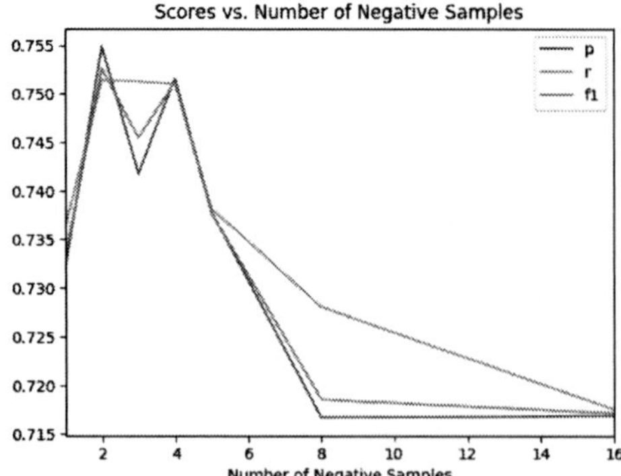

Figure 4: End to end pipeline scores for different negative sampling strategies for SciBERT.

Model	Cond?	P	R	F
BR Pipeline	✓	.784	.777	.780
SB Pipeline	✓	.750	.750	.749
BR Pipeline	✗	.513	.510	.510
SB Pipeline	✗	.489	.486	.486
BR Pipeline abs.	✓	.776	.777	.776
SB Pipeline abs.	✓	.803	.798	.799
Baseline	✓	.526	.516	.514
Diagnostics:				
BR Pipeline 1.0	✓	.762	.764	.763
SB Pipeline 1.0	✓	.749	.761	.753
Baseline 1.0	✓	.531	.519	.520
BR ICO Only		.522	.515	.511
SB ICO Only		.494	.501	.494
BR Oracle Spans	✓	.851	.853	.851
SB Oracle Spans	✓	.840	.840	.838
BR Oracle Sentence	✓	.845	.843	.843
SB Oracle Sentence	✓	.829	.830	.829
BR Oracle Spans	✗	.806	.812	.808
SB Oracle Spans	✗	.786	.789	.787
BR Oracle Sentence	✗	.802	.795	.797
SB Oracle Sentence	✗	.780	.770	.773
BR Oracle Spans abs.	✓	.830	.823	.824
SB Oracle Spans abs.	✓	.866	.862	.863
Baseline Oracle 1.0	✓	.740	.739	.739
Baseline Oracle	✓	.760	.761	.759

Table 7: Replica of Table 1 with both SciBERT and Biomed RoBERTa results. **Classification Scores**. BR Pipeline: Biomed RoBERTa BERT Pipeline, SB Pipeline: SciBERT Pipeline. *abs*: Abstracts only. *Baseline*: model from Lehman et al. (2019). **Diagnostic models**: *Baseline* scores Lehman et al. (2019), BR Pipeline when trained using the Evidence Inference 1.0 data, BR classifier when presented with only the ICO element, an entire human selected evidence span, or a human selected evidence sentence. Full document BR models are trained with four negative samples; abstracts are trained with sixteen; Baseline oracle span results from Lehman et al. (2019). In all cases: *'Cond?'* indicates whether or not the model had access to the ICO elements; P/R/F scores are macro-averaged over classes.

Ev. Cls	ID Acc.	Predicted Class		
		Sig $\ominus$	Sig $\sim$	Sig $\oplus$
Sig $\ominus$	.711	.697	.143	.160
Sig $\sim$	.643	.076	.838	.086
Sig $\oplus$	.635	.146	.141	.713

Table 8: Replica of Table 2 for SciBERT. Breakdown of the conditioned BERT pipeline model mistakes and performance by evidence class. ID Acc. is the "identification accuracy", or percentage of . To the right is a confusion matrix for end-to-end predictions. 'Sig $\ominus$' indicates significantly decreased, 'Sig $\sim$' indicates no significant difference, 'Sig $\oplus$' indicates significantly increased.

Neg. Samples	Cond?	AUROC	Top1 Acc
1	✓	.969	.663
2	✓	.959	.673
4	✓	.968	.659
8	✓	.961	.627
16	✓	.967	.593
1	✗	.894	.094
2	✗	.890	.181
4	✗	.843	.083
8	✗	.862	.170
16	✗	.403	.014

Table 9: Evidence Inference v1.0 evidence identification validation scores varying across negative sampling strategies for SciBERT.

Model	Cond?	P	R	F
BERT Pipeline	✓	.803	.798	.799
BERT Pipeline	✗	.528	.513	.510
Diagnostics:				
ICO Only		.480	.480	.479
Oracle Spans	✓	.866	.862	.863
Oracle Sentence	✓	.848	.842	.844
Oracle Spans	✗	.804	.802	.801
Oracle Sentence	✗	.817	.776	.783

Table 10: **Classification Scores**. SciBERT/Abstract only version of Table 1. All evidence identification models trained with eight negative samples.

Neg. Samples	Cond?	AUROC	Top1 Acc
1	✓	0.980	0.573
2	✓	0.978	0.596
4	✓	0.977	0.623
8	✓	0.950	0.609
16	✓	0.975	0.615
1	✗	0.946	0.340
2	✗	0.939	0.342
4	✗	0.912	0.286
8	✗	0.938	0.313
16	✗	0.940	0.282

Table 11: Abstract only (v1.0) evidence identification validation scores varying across negative sampling strategies for SciBERT.

Ev. Cls	ID Acc.	Conf. Cls		
		Sig $\ominus$	Sig $\sim$	Sig $\oplus$
Sig $\ominus$	.767	.750	.044	.206
Sig $\sim$	.686	.092	.816	.092
Sig $\oplus$	.591	.109	.064	.827

Table 12: Breakdown of the abstract-only conditioned SciBERT pipeline model mistakes and performance by evidence class. ID Acc. is breakdown by final evidence truth. To the right is a confusion matrix for end-to-end predictions.

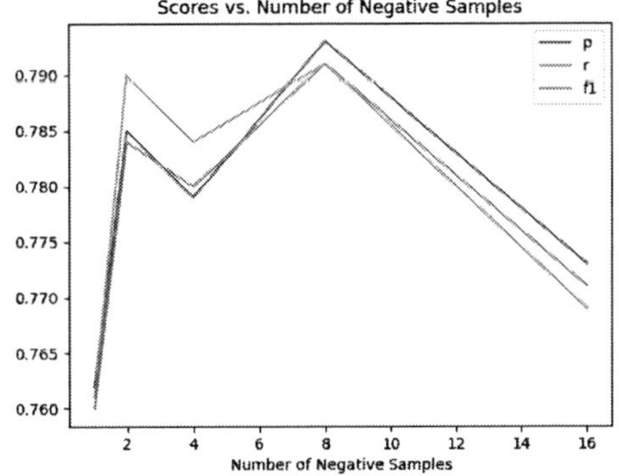

Figure 5: End to end pipeline scores on the abstract-only subset for different negative sampling strategies for SciBERT.

	Train	Dev	Test	Total
Number of prompts	10150	1238	1228	12616
Number of articles	2672	340	334	3346
Label counts (-1 / 0 / 1)	2465 / 4563 / 3122	299 / 544 / 395	295 / 516 / 417	3059 / 5623 / 3934

Table 13: Corpus statistics. Labels -1, 0, 1 indicate *significantly decreased*, *no significant difference* and *significantly increased*, respectively.

Personalized Early Stage Alzheimer's Disease Detection: A Case Study of President Reagan's Speeches

Ning Wang   **Fan Luo**   **Vishal Peddagangireddy**[*]   **K.P. Subbalakshmi**   **R. Chandramouli**

Stevens Institute of Technology

Hoboken, NJ 07030

{nwang7, fluo4, vpeddaga, ksubbala, mouli}@stevens.edu

Paper Number 59

Abstract

Alzheimer's disease (AD)-related global healthcare cost is estimated to be \$1 trillion by 2050. Currently, there is no cure for this disease; however, clinical studies show that early diagnosis and intervention helps to extend the quality of life and inform technologies for personalized mental healthcare. Clinical research indicates that the onset and progression of Alzheimer's disease lead to dementia and other mental health issues. As a result, the language capabilities of patient start to decline.

In this paper, we show that machine learning-based unsupervised clustering of and anomaly detection with linguistic biomarkers are promising approaches for intuitive visualization and personalized early stage detection of Alzheimer's disease. We demonstrate this approach on 10 year's (1980 to 1989) of President Ronald Reagan's speech data set. Key linguistic biomarkers that indicate early-stage AD are identified. Experimental results show that Reagan had early onset of Alzheimer's sometime between 1983 and 1987. This finding is corroborated by prior work that analyzed his interviews using a statistical technique. The proposed technique also identifies the exact speeches that reflect linguistic biomarkers for early stage AD.

1 Introduction

Alzheimer's disease is a serious mental health issue faced by the global population. About 44 million people worldwide are diagnosed with AD. The U.S. alone has 5.5 million AD patients. According to the Alzheimer's association the total cost of care for AD is estimated to be \$1 trillion by 2050. There is no cure for AD yet; however, studies have shown that early diagnosis and intervention can delay the onset.

Regular mental health assessment is a key challenge faced by the medical community. This is due to a variety of reasons including social, economic, and cultural factors. Therefore, Internet based technologies that unobtrusively and continually collect, store, and analyze mental health data are critical. For example, a home smart speaker device can record a subject's speech periodically, automatically extract AD related speech or linguistic features, and present easy to understand machine learning based analysis and visualization. Such a technology will be highly valuable for personalized medicine and early intervention. This may also encourage people to sign-up for such a low cost and home-based AD diagnostic technology. Data and results of such a technology will also instantly provide invaluable information to mental health professionals.

Several studies show that subtle linguistic changes are observed even at the early stages of AD. In (Forbes-McKay and Venneri, 2005), more than 70% of AD patients scored low in a picture description task. Therefore, a critical research question is: can spontaneous temporal language impairments caused by AD be detected at an early stage of the disease? Relation between AD, language functions and language domain are summarized in (Szatloczki et al., 2015). In (Venneri et al., 2008), a significant correlation between the lexical attributes characterising residual linguistic production and the integrity of regions of the medial temporal lobes in early AD patients is observed. Specific lexical and semantic deficiencies of AD patients at early to moderate stages are also detected in verbal communication task in (Boyé et al., 2014). Therefore, in this paper, we explore a machine-learning based clustering and data visualization technique to identify linguistic changes in a subject over a time period. The proposed methodology is also highly personalized since it observes and analyzes the linguistic patterns of each individual separately using only his/her own linguistic biomarkers over a period of time.

Proceedings of the BioNLP 2020 workshop, pages 133–139

Online, July 9, 2020 ©2020 Association for Computational Linguistics

First, we explore a machine learning algorithm called t-distributed stochastic neighbor embedding (t-SNE) (Maaten and Hinton, 2008). t-SNE is useful in dimensionality reduction suitable for visualization of high-dimensional datasets. It calculates the probability that two points in a high-dimensional space are similar, computes the corresponding probability in a low-dimensional space, and minimizes the difference between these two probabilities for mapping or visualization. During this process, the sum of Kullback-Leibler divergences (Liu and Shum, 2003) over all the data points is minimized. Our hypothesis is that high-dimensional AD-related linguistic features when visualized in a low-dimensional space may quickly and intuitively reveal useful information for early diagnosis. Such a visualization will also help individuals and general medical practitioners (who are first points of contact) to assess the situation for further tests. Second, we investigate two unsupervised machine learning techniques, one class support vector machine (SVM) and isolation forest, for temporal detection of linguistic abnormalities indicative of early-stage AD. These proposed approaches are tested on President Reagan's speech dataset and corroborated with results from other research in the literature.

This paper is organized as follows. Background research is discussed in Section 2, Section 3 presents the Reagan speech data set used in this paper, data pre-processing techniques that were applied and AD-related linguistic feature selection rationale and methodology, Section 4 contains the clustering and visualization of the handcrafted linguistic features and anomaly detection to infer the onset of AD and to detect the time period of changes in the linguistic statistical characteristics, and experimental results to demonstrate the proposed method. In Section 5, we describe the machine learning algorithms for detecting anomalies from personalized linguistic biomarkers collected over a period of time to identify early-stage AD. Concluding remarks are given in Section 6.

2 Background

The "picture description" task has been widely studied to differentiate between AD and non-AD or control subjects. In this task, a picture (the "cookie theft picture") is shown and the subject is asked to describe it. It has been observed that subjects with AD usually convey sparse information about the picture ignoring expected facts and inferences (Giles et al., 1996). AD patients have difficulty in naming things and replace target words with simpler semantically neighboring words. Understanding metaphors and sarcasm also deteriorate in people diagnosed with AD (Rapp and Wild, 2011).

Machine learning based classifier design to differentiate between AD and non-AD subjects is an active area of research. A significant correlation between dementia severity and linguistic measures such as confrontation naming, articulation, word-finding ability, and semantic fluency exists. Some studies have reported a 84.8 percent accuracy in distinguishing AD patients from healthy controls using temporal and acoustic features ((Rentoumi et al., 2014); (Fraser et al., 2016)).

Our study differs from the prior work in several ways. Prior work attempt to identify linguistic features and machine learning classifiers for differentiating between AD and non-AD subjects from a corpus (e.g., Dementia Bank (MacWhinney, 2007)) containing text transcripts of interviews with patients and healthy control. In this paper, we first analyze the speech transcripts (over several years) of a single person (President Ronald Reagan) and visualize time-dependent linguistic information using t-SNE. The goal is to identify visual clues for linguistic changes that may indicate the onset of AD. Note that such an easy to understand visual representation depicting differences in linguistic patterns will be useful to both a common person and a general practitioner (who is the first point of contact for majority of patients). Since most general practitioners are not trained mental health professionals the impact of such a visualization tool will be high, especially at the early stages of AD. Sigificant AD-related linguistic biomarkers derived from t-SNE analysis are then used in two unsupervised clustering algorithms for detecting AD-related temporal linguistic anomalies. This provides an estimate for the time when early-stage AD symptoms are beginning to be observed.

3 Reagan Speeches: Data Collection, Pre-processing, and Feature Selection

We describe the data collection, pre-processing, and feature engineering methodologies in this section. Ronald Reagan was the 40th president (served from 1981 to 1989) of the United States of America. He was an extraordinary orator, a radio announcer, actor, and the host for a show called "General Electric Theatre." Clearly, for being successful in these professional domains one needs to have good memory, consciousness, intuition, command over the language, and ease of communicating with a large audience. Reagan officially announced that he had been diagnosed with AD on November 5, 1994. But it was speculated that his cognitive abilities where on the decline even while in office (Gottschalk et al., 1988). Therefore, analyzing his speech transcripts for early signs of AD may reveal interesting patterns, if any.

The Reagan Library is the repository of presidential records for President Reagan's administration. We download his 98 speeches from 1980 to 1989 as shown in Table 1. We removed special characters, tags, and numbers and kept only the words from each speech transcript. The resulting data was then lemmatized and tokenized.

Table 1: President Reagan's speech dataset

Year	No. of speeches
1980	6
1981	8
1982	11
1983	12
1984	14
1985	13
1986	14
1987	10
1988	9
1989	1

Part-of-speech (POS) features: People diagnosed with AD use more pronouns than nouns. Therefore, POS features such as the number of pronouns and the pronouns-to-nouns ratio are important. We identified adverbs, nouns, verbs, and pronouns for each speech transcript with natural language processing (NLP) tools. POS tags having at least 10 occurrences were selected to compute their percentage ratios. Similarly, words that have at least a frequency of 10 were selected and their occurrence percentages were computed.

The full set of *POS features* we used were: (1) number of pronouns, (2) pronoun-noun ratio, (3) number of adverbs, (4) number of nouns, (5) number of verbs, (6) pro-noun frequency rate, (7) noun frequency rate, (8) verb frequency rate, (9) adverb frequency rate, (10) word frequency rate, and (11) word frequency rate without excluding stop words.

Vocabulary Richness: AD patients show a decline in their vocabulary range. Therefore, vocabulary richness metrics: Honore's Statistic (HS), Sichel Measure (SICH), and Brunet's Measure (BM) were calculated for each speech. Higher values of Honore's and Sichel measures indicate greater vocabulary richness. But a higher value corresponds to low vocabulary richness for the Brunet's measure.

Readability Measures: We computed two readability measures, namely, Automated Readability Index (ARI) and Flesch-Kincaid readability (FKR) score. A higher ARI indicates complex speech with rich vocabulary whereas lower Flesch-Kincaid score indicates rich vocabulary.

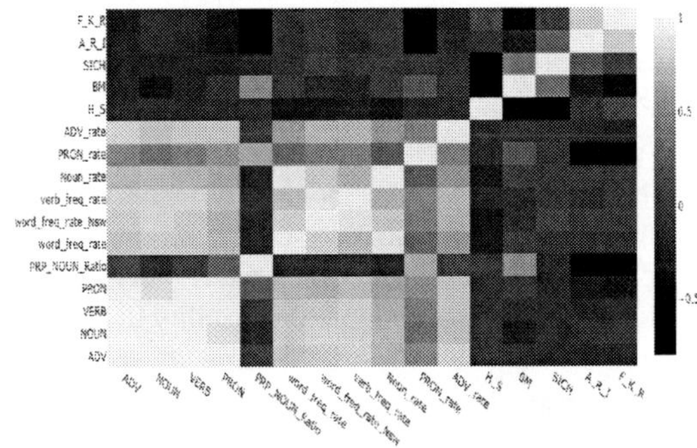

Figure 1: Correlation matrix of the linguistic features.

Figure 1 shows the correlation matrix of the chosen linguistic features computed for the Reagan speech dataset. Note that some of the features are highly correlated, therefore, we pruned the feature set to the following 9 features:

1. pronoun-noun ratio

2. word frequency rate

3. verb frequency rate

4. pronoun frequency rate

5. adverb frequency rate

6. Honore's measure

7. Brunet's measure

8. Sichel measure

9. Automated Readability Index

Any further analysis in this paper uses the above 9 selected linguistic features.

4 Clustering and Visualization of Linguistic Features

We selected the t-SNE machine learning technique for clustering and visulaization of linguistic features extracted from Reagan's speeches. t-SNE is better at creating a single map for revealing structures at many different scales important for high-dimensional data that lie on several different, but related, low-dimensional manifolds. This implies that it can capture much of the local structure of the high-dimensional data very well, while also revealing global structure such as the presence of clusters at several scales (van der Maaten and Hinton 2008). If there are AD-related linguistic patterns then t-SNE may reveal them as clusters.

In t-SNE, the high-dimensional Euclidean distances between datapoints are converted into conditional probabilities representing similarities between them. For example, the similarity of data point x_j with x_i is the conditional probability $p_{j|i}$ that x_i would pick x_j as its neighbor. Neighbors are picked in proportion to their probability density under Student t-distribution centered at x_i in the low-dimensional space. Then the Kullback-Leibler divergence between a joint probability distribution, P, in the high-dimensional space and a joint probability distribution, Q, in the low-dimensional space is then minimized:

$$\min KL(P||Q) = \sum_i \sum_j p_{ij} \log \frac{p_{ij}}{q_{ij}} \quad (1)$$

t-SNE has two tunable hyperparameters: perplexity and learning rate. Perplexity allows us to balance the weighting between local and global relationships of the data. It gives a sense of the number of close neighbors for each point. A perplexity value between 5 and 50 is recommended. The learning rate for t-SNE is usually in the interval [10, 1000]. For a high learning rate, each point will approximately be equidistant from its nearest neighbours. For a low rate, there will be few outliers and therefore the points may look compressed into a dense cloud. Since t-SNE's cost function is not convex different initializations can produce different results. After tuning t-SNE's hyperparameters for the dataset, we chose perplexity value equal to 4 and learning rate equal to 100.

Pronoun-to-noun ratio: Figure 2 shows t-SNE based clustering of the speech transcripts, speeches sharing similar patterns are clustered together. Besides, the radius of each circle is proportional to the pronoun-to-noun ratio. Notice that the cluster on the left side of the graph contains speeches (from 1983 to 1987) have higher values of the ratio. Recall that higher pronoun-to-noun ratio is an indicator of early stage AD.

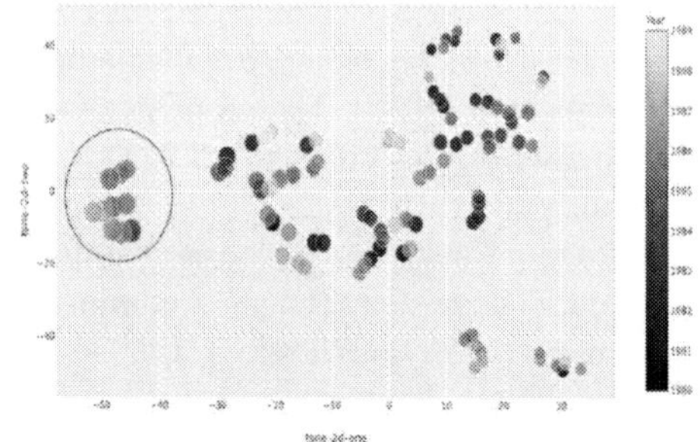

Figure 2: 2-dimensional visualization of speech transcripts where size of each circle in the map is proportional to the pronoun-to-noun ratio.

Pronoun frequency: Fig. 3 indicates clustering and low-dimensional visualization results when the size of each circle is proportional to the pronoun frequency. We again see that the cluster on the left side of the map contains speeches with higher pronoun frequency, another indicator of early stage AD.

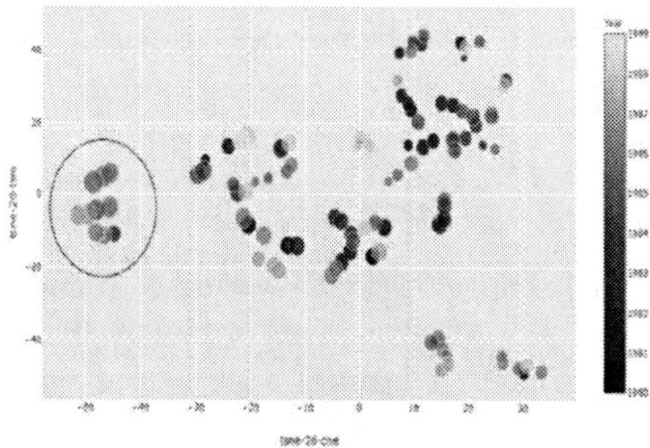

Figure 3: 2-dimensional visualization of speech transcripts where size of each circle in the map is proportional to the pronoun frequency.

Readability score: From Fig. 4 we observe that the speeches from 1983-1987 have lower readability scores.

Word repetition frequency: The word repetition frequency map in Fig. 5 shows that three speeches standout. These three speeches have a higher repetition of high frequency words. Interestingly, two of these speeches have word lengths smaller than the mean length of all the speeches.

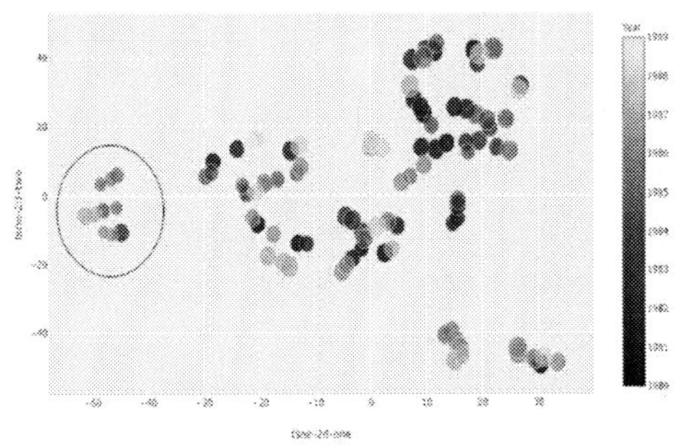

Figure 4: 2-dimensional visualization of speech transcripts where size of each circle in the map is proportional to the readability score.

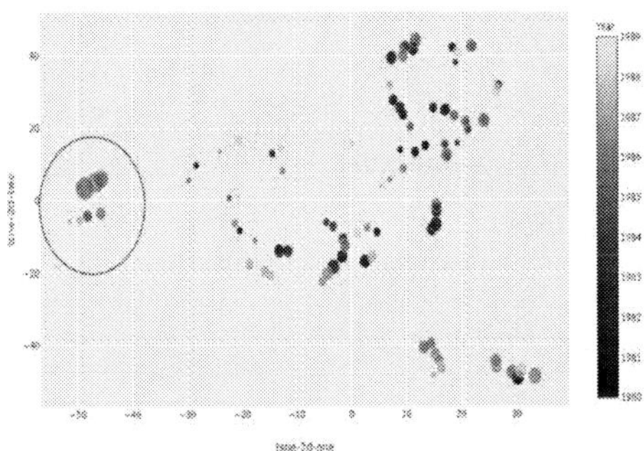

Figure 5: 2-dimensional visualization of speech transcripts where size of each circle in the map is proportional to the word repetition frequency.

From these clustering results the speeches identified as showing early signs of AD are:

- 03-23-1983: Address to the Nation on National Security ("Star Wars" SDI Speech)

- 05-28-1984: Remarks honoring the Vietnam War's Unknown

- 02-06-1985: State of the Union Address, (American Revolution II)

- 04-24-1985: Address to the Nation on the Federal Budget and Deficit Reduction

- 05-05-1985: Speech at Bergen-Belsen Concentration Camp Memorial

- 05-05-1985: Speech at Bitburg Air Base

- 04-14-1986: Address to the Nation on the Air Strike Against Libya

- 06-24-1986: Address to the Nation on Aid to the Contras

- 08-12-1987: Address to the Nation on the Iran-ContraAffair

- 08-12-1987: Address to the Nation on the Iran-ContraAffair

- 09-21-1987: Address to the General Assembly of the United Nations, (INF Agreement and Iran)

Therefore, t-SNE based clustering and low-dimensional visualization of President Reagan's speeches from 1964 to 1989 reveals the following:

- he started showing signs of early AD well before the official announcement in 1994

- it is highly likely that he developed AD sometime between 1983 and 1987

- over time, President Reagan's showed a significant reduction the number of unique words but a significant increase in conversational fillers and non-specific nouns

- the proposed method identifies specific speeches that exhibit linguistic markers for AD

Some of these findings are corroborated by prior research that analyzed his interviews and compared them with President Bursh's public speeches, (Berisha et al., 2015).

5 Linguistic Anomaly Detector for AD

t-SNE-clustering-based approach provides visualization of speeches that are statistically different ("anomalies"). But we need an automated method to identify these anomalies for early signs of AD. Therefore, we investigate a one-class support vector machine (SVM) anomaly detector (Erfani et al., 2016). This method is useful in practice when majority of a subject's speeches over several years would be (statistically) typical of a healthy control ("normal") until he/she begins to exhibit early signs of AD. Our hypothesis is that early stage AD will begin to reveal itself as statistical anomalies in linguistic feature space. In this section we investigate this hypothesis.

We designed a one class SVM with the following hyperparameter values (the choices of these values are not discussed for the sake of clarity and focus), $\nu = 0.5$ (an upper bound on the fraction of training errors and a lower bound of the fraction of support vectors), kernel=rbf (radial basis function) and

$\gamma = \frac{1}{(\text{number of features} \times \text{variance of features})}$ (kernel coefficient). Figure 6 shows that the one class SVM

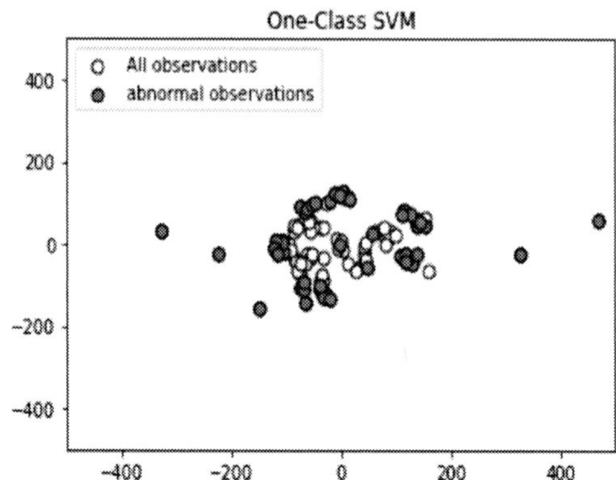

Figure 6: Speeches from 1984 to 1986 are detected as abnormal or anomalies.

detector identified several speeches from 1984 to 1986 as anomalous. That is, President Reagan's speeches in these years are different from the previous years in the linguistic feature space. Therefore, it is likely that:

- he started showing signs of early AD well before the official announcement in 1994

- the onset of AD start from 1984 and became more pronounced in 1985 and 1986, which is corroborated by prior research (e.g., (Berisha et al., 2015)) that analyzed his interviews.

One class SVM learns the profile of non-AD speeches as "normal" over a period of time and detects anomalies to signal the onset of AD. But in many practical instances long historical data may not be available for a subject. In this case, we must identify anomalies explicitly instead of learning what is normal. Isolation forest (Ding and Fei, 2013) is an unsupervised machine learning algorithm that is applicable for this purpose.

We applied the isolation forest algorithm on the speech dataset and tuned its hyperparameters. Figure 7 and Figure 8 show the results. Figure 7 shows the ten speeches that were isolated as anomalies. The corresponding dates of these speeches are seen in Figure 8. We observe that a few speeches between 1984 and 1988 as linguistic anomalies. This time period overlaps significantly with the results of once class SVM and t-SNE clustering.

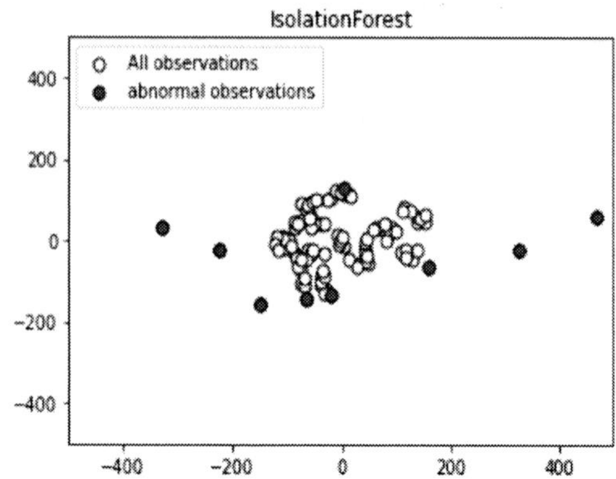

Figure 7: Isolation forest algorithm for AD detection.

```
   Dates   Year     scores   anomaly
1984-06-06  1984  -0.083271       -1
1985-05-05  1985  -0.064989       -1
1985-05-08  1985  -0.092632       -1
1985-10-24  1985  -0.129530       -1
1986-02-26  1986  -0.071642       -1
1986-10-13  1986  -0.240466       -1
1987-12-10  1987  -0.069543       -1
1988-05-31  1988  -0.203001       -1
1988-06-03  1988  -0.060763       -1
1988-12-16  1988  -0.088826       -1
```

Figure 8: Year-wise speeches identified as anomalous.

6 Conclusions

A set of nine linguistic biomarkers for AD was identified. Two complementary unsupervised machine learning methods were applied on the linguistic features extracted from President Reagan's 98 speeches given between 1980 to 1989. The first method, t-SNE, identified and visualized speeches indicating early onset of AD. A higher pronoun usage frequency, lower readability scores, higher repetition of high frequency words were revealed to be the key characteristics of potential AD-related speeches. A subset speeches from 1983 to 1987 were detected to possess these characteristics.

The second machine learning method, one class SVM, learned what is "normal" (i.e., non-AD speech) to detect anomalies in speeches over a period of time. This approach detected several speeches between 1984 and 1986 as potential AD-related. Since normal speech may not be available historically we applied the isolation forest algorithm that explicitly detects anomalies without learning what is normal. This detected 10 speeches from 1983 to 1987 as AD-related.

From the experimental analysis our conclusion

is that President Reagan had signs of AD sometime between 1983 and 1988. This conclusion corroborates results from other studies in the literature. Note that that President Reagan had AD was publicly disclosed only in November 1994.

References

Visar Berisha, Shuai Wang, Amy LaCross, and Julie Liss. 2015. Tracking discourse complexity preceding alzheimer's disease diagnosis: a case study comparing the press conferences of presidents ronald reagan and george herbert walker bush. *Journal of Alzheimer's Disease*, 45(3):959–963.

Maïté Boyé, Natalia Grabar, and Thi Mai Tran. 2014. Contrastive conversational analysis of language production by alzheimer's and control people. In *MIE*, pages 682–686.

Zhiguo Ding and Minrui Fei. 2013. An anomaly detection approach based on isolation forest algorithm for streaming data using sliding window. *IFAC Proceedings Volumes*, 46(20):12–17.

Sarah M Erfani, Sutharshan Rajasegarar, Shanika Karunasekera, and Christopher Leckie. 2016. High-dimensional and large-scale anomaly detection using a linear one-class svm with deep learning. *Pattern Recognition*, 58:121–134.

Katrina E Forbes-McKay and Annalena Venneri. 2005. Detecting subtle spontaneous language decline in early alzheimer's disease with a picture description task. *Neurological sciences*, 26(4):243–254.

Kathleen C Fraser, Jed A Meltzer, and Frank Rudzicz. 2016. Linguistic features identify alzheimer's disease in narrative speech. *Journal of Alzheimer's Disease*, 49(2):407–422.

Elaine Giles, Karalyn Patterson, and John R Hodges. 1996. Performance on the boston cookie theft picture description task in patients with early dementia of the alzheimer's type: missing information. *Aphasiology*, 10(4):395–408.

Louis A Gottschalk, Regina Uliana, and Ronda Gilbert. 1988. Presidential candidates and cognitive impairment measured from behavior in campaign debates. *Public Administration Review*, pages 613–619.

Ce Liu and Hueng-Yeung Shum. 2003. Kullback-leibler boosting. In *2003 IEEE Computer Society Conference on Computer Vision and Pattern Recognition, 2003. Proceedings.*, volume 1, pages I–I. IEEE.

Laurens van der Maaten and Geoffrey Hinton. 2008. Visualizing data using t-sne. *Journal of machine learning research*, 9(Nov):2579–2605.

Brian MacWhinney. 2007. The talkbank project. In *Creating and digitizing language corpora*, pages 163–180. Springer.

Alexander M Rapp and Barbara Wild. 2011. Nonliteral language in alzheimer dementia: a review. *Journal of the International Neuropsychological Society*, 17(2):207–218.

Vassiliki Rentoumi, Ladan Raoufian, Samrah Ahmed, Celeste A de Jager, and Peter Garrard. 2014. Features and machine learning classification of connected speech samples from patients with autopsy proven alzheimer's disease with and without additional vascular pathology. *Journal of Alzheimer's Disease*, 42(s3):S3–S17.

Greta Szatloczki, Ildiko Hoffmann, Veronika Vincze, Janos Kalman, and Magdolna Pakaski. 2015. Speaking in alzheimer's disease, is that an early sign? importance of changes in language abilities in alzheimer's disease. *Frontiers in aging neuroscience*, 7:195.

Annalena Venneri, William J McGeown, Heidi M Hietanen, Chiara Guerrini, Andrew W Ellis, and Michael F Shanks. 2008. The anatomical bases of semantic retrieval deficits in early alzheimer's disease. *Neuropsychologia*, 46(2):497–510.

BIOMRC: A Dataset for Biomedical Machine Reading Comprehension

Petros Stavropoulos[1,2], Dimitris Pappas[1,2], Ion Androutsopoulos[1],
Ryan McDonald[3,1]

[1]Department of Informatics, Athens University of Economics and Business, Greece
[2]Institute for Language and Speech Processing, Research Center 'Athena', Greece
[3]Google Research
{pstav1993, pappasd, ion}@aueb.gr
ryanmcd@google.com

Abstract

We introduce BIOMRC, a large-scale cloze-style biomedical MRC dataset. Care was taken to reduce noise, compared to the previous BIOREAD dataset of Pappas et al. (2018). Experiments show that simple heuristics do not perform well on the new dataset, and that two neural MRC models that had been tested on BIOREAD perform much better on BIOMRC, indicating that the new dataset is indeed less noisy or at least that its task is more feasible. Non-expert human performance is also higher on the new dataset compared to BIOREAD, and biomedical experts perform even better. We also introduce a new BERT-based MRC model, the best version of which substantially outperforms all other methods tested, reaching or surpassing the accuracy of biomedical experts in some experiments. We make the new dataset available in three different sizes, also releasing our code, and providing a leaderboard.

1 Introduction

Creating large corpora with human annotations is a demanding process in both time and resources. Research teams often turn to distantly supervised or unsupervised methods to extract training examples from textual data. In machine reading comprehension (MRC) (Hermann et al., 2015), a training instance can be automatically constructed by taking an unlabeled passage of multiple sentences, along with another smaller part of text, also unlabeled, usually the next sentence. Then a named entity of the smaller text is replaced by a placeholder. In this setting, MRC systems are trained (and evaluated for their ability) to read the passage and the smaller text, and guess the named entity that was replaced by the placeholder, which is typically one of the named entities of the passage. This kind of question answering (QA) is also known as cloze-type questions (Taylor, 1953). Several datasets have

been created following this approach either using books (Hill et al., 2016; Bajgar et al., 2016) or news articles (Hermann et al., 2015). Datasets of this kind are noisier than MRC datasets containing human-authored questions and manually annotated passage spans that answer them (Rajpurkar et al., 2016, 2018; Nguyen et al., 2016). They require no human annotations, however, which is particularly important in biomedical question answering, where employing annotators with appropriate expertise is costly. For example, the BIOASQ QA dataset (Tsatsaronis et al., 2015) currently contains approximately 3k questions, much fewer than the 100k questions of a SQUAD (Rajpurkar et al., 2016), exactly because it relies on expert annotators.

To bypass the need for expert annotators and produce a biomedical MRC dataset large enough to train (or pre-train) deep learning models, Pappas et al. (2018) adopted the cloze-style questions approach. They used the full text of unlabeled biomedical articles from PUBMED CENTRAL,[1] and METAMAP (Aronson and Lang, 2010) to annotate the biomedical entities of the articles. They extracted sequences of 21 sentences from the articles. The first 20 sentences were used as a passage and the last sentence as a cloze-style question. A biomedical entity of the 'question' was replaced by a placeholder, and systems have to guess which biomedical entity of the passage can best fill the placeholder. This allowed Pappas et al. to produce a dataset, called BIOREAD, of approximately 16.4 million questions. As the same authors reported, however, the mean accuracy of three humans on a sample of 30 questions from BIOREAD was only 68%. Although this low score may be due to the fact that the three subjects were not biomedical experts, it is easy to see, by examining samples of BIOREAD, that many examples of the dataset do

---

[1]https://www.ncbi.nlm.nih.gov/pmc/

140

Proceedings of the BioNLP 2020 workshop, pages 140–149
Online, July 9, 2020 ©2020 Association for Computational Linguistics

'question' originating from caption:
"figure 4 htert @entity6 and @entity4 XXXX cell invasion."
'question' originating from reference:
"2004 , 17 , 250 257 .14967013 c samuni y. ; samuni u. ; goldstein s. the use of cyclic XXXX as hno scavengers ."
'passage' containing captions:
"figure 2: distal UNK showing high insertion of rectum into common channel. figure 3: illustration of the cloacal malformation. figure 4: @entity5 showing UNK"

Table 1: Examples of noisy BIOREAD data. XXXX is the placeholder, and UNK is the 'unknown' token.

not make sense. Many instances contain passages or questions crossing article sections, or originating from the references sections of articles, or they include captions and footnotes (Table 1). Another source of noise is METAMAP, which often misses or mistakenly identifies biomedical entities (e.g., it often annotates 'to' as the country Togo).

In this paper, we introduce BIOMRC, a new dataset for biomedical MRC that can be viewed as an improved version of BIOREAD. To avoid crossing sections, extracting text from references, captions, tables etc., we use abstracts and titles of biomedical articles as passages and questions, respectively, which are clearly marked up in PUBMED data, instead of using the full text of the articles. Using titles and abstracts is a decision that favors precision over recall. Titles are likely to be related to their abstracts, which reduces the noise-to-signal ratio significantly and makes it less likely to generate irrelevant questions for a passage. We replace a biomedical entity in each title with a placeholder, and we require systems to guess the hidden entity by considering the entities of the abstract as candidate answers. Unlike BIOREAD, we use PUBTATOR (Wei et al., 2012), a repository that provides approximately 25 million abstracts and their corresponding titles from PUBMED, with multiple annotations.[2] We use DNORM's biomedical entity annotations, which are more accurate than METAMAP's (Leaman et al., 2013). We also perform several checks, discussed below, to discard passage-question instances that are too easy, and we show that the accuracy of experts and non-expert humans reaches 85% and 82%, respectively, on a sample of 30 instances for each annotator type, which is an indication that the new dataset is indeed less noisy, or at least that the task is more feasible for humans. Following Pappas et al. (2018), we release two versions of BIOMRC, LARGE and LITE, containing 812k and 100k instances respectively,

for researchers with more or fewer resources, along with the 60 instances (TINY) humans answered. Random samples from BIOMRC LARGE where selected to create LITE and TINY. BIOMRC TINY is used only as a test set; it has no training and validation subsets.

We tested on BIOMRC LITE the two deep learning MRC models that Pappas et al. (2018) had tested on BIOREAD LITE, namely Attention Sum Reader (AS-READER) (Kadlec et al., 2016) and Attention Over Attention Reader (AOA-READER) (Cui et al., 2017). Experimental results show that AS-READER and AOA-READER perform better on BIOMRC, with the accuracy of AOA-READER reaching 70% compared to the corresponding 52% accuracy of Pappas et al. (2018), which is a further indication that the new dataset is less noisy or that at least its task is more feasible. We also developed a new BERT-based (Devlin et al., 2019) MRC model, the best version of which (SCIBERT-MAX-READER) performs even better, with its accuracy reaching 80%. We encourage further research on biomedical MRC by making our code and data publicly available, and by creating an on-line leaderboard for BIOMRC.[3]

2 Dataset Construction

Using PUBTATOR, we gathered approx. 25 million abstracts and their titles. We discarded articles with titles shorter than 15 characters or longer than 60 tokens, articles without abstracts, or with abstracts shorter than 100 characters, or fewer than 10 sentences. We also removed articles with abstracts containing fewer than 5 entity annotations, or fewer than 2 or more than 20 distinct biomedical entity identifiers. (PUBTATOR assigns the same identifier to all the synonyms of a biomedical entity; e.g., 'hemorrhagic stroke' and 'stroke' have the same identifier 'MESH:D020521'.) We also discarded articles containing entities not linked to any of the ontologies used by PUBTATOR,[4] or entities linked to multiple ontologies (entities with multiple ids), or entities whose spans overlapped with those of other entities. We also removed articles with no entities in their titles, and articles with no entities shared by the title and abstract.[5]

---

[2]Like PUBMED, PUBTATOR is supported by NCBI. Consult: `www.ncbi.nlm.nih.gov/research/pubtator/`

[3]Our code, data, and information about the leaderboard will be available at `http://nlp.cs.aueb.gr/publications.html`.

[4]PUBTATOR uses the Open Biological and Biomedical Ontology (OBO) Foundry, which comprises over 60 ontologies.

[5]A further reason for using the title as the question is that the entities of the titles are typically mentioned in the abstract.

	Passage
Passage	BACKGROUND: Most brain metastases arise from @entity0 . Few studies compare the brain regions they involve, their numbers and intrinsic attributes. METHODS: Records of all @entity1 referred to Radiation Oncology for treatment of symptomatic brain metastases were obtained. Computed tomography (n = 56) or magnetic resonance imaging (n = 72) brain scans were reviewed. RESULTS: Data from 68 breast and 62 @entity2 @entity1 were compared. Brain metastases presented earlier in the course of the lung than of the @entity0 @entity1 (p = 0.001). There were more metastases in the cerebral hemispheres of the breast than of the @entity2 @entity1 (p = 0.014). More @entity0 @entity1 had cerebellar metastases (p = 0.001). The number of cerebral hemisphere metastases and presence of cerebellar metastases were positively correlated (p = 0.001). The prevalence of at least one @entity3 surrounded with > 2 cm of @entity4 was greater for the lung than for the breast @entity1 (p = 0.019). The @entity5 type, rather than the scanning method, correlated with differences between these variables. CONCLUSIONS: Brain metastases from lung occur earlier, are more @entity4 , but fewer in number than those from @entity0 . Cerebellar brain metastases are more frequent in @entity0 .
Candidates	@entity0 : ['breast and lung cancer'] ; @entity1 : ['patients'] ; @entity2 : ['lung cancer'] ; @entity3 : ['metastasis'] ; @entity4 : ['edematous', 'edema'] ; @entity5 : ['primary tumor']
Question	Attributes of brain metastases from XXXX .
Answer	@entity0 : ['breast and lung cancer']

Figure 1: Example passage-question instance of BIOMRC. The passage is the abstract of an article, with biomedical entities replaced by @entity$N$ pseudo-identifiers. The original entity names are shown in square brackets. Both 'edematous' and 'edema' are replaced by '@entity4', because PUBTATOR considers them synonyms. The question is the title of the article, with a biomedical entity replaced by XXXX. @entity0 is the correct answer.

	BIOMRC LARGE				BIOMRC LITE				BIOMRC TINY		
	Training	Development	Test	Total	Training	Development	Test	Total	Setting A	Setting B	Total
Instances	700,000	50,000	62,707	812,707	87,500	6,250	6,250	100,000	30	30	60
Avg candidates	6.73	6.68	6.68	6.72	6.72	6.68	6.65	6.71	6.60	6.57	6.58
Max candidates	20	20	20	20	20	20	20	20	13	11	13
Min candidates	2	2	2	2	2	2	2	2	2	3	2
Avg abstract len.	253.79	257.41	253.70	254.01	253.78	257.32	255.56	254.11	248.13	264.37	256.25
Max abstract len.	543	516	511	543	519	500	510	519	371	386	386
Min abstract len.	57	89	77	57	60	109	103	60	147	154	147
Avg title len.	13.93	14.28	13.99	13.96	13.89	14.22	14.09	13.92	14.17	14.70	14.43
Max title len.	51	46	43	51	49	40	42	49	21	35	35
Min title len.	3	3	3	3	3	3	3	3	6	4	4

Table 2: Statistics of BIOMRC LARGE, LITE, TINY. The questions of the TINY version were answered by humans. All lengths are measured in tokens using a whitespace tokenizer.

Finally, to avoid making the dataset too easy for a system that would always select the entity with the most occurrences in the abstract, we removed a passage-question instance if the most frequent entity of its passage (abstract) was also the answer to the cloze-style question (title with placeholder); if multiple entities had the same top frequency in the passage, the instance was retained. We ended up with approx. 812k passage-question instances, which form BIOMRC LARGE, split into training, development, and test subsets (Table 2). The LITE and TINY versions of BIOMRC are subsets of LARGE.

In all versions of BIOMRC (LARGE, LITE, TINY), the entity identifiers of PUBTATOR are replaced by pseudo-identifiers of the form @entity$N$ (Fig. 1), as in the CNN and Daily Mail datasets (Hermann et al., 2015). We provide all BIOMRC versions in two forms, corresponding to what Pappas et al. (2018) call Settings A and B in BIOREAD.[6] In Setting A, each pseudo-identifier has a global scope, meaning that each biomedical entity has a unique pseudo-identifier in the whole dataset. This allows a system to learn information about the entity represented by a pseudo-identifier from all the occurrences of the pseudo-identifier in the training set. For example after seeing the same pseudo-identifier multiple times a model may learn that it stands for a drug, or that a particular pseudo-identifier tends to neighbor with specific words. Then, much like a language model, a system may guess the pseudo-identifier that should fill in the placeholder even without the passage, or at least it may infer a prior probability for each candidate answer. In contrast, Setting B uses a local scope, i.e., it restarts the numbering of the pseudo-identifiers (from @entity0) anew in each passage-question instance. This forces the models to rely only on information about the entities that can be inferred from the particular passage and question. This corresponds to a non-expert answering the question, who does not have any prior knowledge of the biomedical entities.

Table 2 provides statistics on BIOMRC. In TINY, we use 30 different passage-question instances in Settings A and B, because in both settings we asked the same humans to answer the questions, and we

---

[6]Pappas et al. (2018) actually call 'option a' and 'option b' our Setting B and A, respectively.

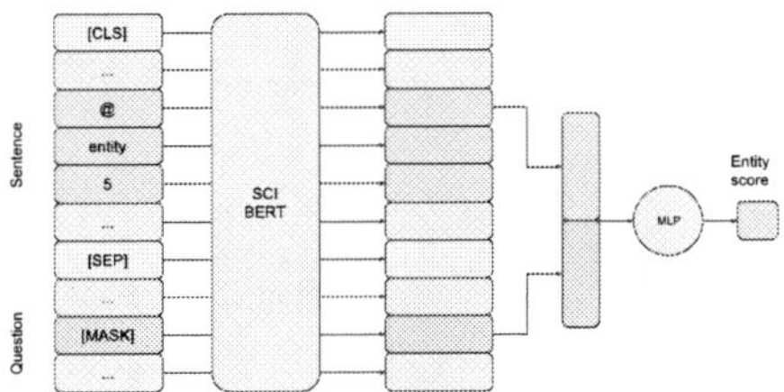

Figure 2: Illustration of our SCIBERT-based models. Each sentence of the passage is concatenated with the question and fed to SCIBERT. The top-level embedding produced by SCIBERT for the first sub-token of each candidate answer is concatenated with the top-level embedding of [MASK] (which replaces the placeholder XXXX) of the question, and they are fed to an MLP, which produces the score of the candidate answer. In SCIBERT-SUM-READER, the scores of multiple occurrences of the same candidate are summed, whereas SCIBERT-MAX-READER takes their maximum.

did not want them to remember instances from one setting to the other. In LARGE and LITE, the instances are the same across the two settings, apart from the numbering of the entity identifiers.

3 Experiments and Results

We experimented only on BIOMRC LITE and TINY, since we did not have the computational resources to train the neural models we considered on the LARGE version of BIOREAD. Pappas et al. (2018) also reported experimental results only on a LITE version of their BIOREAD dataset. We hope that others may be able to experiment on BIOMRC LARGE, and we make our code available, as already noted.

3.1 Methods

We experimented with the four basic baselines (BASE1–4) that Pappas et al. (2018) used in BIOREAD, the two neural MRC models used by the same authors, AS-READER (Kadlec et al., 2016) and AOA-READER (Cui et al., 2017), and a BERT-based (Devlin et al., 2019) model we developed.

Basic baselines: BASE1, 2, 3 return the first, last, and the entity that occurs most frequently in the passage (or randomly one of the entities with the same highest frequency, if multiple exist), respectively. Since in BIOREAD the correct answer is never (by construction) the most frequent entity of the passage, unless there are multiple entities with the same highest frequency, BASE3 performs poorly. Hence, we also include a variant, BASE3+, which randomly selects one of the entities of the

passage with the same highest frequency, if multiple exist, otherwise it selects the entity with the second highest frequency. BASE4 extracts all the token n-grams from the passage that include an entity identifier (@entity$N$), and all the n-grams from the question that include the placeholder (XXXX).[7] Then for each candidate answer (entity identifier), it counts the tokens shared between the n-grams that include the candidate and the n-grams that include the placeholder. The candidate with the most shared tokens is selected. These baselines are used to check that the questions cannot be answered by simplistic heuristics (Chen et al., 2016).

Neural baselines: We use the same implementations of AS-READER (Kadlec et al., 2016) and AOA-READER (Cui et al., 2017) as Pappas et al. (2018), who also provide short descriptions of these neural models, not provided here to save space. The hyper-parameters of both methods were tuned on the development set of BIOMRC LITE.

BERT-based model: We use SCIBERT (Beltagy et al., 2019), a pre-trained BERT (Devlin et al., 2019) model for scientific text. SCIBERT is pre-trained on 1.14 million articles from Semantic Scholar,[8] of which 82% (935k) are biomedical and the rest come from computer science. For each passage-question instance, we split the passage into sentences using NLTK (Bird et al., 2009). For each sentence, we concatenate it (using BERT's [SEP] token) with the question, after replacing the XXXX with BERT's [MASK] token, and we feed the concatenation to SCIBERT (Fig. 2). We collect SCIBERT's top-level vector representations of the entity identifiers (@entity$N$) of the sentence and [MASK].[9] For each entity of the sentence, we concatenate its top-level representation with that of [MASK], and we feed them to a Multi-Layer Perceptron (MLP) to obtain a score for the particular entity (candidate answer). We thus obtain a score for all the entities of the passage. If an entity occurs multiple times in the passage, we take the sum or the maximum of the scores of its occurrences. In both cases, a softmax is then applied to the scores of all the entities, and the entity with the maximum score is selected as the answer. We call

---

[7]We tried $n = 2, \ldots, 6$ and use $n = 3$, which gave the best accuracy on the development set of BIOMRC LARGE.

[8]https://www.semanticscholar.org/

[9]BERT's tokenizer splits the entity identifiers into sub-tokens (Devlin et al., 2019). We use the first one. The top-level token representations of BERT are context-aware, and it is common to use the first or last sub-token of each named-entity.

143

Method	BIOMRC Lite – Setting A							BIOMRC Lite – Setting B						
	Train Acc	Dev Acc	Test Acc	Train Time	All Params	Word Embeds	Entity Embeds	Train Acc	Dev Acc	Test Acc	Train Time	All Params	Word Embeds	Entity Embeds
BASE1	37.58	36.38	37.63	0	0	0	0	37.58	36.38	37.63	0	0	0	0
BASE2	22.50	23.10	21.73	0	0	0	0	22.50	23.10	21.73	0	0	0	0
BASE3	10.03	10.02	10.53	0	0	0	0	10.03	10.02	10.53	0	0	0	0
BASE3+	44.05	43.28	44.29	0	0	0	0	44.05	43.28	44.29	0	0	0	0
BASE4	56.48	57.36	56.50	0	0	0	0	56.48	57.36	56.50	0	0	0	0
AS-READER	**84.63**	62.29	62.38	18 x 0.92 hr	12.87M	12.69M	1.59M	79.64	66.19	66.19	18 x 0.65 hr	6.82M	6.66M	0.60k
AOA-READER	82.51	70.00	69.87	29 x 2.10 hr	12.87M	12.69M	1.59M	**84.62**	71.63	71.57	36 x 1.82 hr	6.82M	6.66M	0.60k
SCIBERT-SUM-READER	71.74	71.73	71.28	11 x 4.38 hr	**154k**	**0**	**0**	68.92	68.64	68.24	6 x 4.38 hr	**154k**	**0**	**0**
SCIBERT-MAX-READER	81.38	**80.06**	**79.97**	19 x 4.38 hr	**154k**	**0**	**0**	81.43	**80.21**	**79.10**	15 x 4.38 hr	**154k**	**0**	**0**

Table 3: Training, development, test accuracy (%) on BIOMRC LITE in Settings A (global scope of entity identifiers) and B (local scope), training times (epochs $\times$ time per epoch), and number of trainable parameters (total, word embedding parameters, entity identifier embedding parameters). In the lower zone (neural methods), the difference from each accuracy score to the next best is statistically significant ($p < 0.02$). We used singe-tailed Approximate Randomization (Dror et al., 2018), randomly swapping the answers to 50% of the questions for 10k iterations.

this model SCIBERT-SUM-READER or SCIBERT-MAX-READER, depending on how it aggregates the scores of multiple occurrences of the same entity.

SCIBERT-SUM-READER is closer to AS-READER and AOA-READER, which also sum the scores of multiple occurrences of the same entity. This summing aggregation, however, favors entities with several occurrences in the passage, even if the scores of all the occurrences are low. Our experiments indicate that SCIBERT-MAX-READER performs better. In all cases, we only update the parameters of the MLP during training, keeping the parameters of SCIBERT frozen to their pre-trained values to speed up training. With more computing resources, it may be possible to improve the scores of SCIBERT-MAX-READER (and SCIBERT-SUM-READER) further by fine-tuning SCIBERT on BIOMRC training data.

3.2 Results on BIOMRC LITE

Table 3 reports the accuracy of all methods on BIOMRC LITE for Settings A and B. In both settings, all the neural models clearly outperform all the basic baselines, with BASE3 (most frequent entity of the passage) performing worst and BASE3+ performing much better, as expected. In both settings, SCIBERT-MAX-READER clearly outperforms all the other methods on both the development and test sets. The performance of SCIBERT-SUM-READER is approximately ten percentage points worse than SCIBERT-MAX-READER's on the development and test sets of both settings, indicating that the superior results of SCIBERT-MAX-READER are to a large extent due to the different aggregation function (max instead of sum) it uses to combine the scores of multiple occurrences of a candidate answer, not to the extensive pre-training of SCIBERT. AOA-READER, which does not employ any pre-training, is competitive to SCIBERT-SUM-READER in Setting A, and performs better than SCIBERT-SUM-

READER in Setting B, which again casts doubts on the value of SCIBERT's extensive pre-training. We expect, however, that the performance of the SCIBERT-based models, could be improved further by fine-tuning SCIBERT's parameters.

The performance of SCIBERT-SUM-READER is slightly better in Setting A than in Setting B, which might suggest that the model manages to capture global properties of the entity pseudo-identifiers from the entire training set. However, the performance of SCIBERT-MAX-READER is almost the same across the two settings, which contradicts the previous hypothesis. Furthermore, the development and test performance of AS-READER and AOA-READER is higher in Setting B than A, indicating that these two models do not capture global properties of entities well, performing better when forced to consider only the information of the particular passage-question instance. Overall, we see no strong evidence that the models we considered are able to learn global properties of the entities.

In both Settings A and B, AOA-READER performs better than AS-READER, which was expected since it uses a more elaborate attention mechanism, at the expense of taking longer to train (Table 3).[10] The two SCIBERT-based models are also competitive in terms of training time, because we only train the MLP (154k parameters) on top of SCIBERT, keeping the parameters of SCIBERT frozen.

The trainable parameters of AS-READER and AOA-READER are almost double in Setting A compared to Setting B. To some extent, this difference is due to the fact that for both models we learn a word embedding for each @entity$N$ pseudo-identifier, and in Setting A the numbering of the identifiers is not reset for each passage-question

---

[10]We trained all models for a maximum of 40 epochs, using early stopping on the dev. set, with patience of 3 epochs.

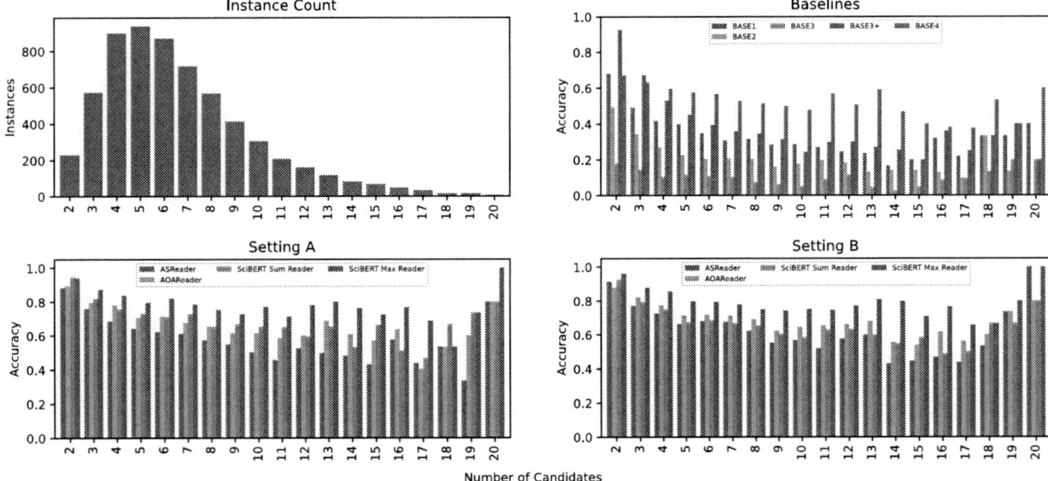

Figure 3: More detailed statistics and results on the development subset of BIOMRC LITE. Number of passage-question instances with 2, 3, ..., 20 candidate answers (top left). Accuracy (%) of the basic baselines (top right). Accuracy (%) of the neural models in Settings A (bottom left) and B (bottom right).

instance, leading to many more pseudo-identifiers (31.77k pseudo-identifiers in the vocabulary of Setting A vs. only 20 in Setting B); this accounts for a difference of 1.59M parameters.[11] The rest of the difference in total parameters (from Setting A to B) is due to the fact that we tuned the hyper-parameters of each model separately for each setting (A, B), on the corresponding development set. Hyper-parameter tuning was performed separately for each model in each setting, but led to the same numbers of trainable parameters for AS-READER and AOA-READER, because the trainable parameters are dominated by the parameters of the word embeddings. Note that the hyper-parameters of the two SCIBERT-based models (of their MLPs) were very minimally tuned, hence these models may perform even better with more extensive tuning.

AOA-READER was also better than AS-READER in the experiments of Pappas et al. (2018) on a LITE version of their BIOREAD dataset, but the development and test accuracy of AOA-READER in Setting A of BIOREAD was reported to be only 52.41% and 51.19%, respectively (cf. Table 3); in Setting B, it was 50.44% and 49.94%, respectively. The much higher scores of AOA-READER (and AS-READER) on BIOMRC LITE are an indication that the new dataset is less noisy, or that the task is at

least more feasible for machines. The results of Pappas et al. (2018) were slightly higher in Setting A than in Setting B, suggesting that AOA-READER was able to benefit from the global scope of entity identifiers, unlike our findings in BIOMRC.[12]

Figure 3 shows how many passage-question instances of the development subset of BIOMRC LITE have 2, 3, ..., 20 candidate answers (top left), and the corresponding accuracy of the basic baselines (top right), and the neural models (bottom). BASE3+ is the best basic baseline for 2 and 3 candidates, and for 2 candidates it is competitive to the neural models. Overall, however, BASE4 is clearly the best basic baseline, but it is outperformed by all neural models in almost all cases, as in Table 3. SCIBERT-MAX-READER is again the best system in both settings, almost always outperforming the other systems. AS-READER is the worst neural model in almost all cases. AOA-READER is competitive to SCIBERT-SUM-READER in Setting A, and slightly better overall than SCIBERT-SUM-READER in Setting B, as can be seen in Table 3.

3.3 Results on BIOMRC TINY

Pappas et al. (2018) asked humans (non-experts) to answer 30 questions from BIOREAD in Setting A, and 30 other questions in Setting B. We mirrored their experiment by providing 30 questions (from

---

[11]Hyper-parameter tuning led to 50- and 30-dimensional word embeddings in Settings A, B, respectively. AS-READER and AOA-READER learn word embeddings from the training set, without using pre-trained embeddings.

[12]For AS-READER, Pappas et al. (2018) report results only for Setting B: 37.90% development and 42.01% test accuracy on BIOREAD LITE. They did not consider BERT-based models.

Passage	The study enrolled 53 @entity1 (29 males, 24 females) with @entity1576 aged 15-88 years. Most of them were 59 years of age and younger. In 1/3 of the @entity1 the diseases started with symptoms of @entity1729, in 2/3 of them–with pulmonary affection. @entity55 was diagnosed in 50 @entity1 (94.3%), acute @entity3617 –in 3 @entity1. ECG changes were registered in about half of the examinees who had no cardiac complaints. 25 of them had alterations in the end part of the ventricular ECG complex; rhythm and conduction disturbances occurred rarely. Mycoplasmosis @entity1 suffering from @entity741 (@entity741) had stable ECG changes while in those free of @entity741 the changes were short. @entity296 foci were absent. @entity299 comparison in @entity1 with @entity1576 and in other @entity1729 has found that cardiovascular system suffers less in acute mycoplasmosis. These data are useful in differential diagnosis of @entity296 .
Candidates	@entity1 : ['patients'] ; @entity1576 : ['respiratory mycoplasmosis'] ; @entity1729 : ['acute respiratory infections', 'acute respiratory viral infection'] ; @entity55 : ['Pneumonia'] ; @entity3617 : ['bronchitis'] ; @entity741 : ['IHD', 'ischemic heart disease'] ; @entity296 : ['myocardial infections', 'Myocardial necrosis'] ; @entity299 : ['Cardiac damage'] .
Question	Cardio-vascular system condition in XXXX .
Expert Human Answers	annotator1: @entity1576; annotator2: @entity1576.
Non-expert Human Answers	annotator1: @entity296; annotator2: @entity296; annotator3: @entity1576.
Systems' Answers	AS-READER: @entity1729; AOA-READER: @entity296; SCIBERT-SUM-READER: @entity1576.

Figure 4: Example from BIOMRC TINY. In Setting A, humans see both the pseudo-identifiers (@entity$N$) and the original names of the biomedical entities (shown in square brackets). Systems see only the pseudo-identifiers, but the pseudo-identifiers have global scope over all instances, which allows the systems, at least in principle, to learn entity properties from the entire training set. In Setting B, humans no longer see the original names of the entities, and systems see only the pseudo-identifiers with local scope (numbering reset per passage-question instance).

BIOMRC LITE) to three non-experts (graduate CS students) in Setting A, and 30 other questions in Setting B. We also showed the same questions of each setting to two biomedical experts. As in the experiment of Pappas et al. (2018), in Setting A both the experts and non-experts were also provided with the original names of the biomedical entities (entity names before replacing them with @entity$N$ pseudo-identifiers) to allow them to use prior knowledge; see the top three zones of Fig. 4 for an example. By contrast, in Setting B the original names of the entities were hidden.

Table 4 reports the human and system accuracy scores on BIOMRC TINY. Both experts and non-experts perform better in Setting A, where they can use prior knowledge about the biomedical entities. The gap between experts and non-experts is three points larger in Setting B than in Setting A, presumably because experts can better deduce properties of the entities from the local context. Turning to the system scores, SCIBERT-MAX-READER is again the best system, but again much of its performance is due to the max-aggregation of the scores of multiple occurrences of entities. With sum-aggregation, SCIBERT-SUM-READER obtains exactly the same scores as AOA-READER, which again performs better than AS-READER. (AOA-READER and SCIBERT-SUM-READER make different mistakes, but their scores just happen to be identical because of the small size of TINY.) Unlike our results on BIOMRC LITE, we now see all systems performing better in Setting A compared to Setting B, which suggests

they do benefit from the global scope of entity identifiers. Also, SCIBERT-MAX-READER performs better than both experts and non-experts in Setting A, and better than non-experts in Setting B. However, BIOMRC TINY contains only 30 instances in each setting, and hence the results of Table 4 are less reliable than those from BIOMRC LITE (Table 3).

In the corresponding experiments of Pappas et al. (2018), which were conducted in Setting B only, the average accuracy of the (non-expert) humans was 68.01%, but the humans were also allowed not to answer (when clueless), and unanswered questions were excluded from accuracy. On average, they did not answer 21.11% of the questions, hence their accuracy drops to 46.90% if unanswered questions are counted as errors. In our experiment, the humans were also allowed not to answer (when clueless), but we counted unanswered questions as errors, which we believe better reflects human performance. Non-experts answered all questions in Setting A, and did not answer 13.33% (4/30) of the questions on average in Setting B. The decrease in the questions non-experts did not answer (from 21.11% to 13.33%) in Setting B (the only one considered in BIOREAD) again suggests that the new dataset is less noisy, or at least that the task is more feasible for humans, even when the names of the entities are hidden. Experts did not answer 2.5% (0.75/30) and 1.67% (0.5/30) of the questions on average in Settings A and B, respectively.

Inter-annotator agreement was also higher for experts than non-experts in our experiment, in both

Method	Setting A	Setting B
Experts (Avg)	**85.00**	**61.67**
Non-Experts (Avg)	81.67	55.56
AS-READER	66.67	46.67
AOA-READER	70.00	56.67
SCIBERT-SUM-READER	70.00	56.67
SCIBERT-MAX-READER	**90.00**	**60.00**

Table 4: Accuracy (%) on BIOMRC TINY. Best human and system scores shown in bold.

Settings A and B (Table 5). In Setting B, the agreement of non-experts was particularly low (47.22%), possibly because without entity names they had to rely more on the text of the passage and question, which they had trouble understanding. By contrast, the agreement of experts was slightly higher in Setting B than Setting A, possibly because without prior knowledge about the entities, which may differ across experts, they had to rely to a larger extent on the particular text of the passage and question.

4   Related work

Several biomedical MRC datasets exist, but have orders of magnitude fewer questions than BIOMRC (Ben Abacha and Demner-Fushman, 2019) or are not suitable for a cloze-style MRC task (Pampari et al., 2018; Ben Abacha et al., 2019; Zhang et al., 2018). The closest dataset to ours is CLICR (Šuster and Daelemans, 2018), a biomedical MRC dataset with cloze-type questions created using full-text articles from BMJ case reports.[13] CLICR contains 100k passage-question instances, the same number as BIOMRC LITE, but much fewer than the 812.7k instances of BIOMRC LARGE. Šuster et al. used CLAMP (Soysal et al., 2017) to detect biomedical entities and link them to concepts of the UMLS Metathesaurus (Lindberg et al., 1993). Cloze-style questions were created from the 'learning points' (summaries of important information) of the reports, by replacing biomedical entities with placeholders. Šuster et al. experimented with the Stanford Reader (Chen et al., 2017) and the Gated-Attention Reader (Dhingra et al., 2017), which perform worse than AOA-READER (Cui et al., 2017).

The QA dataset of BIOASQ (Tsatsaronis et al., 2015) contains questions written by biomedical experts. The gold answers comprise multiple relevant documents per question, relevant snippets from the documents, exact answers in the form of entities, as well as reference summaries, written by the ex-

Annotators (Setting)	Kappa
Experts (A)	70.23
Non Experts (A)	65.61
Experts (B)	72.30
Non Experts (B)	47.22

Table 5: Human agreement (Cohen's Kappa, %) on BIOMRC TINY. Avg. pairwise scores for non-experts.

perts. Creating data of this kind, however, requires significant expertise and time. In the eight years of BIOASQ, only 3,243 questions and gold answers have been created. It would be particularly interesting to explore if larger automatically generated datasets like BIOMRC and CLICR could be used to pre-train models, which could then be fine-tuned for human-generated QA or MRC datasets.

Outside the biomedical domain, several cloze-style open-domain MRC datasets have been created automatically (Hill et al., 2016; Hermann et al., 2015; Dunn et al., 2017; Bajgar et al., 2016), but have been criticized of containing questions that can be answered by simple heuristics like our basic baselines (Chen et al., 2016). There are also several large open-domain MRC datasets annotated by humans (Kwiatkowski et al., 2019; Rajpurkar et al., 2016, 2018; Trischler et al., 2017; Nguyen et al., 2016; Lai et al., 2017). To our knowledge the biggest human annotated corpus is Google's Natural Questions dataset (Kwiatkowski et al., 2019), with approximately 300k human annotated examples. Datasets of this kind require extensive annotation effort, which for open-domain datasets is usually crowd-sourced. Crowd-sourcing, however, is much more difficult for biomedical datasets, because of the required expertise of the annotators.

5   Conclusions and Future Work

We introduced BIOMRC, a large-scale cloze-style biomedical MRC dataset. Care was taken to reduce noise, compared to the previous BIOREAD dataset of Pappas et al. (2018). Experiments showed that BIOMRC's questions cannot be answered well by simple heuristics, and that two neural MRC models that had been tested on BIOREAD perform much better on BIOMRC, indicating that the new dataset is indeed less noisy or at least that its task is more feasible. Human performance was also higher on a sample of BIOMRC compared to BIOREAD, and biomedical experts performed even better. We also developed a new BERT-based model, the best version of which outperformed all other meth-

---

[13]https://casereports.bmj.com/

ods tested, reaching or surpassing the accuracy of biomedical experts in some experiments. We make BIOMRC available in three different sizes, also releasing our code, and providing a leaderboard.

We plan to tune more extensively the BERT-based model to further improve its efficiency, and to investigate if some of its techniques (mostly its max-aggregation, but also using sub-tokens) can also benefit the other neural models we considered. We also plan to experiment with other MRC models that recently performed particularly well on open-domain MRC datasets (Zhang et al., 2020). Finally, we aim to explore if pre-training neural models on BIOREAD is beneficial in human-generated biomedical datasets (Tsatsaronis et al., 2015).

Acknowledgments

We are most grateful to I. Almirantis, S. Kotitsas, V. Kougia, A. Nentidis, S. Xenouleas, who participated in the human evaluation with BIOMRC TINY.

References

Alan R Aronson and François-Michel Lang. 2010. An overview of MetaMap: historical perspective and recent advances. *Journal of the American Medical Informatics Association*, 17(3):229–236.

Ondrej Bajgar, Rudolf Kadlec, and Jan Kleindienst. 2016. Embracing data abundance: BookTest Dataset for Reading Comprehension. *CoRR*.

Iz Beltagy, Kyle Lo, and Arman Cohan. 2019. SciBERT: Pretrained Language Model for Scientific Text. In *EMNLP*.

Asma Ben Abacha and Dina Demner-Fushman. 2019. A question-entailment approach to question answering. *BMC Bioinformatics*, 20:511.

Asma Ben Abacha, Chaitanya Shivade, and Dina Demner-Fushman. 2019. Overview of the MEDIQA 2019 Shared Task on Textual Inference, Question Entailment and Question Answering. In *Proceedings of the 18th BioNLP Workshop and Shared Task*, pages 370–379, Florence, Italy.

Steven Bird, Loper Edward, and Ewan Klein. 2009. *Natural Language Processing with Python*. O'Reilly Media Inc.

Danqi Chen, Jason Bolton, and Christopher D. Manning. 2016. A Thorough Examination of the CNN/Daily Mail Reading Comprehension Task. In *Proceedings of the 54th Annual Meeting of the Association for Computational Linguistics (Volume 1: Long Papers)*, pages 2358–2367, Berlin, Germany.

Danqi Chen, Adam Fisch, Jason Weston, and Antoine Bordes. 2017. Reading Wikipedia to Answer Open-Domain Questions. In *Proceedings of the 55th Annual Meeting of the Association for Computational Linguistics (Volume 1: Long Papers)*, pages 1870–1879, Vancouver, Canada.

Yiming Cui, Zhipeng Chen, Si Wei, Shijin Wang, Ting Liu, and Guoping Hu. 2017. Attention-over-Attention Neural Networks for Reading Comprehension. In *Proceedings of the 55th Annual Meeting of the Association for Computational Linguistics (Volume 1: Long Papers)*, pages 593–602, Vancouver, Canada.

Jacob Devlin, Ming-Wei Chang, Kenton Lee, and Kristina Toutanova. 2019. BERT: Pre-training of Deep Bidirectional Transformers for Language Understanding. In *NAACL-HLT*.

Bhuwan Dhingra, Hanxiao Liu, Zhilin Yang, William Cohen, and Ruslan Salakhutdinov. 2017. Gated-Attention Readers for Text Comprehension. In *Proceedings of the 55th Annual Meeting of the Association for Computational Linguistics (Volume 1: Long Papers)*, pages 1832–1846, Vancouver, Canada.

Rotem Dror, Gili Baumer, Segev Shlomov, and Roi Reichart. 2018. The Hitchhiker's Guide to Testing Statistical Significance in Natural Language Processing. In *Proceedings of the 56th Annual Meeting of the Association for Computational Linguistics (Volume 1: Long Papers)*, pages 1383–1392, Melbourne, Australia.

Matthew Dunn, Levent Sagun, Mike Higgins, V. Ugur Güney, Volkan Cirik, and Kyunghyun Cho. 2017. SearchQA: A New Q&A Dataset Augmented with Context from a Search Engine. *CoRR*.

Karl Moritz Hermann, Tomáš Kočiský, Edward Grefenstette, Lasse Espeholt, Will Kay, Mustafa Suleyman, and Phil Blunsom. 2015. Teaching Machines to Read and Comprehend. In *Proceedings of the 28th International Conference on Neural Information Processing Systems - Volume 1*, page 1693–1701, Cambridge, MA, USA.

Felix Hill, Antoine Bordes, Sumit Chopra, and Jason Weston. 2016. The Goldilocks Principle: Reading Children's Books with Explicit Memory Representations. *CoRR*.

Rudolf Kadlec, Martin Schmid, Ondrej Bajgar, and Jan Kleindienst. 2016. Text Understanding with the Attention Sum Reader Network. In *Proceedings of the 54th Annual Meeting of the Association for Computational Linguistics (Volume 1: Long Papers)*, pages 908–918, Berlin, Germany.

Tom Kwiatkowski, Jennimaria Palomaki, Olivia Redfield, Michael Collins, Ankur Parikh, Chris Alberti, Danielle Epstein, Illia Polosukhin, Jacob Devlin, Kenton Lee, Kristina Toutanova, Llion Jones, Matthew Kelcey, Ming-Wei Chang, Andrew M. Dai,

Jakob Uszkoreit, Quoc Le, and Slav Petrov. 2019. Natural Questions: A Benchmark for Question Answering Research. *Transactions of the Association for Computational Linguistics*, 7:453–466.

Guokun Lai, Qizhe Xie, Hanxiao Liu, Yiming Yang, and Eduard Hovy. 2017. RACE: Large-scale ReAding Comprehension Dataset From Examinations. In *Proceedings of the 2017 Conference on Empirical Methods in Natural Language Processing*, pages 785–794, Copenhagen, Denmark.

Robert Leaman, Rezarta Islamaj Doğan, and Zhiyong Lu. 2013. DNorm: disease name normalization with pairwise learning to rank. *Bioinformatics*, 29(22):2909–2917.

Donald A. B. Lindberg, Betsy L. Humphreys, and Alexa T. McCray. 1993. The Unified Medical Language System. *Yearbook of medical informatics*, 1:41–51.

Tri Nguyen, Mir Rosenberg, Xia Song, Jianfeng Gao, Saurabh Tiwary, Rangan Majumder, and Li Deng. 2016. MS MARCO: A Human Generated MAchine Reading COmprehension Dataset. *ArXiv*.

Anusri Pampari, Preethi Raghavan, Jennifer Liang, and Jian Peng. 2018. emrQA: A Large Corpus for Question Answering on Electronic Medical Records. In *Proceedings of the 2018 Conference on Empirical Methods in Natural Language Processing*, pages 2357–2368, Brussels, Belgium.

Dimitris Pappas, Ion Androutsopoulos, and Haris Papageorgiou. 2018. BioRead: A New Dataset for Biomedical Reading Comprehension. In *Proceedings of the Eleventh International Conference on Language Resources and Evaluation (LREC 2018)*, Miyazaki, Japan.

Pranav Rajpurkar, Robin Jia, and Percy Liang. 2018. Know What You Don't Know: Unanswerable Questions for SQuAD. In *Proceedings of the 56th Annual Meeting of the Association for Computational Linguistics (Volume 2: Short Papers)*, pages 784–789, Melbourne, Australia.

Pranav Rajpurkar, Jian Zhang, Konstantin Lopyrev, and Percy Liang. 2016. SQuAD: 100,000+ Questions for Machine Comprehension of Text. In *Proceedings of the 2016 Conference on Empirical Methods in Natural Language Processing*, pages 2383–2392, Austin, Texas.

Ergin Soysal, Jingqi Wang, Min Jiang, Yonghui Wu, Serguei Pakhomov, Hongfang Liu, and Hua Xu. 2017. CLAMP – a toolkit for efficiently building customized clinical natural language processing pipelines. *Journal of the American Medical Informatics Association*, 25(3):331–336.

Simon Šuster and Walter Daelemans. 2018. CliCR: a Dataset of Clinical Case Reports for Machine Reading Comprehension. In *Proceedings of the 2018 Conference of the North American Chapter of the Association for Computational Linguistics: Human Language Technologies, Volume 1 (Long Papers)*, pages 1551–1563, New Orleans, Louisiana.

Wilson L. Taylor. 1953. "Cloze Procedure": A New Tool for Measuring Readability. *Journalism Quarterly*, 30(4):415–433.

Adam Trischler, Tong Wang, Xingdi Yuan, Justin Harris, Alessandro Sordoni, Philip Bachman, and Kaheer Suleman. 2017. NewsQA: A Machine Comprehension Dataset. In *Proceedings of the 2nd Workshop on Representation Learning for NLP*, pages 191–200, Vancouver, Canada.

G. Tsatsaronis, G. Balikas, P. Malakasiotis, I. Partalas, M. Zschunke, M.R. Alvers, D. Weissenborn, A. Krithara, S. Petridis, D. Polychronopoulos, Y. Almirantis, J. Pavlopoulos, N. Baskiotis, P. Gallinari, T. Artieres, A. Ngonga, N. Heino, E. Gaussier, L. Barrio-Alvers, M. Schroeder, I. Androutsopoulos, and G. Paliouras. 2015. An Overview of the BioASQ Large-Scale Biomedical Semantic Indexing and Question Answering Competition. *BMC Bioinformatics*, 16(138).

Chih-Hsuan Wei, Bethany R. Harris, Donghui Li, Tanya Z. Berardini, Eva Huala, Hung-Yu Kao, and Zhiyong Lu. 2012. Accelerating literature curation with text-mining tools: a case study of using PubTator to curate genes in PubMed abstracts. *Database*, 2012.

Xiao Zhang, Ji Wu, Zhiyang He, Xien Liu, and Ying Su. 2018. Medical Exam Question Answering with Large-scale Reading Comprehension. *ArXiv*.

Zhuosheng Zhang, Jun jie Yang, and Hai Zhao. 2020. Retrospective Reader for Machine Reading Comprehension. *ArXiv*.

Neural Transduction of Letter Position Dyslexia using an Anagram Matrix Representation

Avi Bleiweiss

BShalem Research

Sunnyvale, CA, USA

avibleiweiss@bshalem.onmicrosoft.com

Abstract

Research on analyzing reading patterns of dyslectic children has mainly been driven by classifying dyslexia types offline. We contend that a framework to remedy reading errors in-line is more far-reaching and will help to further advance our understanding of this impairment. In this paper, we propose a simple and intuitive neural model to reinstate migrating words that transpire in letter position dyslexia, a visual analysis deficit to the encoding of character order within a word. Introduced by the anagram matrix representation of an input verse, the novelty of our work lies in the expansion from one to a two dimensional context window for training. This warrants words that only differ in the disposition of letters to remain interpreted semantically similar in the embedding space. Subject to the apparent constraints of the self-attention transformer architecture, our model achieved a unigram BLEU score of 40.6 on our reconstructed dataset of the Shakespeare sonnets.

1 Introduction

Dyslexia is a reading disorder that is perhaps the most studied of learning disabilities, with an estimated prevalence rate of 5 to 17 percentage points of school-age children in the US (Shaywitz and Shaywitz, 2005; Made by Dyslexia, 2019). Counter to popular belief, dyslexia is not only tied to the visual analysis system of the brain, but also presents a linguistic problem and hence its relevance to natural language processing (NLP). Dyslexia manifests itself in several forms as this work centers on Letter Position Dyslexia (LPD), a selective deficit to encoding the position of a letter within a word while sustaining both letter identification and character binding to words (Friedmann and Gvion, 2001).

A growing body of research advocates heterogeneity of dyslexia causes to poor non-word and irregular-word reading (McArthur et al., 2013).

Along the same lines Kezilas et al. (2014) suggest that character transposition effects in LPD are most likely caused by a deficit specific to coding the letter position and is evidenced by an interaction between the orthographic and visual analysis stages of reading. To this end, more recently Marcet et al. (2019) managed to significantly reduce migration errors by either altering letter contrast or presenting letters to the young adult sequentially.

To dyslectic children not all letter positions are equally impaired as medial letters in a word are by far more vulnerable to reading errors compared to the first and last characters of the word (Friedmann and Gvion, 2001). Children with LPD have high migration errors where the transposition of letters in the middle of the word leads to another word, for example, *slime–smile* or *cloud–could*. On the other hand, not all reading errors in cases of selective LPD are migratable and are evidenced by words read without a lexical sense e.g., *slime–silme*. Intriguingly, increasing the word length does not elevate the error rate, and moreover, shorter words that have lexical anagrams are prone to a larger proportion of migration errors compared to longer words that possess no-anagram words. A key observation for LPD is that although words read may share all letters in most of the positions, they still remain semantically unrelated.

Machine learning tools to classify dyslexia use a large corpus of reading errors for training and mainly aim to automate and substitute diagnostic procedures expensively managed by human experts. Lakretz et al. (2015) used both LDA and Naive Bayes models and showed an area under curve (AUC) performance of about 0.8 that exceeded the quality of clinician-rendered labels. In their study, Rello and Ballesteros (2015) proposed a statistical model that predicts dyslectic readers using eye tracking measures. Employing an SVM-based binary classifier, they achieved about 80% accuracy.

Proceedings of the BioNLP 2020 workshop, pages 150–155

Instead, our approach applies deep learning to the task of restoring LPD inline that we further formulate as a sequence transduction problem. Thus, given an input verse that contains shuffled-letter words identified as transpositional errors, the objective of our neural model is to predict the originating unshuffled words. We use language similarity between predicted verses and ground-truth target text-sequences to quantitatively evaluate our model. Our main contribution is a concise representation of the input verse that scales up to moderate an exhaustive set of LPD permutable data.

2 Anagram Matrix

Using a colon notation, we denote an input verse to our model as a text sequence $w_{1:n} = (w_1, \ldots, w_n)$ of n words interchangeably with n collections of letters $l_{1:n} = (l^{(1)}_{1:|w_1|}, \ldots, l^{(n)}_{1:|w_n|})$. We generate migrated word patterns synthetically by anchoring the first and last character of each word and randomly permuting the position of the inner letters $(l^{(1)}_{2:|w_1|-1}, \ldots, l^{(n)}_{2:|w_n|-1})$. Thus given a word with a character length $|l^{(i)}|$, the number of possible unique transpositions for each word follows $t_{1:n} = (|l^{(1)}|!, \ldots, |l^{(n)}|!)$. Next, we extract a migration amplification factor $k = \mathrm{argmax}^n_{i=1}\, t_i$ that we apply to each word in an input verse independently and form the sequence $m_{1:k} = (m_1, \ldots, m_k)$. Word length commonly used in experiments of previous LPD studies averages five letters and ranges from four to seven letters long, hence migrating to feasible 2, 6, 24, and 120 letter substitutions, respectively. We note that words with 1, 2, or 3 letters are held intact and are not migratable.

when	forty	winters	shall	besiege	thy	brow
wehn	fotry	wenitrs	sahll	bseeige	thy	borw
when	froty	winrtes	slhal	begseie	thy	borw
wehn	fotry	wrenits	slahl	begisee	thy	borw
when	forty	wtenirs	shall	begeise	thy	brow
when	froty	wtneirs	shall	bigeese	thy	brow
when	ftory	weinrts	sahll	bgiesee	thy	borw
wehn	frtoy	wirtens	slhal	bisgeee	thy	borw
wehn	froty	wterins	slahl	beeisge	thy	brow
when	froty	wtiners	shlal	beesgie	thy	borw
wehn	frtoy	wnetris	shall	beisege	thy	borw

Table 1: A snapshot of letter-position migration patterns in the form of an anagram matrix. The unedited version of the text sequence is highlighted on top.

To address the inherent semantic unrelatedness between transpositioned words, we define a two-dimensional migration-verse array in the form of an anagram matrix $A = [m^{(1)}_{1:k}; \ldots; m^{(n)}_{1:k}] \in \mathbb{R}^{k \times n}$, where $m^{(i)}$ are column vectors, $[\cdot; \cdot]$ is column-bound matrix concatenation, and k and n are the transposition and input verse dimensions, respectively. In Table 1, we render a subset of an anagram matrix drawn from a target verse with a maximal word length of seven letters. The anagram matrix founds an effective context structure for a two-pass embedding training, and our training dataset thus reconstructs on the basis of a collection of anagram matrices with varying dimensions.

3 LPD Embeddings

Models for learning word vectors train locally on a one-dimensional context window by scanning the entire corpus (Mikolov et al., 2013). Through evaluation on a word analogy task, these models capture linguistic regularities as linear relationships between word embeddings. Mikolov et al. (2013) proposed the skip-gram and continuous-bag-of-words (CBOW) neural architectures with the objective to predict the context of the target word and the target word given its context, respectively. Notably LPD migrating words tend mostly outside the English vocabulary and thus pretrained word embeddings on large corpora are of limited use in our system. [1]

$w_{t-2,t-2}$	$w_{t-1,t-2}$	$w_{t,t-2}$	$w_{t+1,t-2}$	$w_{t+2,t-2}$
$w_{t-2,t-1}$	$w_{t-1,t-1}$	$w_{t,t-1}$	$w_{t+1,t-1}$	$w_{t+2,t-1}$
$w_{t-2,t}$	$w_{t-1,t}$	$w_{t,t}$	$w_{t+1,t}$	$w_{t+2,t}$
$w_{t-2,t+1}$	$w_{t-1,t+1}$	$w_{t,t+1}$	$w_{t+1,t+1}$	$w_{t+2,t+1}$
$w_{t-2,t+2}$	$w_{t-1,t+2}$	$w_{t,t+2}$	$w_{t+1,t+2}$	$w_{t+2,t+2}$

Figure 1: A two-dimensional context window of size two drawn from outside context cells of an anagram matrix. The center words are shown in gray for both the normal by-row $\{w_{t,t-2}, \ldots, w_{t,t+2}\}$ and transposed column-wise $\{w_{t-2,t}, \ldots, w_{t+2,t}\}$ forms of feeding our neural network.

While the essence of our task is formalized as verse simplification, mending LPD relies on robust discovery of word similarities along both the migration and verse axes of the anagram matrix. To this extent, we reshape the context window to train word embeddings from one to a two-dimensional array. In Figure 1, we show a bi-dimensional con-

____
[1] https://nlp.stanford.edu/projects/glove/

text window of size two that is a visible subset drawn from outside context cells of an anagram matrix. Learning word vectors for LPD is a two-pass process in our model. First, the context window W feeds our neural network row-by-row for each transpositioned verse, and then follows by iterating migration vectors $m^{(i)}$ in W^T as inputs.

4 Model

Our task is inspired by recent advances in neural machine translation (NMT). NMT architectures have shown state-of-the-art results in both the form of a powerful sequence model (Sutskever et al., 2014; Cho et al., 2014; Bahdanau et al., 2015) and more recently, the cross-attention ConvS2S (Elbayad et al., 2018) and the self-attention based transformer (Vaswani et al., 2017) networks. Given an unintelligible diction of shuffled-letter words, our model aims to output a verse that preserves the semantics of the input, and uses the transformer that outperforms both recurrent and convolutional configurations on many language translation tasks.

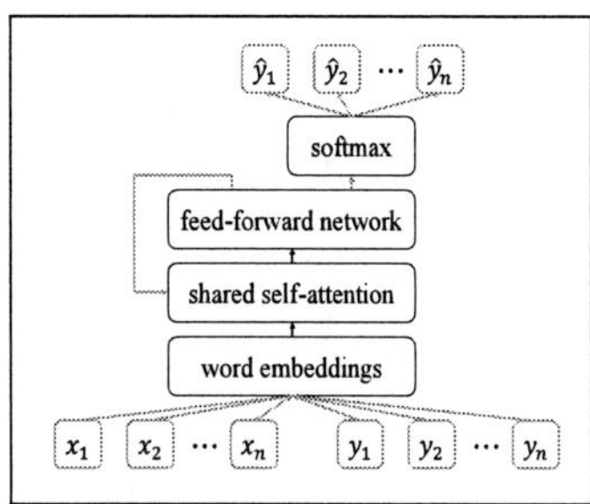

Figure 2: Transformer architecture overview (encoder path shown in blue, decoder in brown).

Stacked with several network layers, the transformer architecture only relies on attention mechanisms and entirely dispensing with recurrence (Hochreiter and Schmidhuber, 1997; Chung et al., 2014). In Figure 2, we show a synoptic rendition of the transformer. Its inputs consist of a source verse with potentially letter-transpositioned words x_i, and a ground-truth target verse of words with unshuffled letters y_i. The transformer encoder and decoder modules largely operate in parallel and provide for a source-to-target attention communication, and a softmax layer operates on the decoder hidden-state outputs to produce predicted words $\hat{y}_i$. In LPD, source and target verses are consistently of the same word count n, however, copying tokens from the source over to predictions is inconsequen-

tial to the quality of repairing reading errors due to extensive out-of-vocabulary non-migrating words.

5 Setup

To quantitatively evaluate our LPD transduction approach, we chose to mainly report n-gram BLEU precision (Papineni et al., 2002) that defines the language similarity between a predicted text sequence and the ground-truth reference verse. In the BLEU metric, higher scores indicate better performance.

5.1 Corpus

Rather than clinical reading tests, we used the Sonnets by William Shakespeare (Shakespeare, 1997). This is motivated by the apostrophe-rich data that forces left-out letters. The raw dataset comprises 2,154 verses that range from four to fifteen word sequences. In Figure 3, we show the distribution of word length across the dataset, as 18,858 unique tokens are of up to seven-letter long inclusive and take about 62 percentage points of the entire corpus words. To conform to preceding LPD research, we conducted a cleanup step that removes all words of eight letters or more from the dataset. We hypothesize that evaluating LPD on a single word basis lets us perform this step without loss of generality.

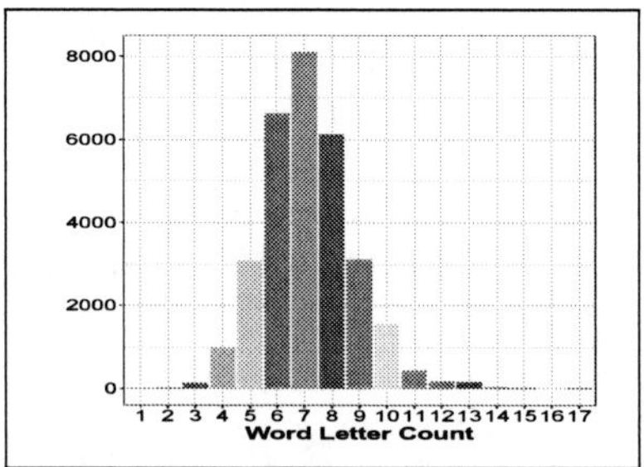

Figure 3: Distribution of word letter count across unique tokens of the Shakespeare Sonnets dataset.

We then transform each verse of the Sonnets to an anagram matrix representation A. The verse word with the maximal letters has a set of distinct traspositions while words of lesser letters are shuffled with repetition (Table 1). In Figure 4, we show the distribution of anagram matrices across the entire Shakespeare Sonnets dataset, with a migration amplification factor $k \in \{1, 2, 6, 24, 120\}$ and a cleaned up verse that spans two to thirteen words. Evidently most prominent tiles are of words with seven letters and consist of verse sizes between seven to nine words. Concatenating the rows of all the anagram matrices presents a sixtyfold extended shape of our LPD training dataset that has 130,021 text sequences, along with source and target vocabularies of 173,575 and 3,147 tokens, respectively.

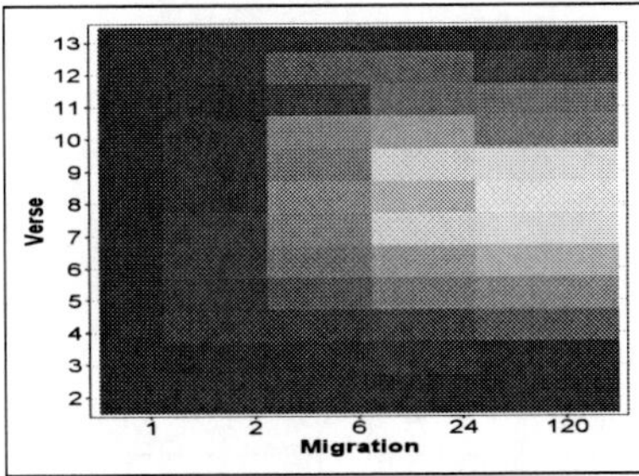

Figure 4: Distribution of anagram matrices across the verse collection of the Shakespeare Sonnets dataset.

5.2 Training

We used PyTorch (Paszke et al., 2017) version 1.0 as our deep learning platform for training and inference. PyTorch supports the building of effective neural architectures for NLP task development.

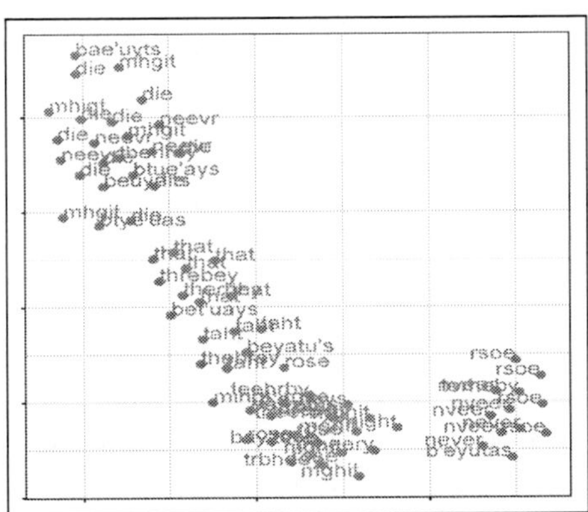

Figure 5: Aided by using our anagram matrix approach, migrated and non-migrated embeddings shown to preserve unedited input similarity. Presented in seven clusters produced by k-means (R Core Team, 2013).

We incorporated in our framework the annotated PyTorch implementation of the transformer (Rush, 2018) and modified it to accommodate our LPD dataset. Multi-head attention was configured with $h = 8$ layers and a model size $d_{model} = 512$, and the query, key, and value vectors were set uniformly to $d_{model}/h = 64$. The inner layer of the encoder and the decoder had dimensionality $d_{ff} = 2,048$. In Figure 5, we show permuted embeddings retaining input semantics by using our anagram matrix

concept. The presence of replicated words in vector space owes to the transformer built-in learned positions of input embeddings. We chose the Adam optimizer (Kingma and Ba, 2014) with a varied learning rate and a fixed model dropout of 0.1, using cross-entropy loss and label smoothing for regularization. Figure 6 reviews epoch-loss progression in training and validating our model.

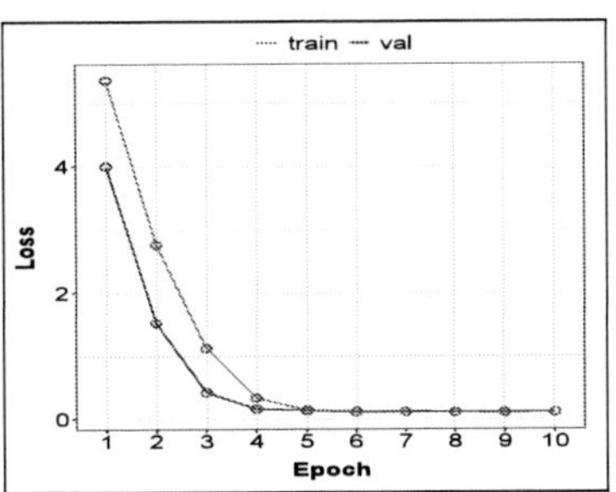

Figure 6: Epoch-loss progression in training and validation. Loss descent subsides near the seventh epoch.

6 Results

We ran our model inference on a split test set that comprises randomly selected rows sampled from the entire collection of anagram matrices and further excluded from the train set. We postulate that the use of matrix columns along the migration axis are only beneficial for embedding training.

Context Window	BLEU-1	BLEU-2	BLEU-3	BLEU-4
one-dimensional	36.8	20.9	13.0	8.3
two-dimensional	**40.6**	**23.7**	**14.7**	**8.9**

Table 2: Model performance using n-gram BLEU measures at a corpus level on our augmented Sonnets testset for repairing letter transpositions. Scores shown are contrasted between the use of one and two dimensional context window for training word embeddings.

In Table 2, we report corpus-level n-gram BLEU scores of our transformer-based model for inline transduction of LPD reading patterns. Uniformly a two-dimensional context window for training embeddings boosts our performance by about ten percentage points on average compared to the one-dimensional window. As expected, BLEU scores decline exponentially when we increase n-gram, from 40.6 for BLEU-1 down to 8.9 for BLEU-4.

While BLEU scores the output by counting n-gram matches with the reference, we also evaluated our model using SARI (Xu et al., 2016), a novel metric that correlates with human judgments and designed to specifically analyze text simplification models. SARI principally compares system output against both the reference and input verse and returns an arithmetic average of n-gram precisions and recalls of addition, copy, and delete rewrite operations. [2] Table 3 summarizes SARI and average BLEU measures of our model. Scores appear fairly correlated with a slight edge in favor of SARI that correctly rewards models like ours which make changes that simplify input verses.

Context Window	SARI	BLEU
one-dimensional	21.2	19.8
two-dimensional	**23.7**	**22.0**

Table 3: Model performance using automatic evaluation measures of SARI and BLEU at a corpus level on our augmented Sonnets test-set. Scores are contrasted between the use of one and two dimensional context window for training word embeddings.

The transformer is known to be bound by a fixed-length context and thus tends to split a long context to segments that often ignore semantic boundaries. This led to the conjecture that context fragmentation may impact our model performance adversely. The novel transformer-xl network (Dai et al., 2019) that learns dependencies across subsequences using recurrence, might be the more effective architecture to perform our task.

7   Discussion

To conduct a baseline evaluation of our model, we hand curated a corpus made of LPD screening tests. Targeted screeners are brief performance measures intended to classify at-risk individuals. To the extent of our knowledge, Lakretz et al. (2015) used for their experiments the largest known screener dataset to date that consisted of 196 loose target words in Hebrew. Correspondingly, we assembled a screening corpus of 196 English words that are prone to erroneous reading. In our system, these words are recast into a set of anagram matrices, each however reduced to a vector $\in \mathbb{R}^{k \times 1}$. Further downstream, we represented context-less words as one-hot vectors. As expected, on the task of reinstating screener data our sequence model achieved a fairly low 1-gram BLEU score of 9.2. Counter to nearly 4.4X improvement on the Sonnets dataset, when trained using a 2D context window.

Compared to almost two orders of magnitude larger Sonnets dataset, the screening corpus was too small and thus overfitting our transformer-based neural model. In addition, to effectively exploit our proposed anagram matrix representation, rather than disjoint words we require to train our sequence model on a dataset comprised of verses or sentences that provides essential context for learning embeddings.

In a practical application framework, our proposed model is rated on successful recovery from LPD reading errors that transpire in a text sequence. We envision our model already pretrained on multiple corpora, each extended to a collection of anagram matrices. Every editing instance follows with a dyslectic individual who reads and utters a verse at a time from a text document. Fed to the network, the verse is then inferred by our model that returns an amended text sequence the user can compare side-by-side on his display. It is key for the system we presented to perform responsively.

8   Conclusions

In this paper, we presented word-level neural sentence simplification to aid letter-position dyslectic children. We modeled the task after a monolingual machine translation and showed the representation effectiveness of a two-dimensional context window to boost our model performance. Future avenues of research include using our model in real-world restoration scenarios of LPD, and exploring the efficacy of the transformer-xl architecture to a non language modeling task like ours. We look forward to leverage the exceptional ability of transformer-xl to perform character-level language modeling and improve mending LPD.

Acknowledgments

We would like to thank the anonymous reviewers for their insightful suggestions and feedback.

References

Dzmitry Bahdanau, Kyunghyun Cho, and Yoshua Bengio. 2015.   Neural machine translation by jointly learning to align and translate. In *International Conference on Learning Representations, (ICLR)*, San Diego, California.

Kyunghyun Cho, Bart van Merrienboer, Caglar Gulcehre, Dzmitry Bahdanau, Fethi Bougares, Holger Schwenk, and Yoshua Bengio. 2014.   Learning phrase representations using RNN encoder–decoder

---

[2]https://github.com/cocoxu/simplification

for statistical machine translation. In *Empirical Methods in Natural Language Processing (EMNLP)*, pages 1724–1734, Doha, Qatar.

Junyoung Chung, Çaglar Gülçehre, KyungHyun Cho, and Yoshua Bengio. 2014. Empirical evaluation of gated recurrent neural networks on sequence modeling. *CoRR*, abs/1412.3555. Http://arxiv.org/abs/1412.3555.

Zihang Dai, Zhilin Yang, Yiming Yang, Jaime Carbonell, Quoc Le, and Ruslan Salakhutdinov. 2019. Transformer-XL: Attentive language models beyond a fixed-length context. In *Annual Meeting of the Association for Computational Linguistics (ACL)*, pages 2978–2988, Florence, Italy.

Maha Elbayad, Laurent Besacier, and Jakob Verbeek. 2018. Pervasive attention: {2D} convolutional neural networks for sequence-to-sequence prediction. In *Proceedings of the 22nd Conference on Computational Natural Language Learning (CONLL)*, pages 97–107, Brussels, Belgium.

Naama Friedmann and Aviah Gvion. 2001. Letter position dyslexia. *Journal of Cognitive Neuropsychology*, 18(8):673–696.

Sepp Hochreiter and Jürgen Schmidhuber. 1997. Long short-term memory. *Neural Computation*, 9(8):1735–1780.

Yvette Kezilas, Saskia Kohnen, Meredith Mckague, and Anne Castles. 2014. The locus of impairment in english developmental letter position dyslexia. *Frontiers in human neuroscience*, 8:1–14.

Diederik P. Kingma and Jimmy Ba. 2014. Adam: A method for stochastic optimization. *CoRR*, abs/1412.6980. http://arxiv.org/abs/1412.6980.

Yair Lakretz, Gal Chechik, Naama Friedmann, and Michal Rosen-Zvi. 2015. Probabilistic graphical models of dyslexia. In *Knowledge Discovery and Data Mining (KDD)*, pages 1919–1928, Sydney, Australia.

Made by Dyslexia. 2019. Dyslexia in schools: A survey. http://madebydyslexia.org/assets/downloads/ Dyslexia_In_Schools_2019.pdf (2019/9/8).

Ana Marcet, Manuel Perea, Ana Baciero, and Pablo Gómez. 2019. Can letter position encoding be modified by visual perceptual elements? *Quarterly Journal of Experimental Psychology*, 72(6):1344–1353.

Genevieve McArthur, Saskia Kohnen, Linda Larsen, Kristy Jones, Thushara Anandakumar, Erin Banales, and Anne Castles. 2013. Getting to grips with the heterogeneity of developmental dyslexia. *Cognitive neuropsychology*, 30:1–24.

Tomas Mikolov, Ilya Sutskever, Kai Chen, Greg S Corrado, and Jeff Dean. 2013. Distributed representations of words and phrases and their compositionality. In C. J. C. Burges, L. Bottou, M. Welling, Z. Ghahramani, and K. Q. Weinberger, editors, *Advances in Neural Information Processing Systems (NIPS)*, pages 3111–3119. Curran Associates, Inc.

Kishore Papineni, Salim Roukos, Todd Ward, and Wei-Jing Zhu. 2002. BLEU: a method for automatic evaluation of machine translation. In *Annual Meeting of the Association for Computational Linguistics (ACL)*, pages 311–318, Philadelphia, Pennsylvania.

Adam Paszke, Sam Gross, Soumith Chintala, Gregory Chanan, Edward Yang, Zachary DeVito, Zeming Lin, Alban Desmaison, Luca Antiga, and Adam Lerer. 2017. Automatic differentiation in pytorch. In *Workshop on Autodiff, Advances in Neural Information Processing Systems (NIPS))*, Long Beach, California.

R Core Team. 2013. *R: A Language and Environment for Statistical Computing*. R Foundation for Statistical Computing, Vienna, Austria. http://www.R-project.org/.

Luz Rello and Miguel Ballesteros. 2015. Detecting readers with dyslexia using machine learning with eye tracking measures. In *Web for All Conference*, pages 16:1–16:8, Florence, Italy.

Alexander Rush. 2018. The annotated transformer. In *Proceedings of Workshop for NLP Open Source Software (NLP-OSS)*, pages 52–60, Melbourne, Australia.

William Shakespeare. 1997. Shakespeare's sonnets. http://www.gutenberg.org/ebooks/1041 (2019/10/24).

Sally E. Shaywitz and Bennett A. Shaywitz. 2005. Dyslexia (specific reading disability). *Biological Psychiatry*, 57(11):1301–1309.

Ilya Sutskever, Oriol Vinyals, and Quoc V. Le. 2014. Sequence to sequence learning with neural networks. In *Advances in Neural Information Processing Systems (NIPS)*, pages 3104–3112. Curran Associates, Inc., Red Hook, NY.

Ashish Vaswani, Noam Shazeer, Niki Parmar, Jakob Uszkoreit, Llion Jones, Aidan N Gomez, Ł ukasz Kaiser, and Illia Polosukhin. 2017. Attention is all you need. In *Advances in Neural Information Processing Systems (NIPS)*, pages 5998–6008. Curran Associates, Inc., Red Hook, NY.

Wei Xu, Courtney Napoles, Ellie Pavlick, Quanze Chen, and Chris Callison-Burch. 2016. Optimizing statistical machine translation for text simplification. *Transactions of the Association for Computational Linguistics*, 4:401–415.

Domain Adaptation and Instance Selection for Disease Syndrome Classification over Veterinary Clinical Notes

Brian Hur[1,2] **Timothy Baldwin**[2] **Karin Verspoor**[2,3]
Laura Hardefeldt[1] **James Gilkerson**[1]

[1]Asia-Pacific Centre for Animal Health
[2]School of Computing and Information Systems
[3]Centre for the Digital Transformation of Health
The University of Melbourne, Australia

{b.hur,tbaldwin,karin.verspoor,laura.hardefeldt,jrgilk}@unimelb.edu.au

Abstract

Identifying the reasons for antibiotic administration in veterinary records is a critical component of understanding antimicrobial usage patterns. This informs antimicrobial stewardship programs designed to fight antimicrobial resistance, a major health crisis affecting both humans and animals in which veterinarians have an important role to play. We propose a document classification approach to determine the reason for administration of a given drug, with particular focus on domain adaptation from one drug to another, and instance selection to minimize annotation effort.

1 Introduction

Microorganisms — such as bacteria, fungi, and viruses — were a major cause of death until the discovery of antibiotics (Demain and Sanchez, 2009). However, antimicrobial resistance ("AMR") to these drugs has been detected since their introduction to clinical practice (Rollo et al., 1952), and risen dramatically over the last decade to be considered an emergent global phenomenon and major public health problem (Roca et al., 2015). Companion animals are capable of acquiring and exchanging multidrug-resistant pathogens with humans, and may serve as a reservoir of AMR (Lloyd, 2007; Guardabassi et al., 2004; Allen et al., 2010; Graveland et al., 2010). In addition, AMR is associated with worse animal health and welfare outcomes in veterinary medicine (Duff et al.; Johnston and Lumsden). "Antimicrobial Stewardship" is broadly used to refer to the implementation of a program for responsible antimicrobial usage, and has been demonstrated to be an effective means of reducing AMR in hospital settings (Arda et al., 2007; Pulcini et al., 2014; Baur et al., 2017; Cisneros et al., 2014). A key part of antimicrobial stewardship is having the ability to monitor antimicrobial usage patterns, including which antibiotic

History: Examination: Still extremely pruritic. There is no frank blood visible. And does not appear to be overt inflammation of skin inside EAC. Laboratory: Assessment: Much improved but still concnered there might be some residual pain/infection. This may be exac by persistent oilinesss from PMP over the last week. Treatment: Cefovecin 1mg/kg sc Owner will also use advocate; Advised needs to lose weight. To be 7kg Plan: Owner may return to recheck in ten days at completion of cefo duration.

Figure 1: Sample clinical note, in which the indication of use for *cefovecin* would be EAR DISORDER

is given and the reason — or "indication" — for its use. This data is generally captured within free text clinical records created at the time of consult. The primary objective of this paper is to develop text categorization methods to automatically label clinical records with the indication for antibiotic use.

We perform this research over the VetCompass Australia corpus, a large dataset of veterinary clinical records from over 180 of the 3,222 clinical practices in Australia which contains over 15 million clinical records and 1.3 billion tokens (McGreevy et al., 2017). An example of a typical clinical note is shown in Figure 1. We aim to map the indication for an antimicrobial into a standardized format such as Veterinary Nomenclature (VeNom) (Brodbelt, 2019), and in doing so, facilitate population-scale quantification of antimicrobial usage patterns.

As illustrated in Figure 1, the data is domain specific, and expert annotators are required to label the training data. This motivates the use of

156

Proceedings of the BioNLP 2020 workshop, pages 156–166
Online, July 9, 2020 ©2020 Association for Computational Linguistics

approaches to minimize the amount of annotation effort required, with specific interest in adapting models developed for one drug to a novel drug.

Previous analysis of this dataset has focused on labeling the antibiotic associated with each clinical note (Hur et al., 2020). In that study, it was found that *cefovecin* along with *amoxycillin clavulanate* and *cephalexin* were the top 3 antibiotics used. As *cefovecin* was the most commonly used antimicrobial with the most critical significance for the development of AMR, it was targeted for additional studies to understand the specific indications of use. The indication of use was manually labeled in 5,008 records. However, there were still over 79,000 clinical records with instances of *cefovecin* administration that did not have labels, in addition to over 1.1 million other clinical records involving other antimicrobial drug administrations missing labels.

Having only a single type of antimicrobial agent labeled causes challenges for training a model to classify the indication of use for other antimicrobials, as antimicrobials vary in how and why they are used, with the form of drug administration (oral, injected, etc.) and different indications of use creating distinct contexts that can be seen as sub-domains. Therefore, models that allow for the transfer of knowledge between the sub-domains of the various antimicrobials are required to effectively label the indication of use.

To explore the interaction between learning methods and the resource constraints on labeling, we develop models using the complete set of labels we had available, but also models derived using only labels that can be created within two hours, following the paradigm of Garrette and Baldridge (2013).

Specifically, our work explores methods to improve the performance of classifying the indication for an antibiotic administration in veterinary records of dogs and cats. In addition to classifying the indication of use, we explore how data selection can be used to improve the transfer of knowledge derived from labeled data of a single antimicrobial agent to the context of other agents. We also release our code, and select pre-trained models used in this study at: `https://github.com/havocy28/VetBERT`.

2   Related Work

Clinical coding of medical documents has been previously done with a variety of methods (Kiritchenko and Cherry, 2011; Goldstein et al., 2007; Li et al., 2018a). Additionally, classifying diseases and medications in clinical text has been addressed in shared tasks for human texts (Uzuner et al., 2010). Previous methods have also been explored for extracting the antimicrobials used, out of veterinary prescription labels, associated with the clinical records (Hur et al., 2019), and labeling of diseases in veterinary clinical records (Zhang et al., 2019; Nie et al., 2018) as well exploring methods for negation of diseases for addressing false positives (Cheng et al., 2017; Kennedy et al., 2019). Our work expands on this work by linking the indication of use to an antimicrobial being administered for that diagnosis.

Contextualized language models have recently gained much popularity due to their ability to greatly improve the representation of texts with fewer training instances, thereby transferring more efficiently between domains (Devlin et al., 2018; Howard and Ruder, 2018). Pre-training these language models on large amounts of text data specific to a given domain, such as clinical records or biomedical literature, has also been shown to further improve the performance in biomedical domains with unique vocabularies (Alsentzer et al., 2019; Lee et al., 2019). These models can also accomplish many tasks in an unsupervised manner. For example, Radford et al. (2019) showed that free text questions could be fed through a language model and generate the correct answer in many cases. In our experiments, we demonstrate the usefulness of contextualized language models by pre-training BERT on a large set of veterinary clinical records, and further explore its usefulness for domain adaptation through instance selection.

Domain adaptation is a task which has a long history in NLP (Blitzer et al., 2006; Jiang and Zhai, 2007; Agirre and De Lacalle, 2008; Daumé III, 2007). There has been further work demonstrating the usefulness of reducing the covariance between domains through adversarial learning (Li et al., 2018b). More recently, it has been shown that domain adversarial training can be effectively done using contextualized models, such as BERT, through using a two-step domain-discriminative data selection (Ma et al., 2019). We adapt these methods to our task to create a more generalizable

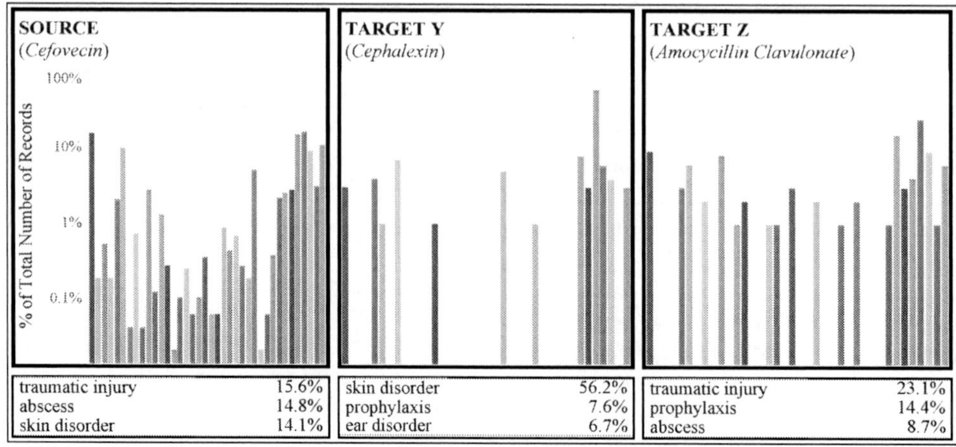

SOURCE (*Cefovecin*)		TARGET Y (*Cephalexin*)		TARGET Z (*Amocycillin Clavulonate*)	
traumatic injury	15.6%	skin disorder	56.2%	traumatic injury	23.1%
abscess	14.8%	prophylaxis	7.6%	prophylaxis	14.4%
skin disorder	14.1%	ear disorder	6.7%	abscess	8.7%

Figure 2: Distribution of labels from the SOURCE and TARGET domains (log scale). The Top-3 labels are noted below each chart.

model that can adapt between domains more effectively.

Previous experiments have used active learning to improve clinical concept extraction with weak supervision (Kholghi et al., 2016). Our work expands on this work through combining approaches to domain adaptation and the effective use of a small number of labels through the development of additional instance selection methods.

3 Dataset

3.1 Creating a set of terms

Standardized terminologies such as VeNom and SNOMED (NIH, 2019) have been created for medical diagnosis codes. While SNOMED has a veterinary extension, VeNom was created specifically for veterinary clinical text and can be mapped back to SNOMED, and is also part of the Unified Medical Language System (UMLS) (Bodenreider, 2004) used widely within human medicine. Therefore, VeNom codes are used here to create labels for the indication of drug administration (Brodbelt, 2019).

The VeNom codes we adopt are not fully comprehensive; they are a subset of the codes used by (O'Neill et al., 2019) which map specific VeNom codes to more generalized codes. These codes were provided by the Royal College of Veterinary Medicine for this study. In this subset of terms, specific labels such as EXTRACTION OF UPPER LEFT PREMOLAR 4 are simply mapped to DENTAL DISORDER. There were a total of 52 of these terms,

of which 38 actually occur in our target dataset.

3.2 Data sub-domains

We consider the individual antibiotic agents in our dataset to be sub-domains, as they are administered differently (e.g. orally vs. injectable), and in response to different indications. In our experiments, we target *cefovecin* (injectable), *amoxycillin clavulanate* (oral or injectable), and *cephalexin* (oral). In addition, *cefovecin* and *amoxycillin clavulanate* are used broadly for many indications, while *cephalexin* is primarily used for skin infections.

3.3 Extracting and labeling the data

A corpus of 5,008 clinical records, where patients had been given *cefovecin*, were sourced from Vet-Compass Australia using methods previously described in Hur et al. (2019). The indication of use for *cefovecin* was then labeled by a veterinarian.

A subset of 100 of these annotations were labeled by another veterinarian and used to calculate agreement, which was measured as Cohen's Kappa = 0.78, with raw agreement of 0.80. An additional 105 and 104 records were randomly selected for each of *cephalexin* and *amoxycillin clavulanate*, respectively, and annotated by the same two veterinarians.

The variance between the distribution of indications for *cefovecin*, *cephalexin*, and *amoxycillin clavulanate* is presented in Figure 2.

An additional set of 3000 unannotated clinical notes was sampled, comprising 1000 clinical notes for each of the three antibiotics of interest. We use these to train a domain classification filter (to identify which antimicrobial is administered), and for data selection. Any notes with fewer than 5 tokens were removed from the corpus.

3.4 Training and development sets

The training of the indication-of-use classifier was performed using the dataset pertaining to *cefovecin*, based on a 90:10 split of train and development data. In evaluation, we will refer to the development set as "SOURCE".

The labeled datasets for *amoxycillin clavulanate* and *cephalexin* are used to test cross-domain accuracy, and are referred to as "TARGET Y" for *cephalexin* and "TARGET Z" for *amoxycillin clavulanate*. The test data used for "TARGET Y" and "TARGET Z" were fixed in all tests and strictly disjoint from any training.

The estimated number of records that could be annotated within two hours was 250, based on the annotation of the three datasets. To assess the setting of having only two hours of annotation time, a subset of 250 records was sampled and annotated for for training and taken only from the "SOURCE" data according to one of the various instance selection methods described in the Approach section.

4 Approach

In this section we detail our approach, as illustrated in Figure 3.

Pretraining

In order to fine-tune our model to veterinary clinical notes, we took ClinicalBERT (Alsentzer et al., 2019) and repeated the pretraining steps as described by Devlin et al. (2018) using the entire corpus of 15 million clinical notes from VetCompass Australia. We refer to this model as "VetBERT".

Training classifiers

A baseline classifier for indication of antibiotic administration was trained using an LSTM ("LSTM": Gers et al. (1999)) with a 100 dimension embedding layer with 0.3 dropout, implemented in keras (Chollet et al., 2015). We also use a baseline BERT model using BERT-Base ("BERT"), in addition to a model based on VetBERT. Both the BERT and VetBERT classifiers were trained using an Adam optimizer, maximum of 512 word pieces, batch size of 32, softmax loss, and Learning Rate of 2e-5. Models trained on the full training set were

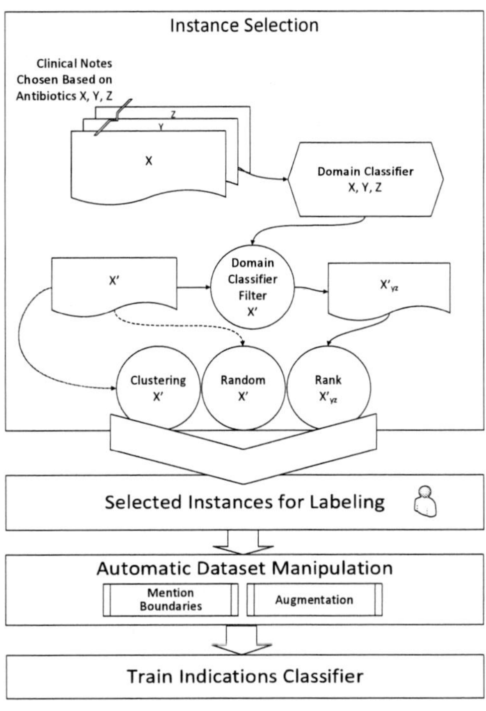

Figure 3: Outline of the proposed approach.

trained for 3 epochs, while models based on limited training data (see Section 4.3) were trained for 60 epochs. All models were tested with 5 different random seeds, and results averaged across them.

Table 1 shows the performance of VetBERT and the two baseline methods both in-domain ("SOURCE"), and for the two out-of-domain antimicrobials using the training data from SOURCE.

While the performance of VetBERT exceeded the interannotator agreement of 78% in-domain, the out-of-domain performance over TARGET Z in particular was substantially less, at 65.4% accuracy. To improve cross-domain performance, we add instance selection and dataset manipulation methods, as described below.

4.1 Instance selection

We hypothesize that filtering out training data that is dissimilar to the target domain will improve performance, despite the lower volume of training data. To this end, we experiment with domain-based instance selection.

We model domain similarity using a domain classification model, trained on the domain (i.e. administered antimicrobial) associated with a given medical record. Note that this is directly avail-

	SOURCE	TARGET Y	TARGET Z
LSTM	51.5±3.5	47.4±3.4	29.2±5.0
BERT	73.4±1.1	71.0±1.3	58.1±1.4
VetBERT	80.1±0.7	81.5±2.1	65.4±1.5
VetBERT+A	80.9±0.6	**83.4**±1.5	68.1±2.1
VetBERT+M	80.5±0.6	80.2±1.3	65.8±2.6
VetBERT+M+A	**81.2**±0.6	82.1±1.8	66.5±1.7
VetBERT+D	78.3±0.7	79.1±2.6	66.7±1.5
VetBERT+D+A	80.5±0.5	83.1±1.4	**68.5**±2.2
VetBERT+D+M	78.5±0.8	78.3±3.5	66.7±0.9
VetBERT+D+M+A	80.3±0.4	82.7±2.1	67.5±2.2

Table 1: Predictive accuracy (%) of reason for antimicrobial administration in the SOURCE and TARGET domains, trained on all available source-domain training data. Notation: +D = domain-based instance selection, +M = mention boundary tagging, +A = data augmentation

able as an artefact of the dataset construction, and doesn't require any manual annotation. Specifically, we identify instances of source domain X (*cefovecin*) for which we have labeled data, which are most similar to instances from target domains Y and Z, i.e., records in which *cephalexin* or *amoxycillin clavulanate*, respectively, were administered. Determination of similarity is based on the probabilistic output of a domain classifier over the three domains. In Figure 3, we label this subset of the training data "X'_{YZ}", reflecting the fact that it is a subset of X similar to Y and Z. This subset of X is then used to train a second classifier focused on the primary task, namely the reason for administering an antibiotic.

To build the domain classification model, we follow the procedure of Ma et al. (2019), first training a domain classifier for 1 epoch, based on the datasets of 1000 instances each of the three domains. We used the same model architecture as the VetBERT model, with a softmax classification layer. This model was then applied to the 5,008 training instances for *cefovecin*, which were sorted in increasing score over domain X (i.e. in decreasing order of similarity to the target domains), the Top-1000, 2000, 3000, or 4000 records were selected, and the VetBERT model was trained over that subset of the training data. The best results were found to occur for 3000 samples. Models with domain-based instance selection are indicated with "+D" in Table 1.

The domain classifier filtering method results in an improvement for TARGET Z (66.7%), but drop in accuracy for TARGET Y (79.1%).

4.2 Automated dataset manipulation

We also explore the use of dataset manipulation, in two forms: (1) mention boundary tagging; and (2) data augmentation.

4.2.1 Mention Boundary Tagging

To sensitize the model to the specific drug of interest, we add special learnable embedding vectors to the start and end of each antibiotic mention, based on the findings of Logeswaran et al. (2019) and Wu et al. (2019). Similar to Wu et al. (2019), we used special tokens to mark the boundaries of the tokens that contained a partial string match for the antibiotic of interest. This allows for the model to attend to these tokens at every layer of the network while training the classifier, and ideally better generalize across antimicrobials. The partial string matches were created by identifying strings that contained the prefixes *clav* or *amoxyclav* for *amoxycillin clavulanate*, *ceph*, *rilex* or *kflex* for *cephalexin*, and *conv* or *cefov* for *cefovecin*. These prefixes were sourced from a previous study exploring mention detection of antimicrobials (Hur et al., 2019). We signal the use of mention boundary embeddings with "+M" in the results tables.

4.2.2 Data augmentation

Synonym-based data augmentation has been successfully applied to contexts including word sense disambiguation (Leacock and Chodorow, 1998), sentiment analysis (Li et al., 2017), text classification (Wei and Zou, 2019), and argument analysis (Joshi et al., 2018).

We perform data augmentation on clinical notes by replacing synonyms using WordNet (Fellbaum,

	SOURCE	TARGET Y	TARGET Z
VetBERT+rank[linear]	74.3±0.2	76.6±3.0	**66.9**±2.2
VetBERT+rank[linear]+A	75.8±1.3	**81.0**±2.6	63.7±1.4
VetBERT+rank[linear]+M	73.4±0.9	77.1±1.9	65.9±2.4
VetBERT+rank[linear]+M+A	75.7±0.8	**81.0**±2.8	63.8±3.5
VetBERT+rank[exp]	68.3±2.1	66.5±2.1	58.1±1.5
VetBERT+rank[exp]+A	76.6±0.3	76.7±2.4	65.4±1.0
VetBERT+rank[exp]+M	68.9±2.0	66.7±1.5	57.9±2.1
VetBERT+rank[exp]+M+A	**76.9**±0.2	77.3±2.3	64.4±1.5
VetBERT+rank[rand]	73.5±1.8	75.4±2.3	61.9±2.8
VetBERT+rank[rand]+A	74.8±1.3	78.9±3.1	64.2±2.5
VetBERT+rank[rand]+M	73.9±1.2	76.2±2.8	62.1±1.1
VetBERT+rank[rand]+M+A	74.9±0.4	80.6±1.3	63.1±2.6

Table 2: Predictive accuracy (%) of reason for antimicrobial administration over the SOURCE and TARGET domains, trained on 2-hours' worth of labeled data with the three domain similarity selection methods over the top-3000 from X'_{YZ} of random sampling ("+rank[rand]"), modified exponential sampling ("+rank[exp]"), and linear step-wise sampling ("+rank[linear]").

	SOURCE	TARGET Y	TARGET Z
VetBERT+rand	70.9±1.5	76.2±1.6	58.0±2.0
VetBERT+rand+A	69.7±0.4	75.8±1.1	59.6±0.0
VetBERT+rand+M	70.5±0.1	77.4±0.6	57.4±2.4
VetBERT+rand+M+A	69.9±0.9	77.4±0.6	59.6±1.7
VetBERT+rank[linear]	74.3±0.2	76.6±3.0	66.9±2.2
VetBERT+rank[linear]+A	**75.8**±1.3	**81.0**±2.6	63.7±1.4
VetBERT+rank[linear]+M	73.4±0.9	77.1±1.9	65.9±2.4
VetBERT+rank[linear]+M+A	75.7±0.8	**81.0**±2.8	63.8±3.5
VetBERT+cluster	73.4±1.1	68.6±1.3	63.0±2.1
VetBERT+cluster+A	73.9±0.1	75.2±2.7	**67.8**±0.7
VetBERT+cluster+M	73.3±0.5	66.2±0.7	62.5±1.4
VetBERT+cluster+M+A	72.8±0.6	75.2±0.0	63.5±5.4

Table 3: Predictive accuracy (%) of reason for antimicrobial administration in the SOURCE and TARGET domains, trained on 2-hours' worth of labeled data, with random selection ("+rand"), linear step-wise sampling ("+rank[linear]"; results duplicated from Table 2), and clustering ("+cluster").

2012), based on random sampling. In this way, we create up to two additional training instances[1] in addition to the original instance, potentially tripling the amount of training data. We signal the use of data augmentation with "+A" in the results tables.

4.2.3 Results for dataset augmentation methods

Mention boundary tagging and data augmentation generally led to improvements in results both in-

and out-of-domain, as seen in Table 1. The highest accuracy over the source domain 81.2% was obtained with both mention boundary tagging and data augmentation (without instance selection), while the best out-of-domain results were obtained with data augmentation (with or without instance selection).

4.3 Instance selection under two annotation-hour constraint

All results to date have been based on the generous supervision setting of 3000 instances, or ap-

---

[1]In the instance of there being no synonym substitutes for any words in the original clinical note, no additional training instances are generated.

proximately 24 hours' annotation time. One natural question, inspired by the work of Garrette and Baldridge (2013) in the context of part-of-speech tagging in low-resource languages, is whether similar results can be achieved with a more realistic budget of expert annotation time. Specifically, we assume access to only 2 hours of expert annotator time, which translates to the annotation of 250 clinical notes. We propose three approaches to instance selection under this constraint: (1) domain similarity selection; and (2) clustering. We contrast these with a random selection baseline ("+rand" in our results tables).

4.3.1 Domain similarity selection

Our first approach is based on the instance selection method from Section 4.1, except that we now select only 250 instances from SOURCE for annotation, based on their similarity with the target domain (as distinct from the top-3000 instances in Table 1). That is, we take the top-3000 instances from X'_{YZ} and perform additional sub-sampling, in the form of: (a) random sampling ("+rank[rand]");[2] (b) modified exponential sampling ("+rank[exp]"); or (c) linear step-wise sampling ("+rank[linear]").

Modified exponential sampling is implemented by mapping 3000 onto an exponential scale of 250 steps over the 3000 results, rounding to the nearest integer, and additionally rounding up in the case that there is a collision with a value earlier in the series. That is, instead of the (rounded) series being $0, 0, 0, ..., 2879, 2938, 2999$ it becomes $0, 1, 2, ..., 2879, 2938, 2999$.

Linear step-wise sampling involves separating the domain space evenly, and taking every nth sample where $n = \lfloor \text{len}(N)/x \rfloor$ where x is the number of labeled instances (= 250) and N is the total number of samples (= 3000).

Results for the different instance selection methods are presented in Table 2. The best-performing method is step-wise sampling, achieving out-of-domain accuracy which is competitive with the results from Table 1 over 12 times the amount of training data.

4.3.2 Clustering-based instance selection

Our second approach is based on the intuition that the diversity in the training data will optimize performance. We achieve this by clustering the source

domain instances, and selecting prototypical instances from each cluster.

First, we generate a representation of each source-domain clinical note using the pretrained VetBERT model, based on the [CLS] token in the second-last layer of the model. Next, we cluster the instances into 250 clusters using k-means++ (Arthur and Vassilvitskii, 2006), and select the instance closest to the centroid for each cluster. This method is labeled "+cluster" in Table 3.

Clustering results in the highest accuracy for TARGET Z of 67.8%, but weaker results for TARGET Y.

5 Discussion

5.1 Pretraining Improvements

Pretraining BERT to the veterinary domain using the VetCompass Australia corpus showed the most dramatic improvement in our experiments. This was demonstrated by marked improvement over other baselines, without any additional steps (Table 1: VetBERT vs. BERT and LSTM). However, even with the pretraining used to create VetBERT, there was significant degradation in performance across the domains where there were fewer training instances (VetBERT in Table 1 vs. VetBERT+rand in Table 3).

5.2 Sub-domain transfer performance

The relative performance over TARGET Z as compared to TARGET Y when transferring from SOURCE was generally poor (Tables 1 and 3). This could be due to TARGET Y sharing more similarities with SOURCE, along with the more skewed class distribution in TARGET Y (Figure 2), potentially making it an easier classification task. More analysis is needed to understand this effect.

5.3 Optimizing for two hours of annotation time

When optimizing for two hours of annotation time, there were consistent improvements with the instance selection methods, compared to random selectin (Table 3: VetBERT+rand vs. others).

5.4 Dataset manipulation methods

The results for data augmentation and the addition of mention boundary embeddings were not as clear, in that they sometimes resulted in improvements and sometimes did not (Table 2 and 3: +A and +M vs. others). The clustering

---

[2] Note that this differs from +rand in that it is over the top-3000 instances, whereas +rand is over all 5008 annotated instances.

method generally performed better with data augmentation and worst with mention boundary embeddings (Table 3: `VetBERT+cluster+A` vs. `VetBERT+cluster+M+A` and `VetBERT+cluster+M`).

5.5 Limitations

The primary limiting factor was also the motivation of this study, namely the difficulty in obtaining sufficient high-quality annotations to perform accurate analysis of the model performance. We were also limited in that the instance selection was performed retrospectively over the 5008 annotated instances, and we were limited to the instances provided for the SOURCE domain, rather than a larger sample that could be obtained from VetCompass. There are also additional domains of data within this corpus that should be evaluated, such as records from specialty practices vs. records from general practices. This was shown to result in significant degradation of performance by Nie et al. (2018), and is a potential area for future research.

6 Conclusions and future work

In conclusion, we proposed a range of methods to transfer knowledge derived from labeled data for one antimicrobial agent to other agents, considering the additional constraint of a limited annotation resource time of two hours. While the in-domain accuracy of 83% exceeds the raw inter-annotator agreement of 80% (Cohen's Kappa = 0.78) on the source domain, transfer to other classes is still substantially lower with an average of 76% between the two classes. This shows that while the accuracy on classifying diseases is on par with human classifications for a single disease, there is still room for improvement on transferability to new data subdomains.

The primary question is whether the labels created are good enough to report the reason for antibiotic administration in epidemiological reporting and antimicrobial stewardship guidelines. While the labels for why *cefovecin* was administered were better than the current standard of using expert annotations, our results indicate that accuracy varies substantially depending on the antibiotic being administered, and testing of the accuracy for each individual antibiotic should be evaluated prior to reporting the results based on labels generated by any model.

In future research, these methods could be im-

proved through utilization of available resources such as UMLS or Drugbank to identify clinical use guidelines for antibiotics, to allow for training or adapting a model with few or no annotations. Additionally, further work is required to apply these models into a data pipeline to create labels for VetCompass data to enable analysis of the key reasons for antimicrobial administration in veterinary hospitals across Australia.

Acknowledgments

We thank Simon Süster, Afshin Rahimi, and the anonymous reviewers for their insightful comments and valuable suggestions.

This research was undertaken with the assistance of information and other resources from the VetCompass Australia consortium under the project "VetCompass Australia: Big Data and Real-time Surveillance for Veterinary Science", which is supported by the Australian Government through the Australian Research Council LIEF scheme (LE160100026).

References

Eneko Agirre and Oier Lopez De Lacalle. 2008. On robustness and domain adaptation using SVD for word sense disambiguation. In *Proceedings of the 22nd International Conference on Computational Linguistics (Coling 2008)*, pages 17–24.

Heather K. Allen, Justin Donato, Helena Huimi Wang, Karen A. Cloud-Hansen, Julian Davies, and Jo Handelsman. 2010. Call of the wild: antibiotic resistance genes in natural environments. *Nature Reviews Microbiology*, 8(4):251–259.

Emily Alsentzer, John R. Murphy, Willie Boag, Wei-Hung Weng, Di Jin, Tristan Naumann, and Matthew B. A. McDermott. 2019. Publicly Available Clinical BERT Embeddings.

Bilgin Arda, Oguz Resat Sipahi, Tansu Yamazhan, Meltem Tasbakan, Husnu Pullukcu, Alper Tunger, Cagri Buke, and Sercan Ulusoy. 2007. Short-term effect of antibiotic control policy on the usage patterns and cost of antimicrobials, mortality, nosocomial infection rates and antibacterial resistance. *Journal of Infection*, 55(1):41–48.

David Arthur and Sergei Vassilvitskii. 2006. k-means++: The advantages of careful seeding. Technical report, Stanford.

David Baur, Beryl Primrose Gladstone, Francesco Burkert, Elena Carrara, Federico Foschi, Stefanie Döbele, and Evelina Tacconelli. 2017. Effect of antibiotic stewardship on the incidence of infection and colonisation with antibiotic-resistant bacteria and Clostridium difficile infection: a systematic review and meta-analysis. *The Lancet Infectious Diseases*, 17(9):990–1001.

John Blitzer, Ryan McDonald, and Fernando Pereira. 2006. Domain adaptation with structural correspondence learning. In *Proceedings of the 2006 Conference on Empirical Methods in Natural Language Processing*, pages 120–128.

Olivier Bodenreider. 2004. The unified medical language system (UMLS): integrating biomedical terminology. *Nucleic acids research*, 32(suppl_1):D267–D270.

David Brodbelt. 2019. VeNom Coding – VeNom Coding Group. http://venomcoding.org/.

Katherine Cheng, Timothy Baldwin, and Karin Verspoor. 2017. Automatic Negation and Speculation Detection in Veterinary Clinical Text. In *Proceedings of the Australasian Language Technology Association Workshop 2017*, pages 70–78, Brisbane, Australia.

François Chollet et al. 2015. Keras. https://keras.io.

J. M. Cisneros, O. Neth, M. V. Gil-Navarro, J. A. Lepe, F. Jiménez-Parrilla, E. Cordero, M. J. Rodríguez-Hernández, R. Amaya-Villar, J. Cano, A. Gutiérrez-Pizarraya, E. García-Cabrera, and J. Molina. 2014. Global impact of an educational antimicrobial stewardship programme on prescribing practice in a tertiary hospital centre. *Clinical Microbiology and Infection*, 20(1):82–88.

Hal Daumé III. 2007. Frustratingly easy domain adaptation. In *Proceedings of the 45th Annual Meeting of the Association of Computational Linguistics*, pages 256–263, Prague, Czech Republic.

Arnold L. Demain and Sergio Sanchez. 2009. Microbial drug discovery: 80 years of progress. *The Journal of Antibiotics*, 62(1):5–16.

Jacob Devlin, Ming-Wei Chang, Kenton Lee, and Kristina Toutanova. 2018. BERT: Pre-training of Deep Bidirectional Transformers for Language Understanding. *arXiv:1810.04805 [cs]*. ArXiv: 1810.04805.

A Duff, SK Keane, and Laura Y. Hardefeldt. Descriptive study of antimicrobial susceptibilty patterns from equine septic synovial structures. In *Proceedings of the 39th Bain Fallon Memorial Lectures*, volume 2017:11. Equine Veterinarians Australia.

Christiane Fellbaum. 2012. WordNet. *The Encyclopedia of Applied Linguistics*.

Dan Garrette and Jason Baldridge. 2013. Learning a Part-of-Speech Tagger from Two Hours of Annotation. In *Proceedings of the 2013 Conference of the*

North American Chapter of the Association for Computational Linguistics: Human Language Technologies, pages 138–147, Atlanta, Georgia.

Felix A Gers, Jürgen Schmidhuber, and Fred Cummins. 1999. Learning to forget: Continual prediction with LSTM. In *Proceedings of the 9th International Conference on Artificial Neural Networks: ICANN '99*, pages 850–855.

Ira Goldstein, Anna Arzumtsyan, and Özlem Uzuner. 2007. Three approaches to automatic assignment of ICD-9-CM codes to radiology reports. In *AMIA Annual Symposium Proceedings*, volume 2007, page 279. American Medical Informatics Association.

Haitske Graveland, Jaap A. Wagenaar, Hans Heesterbeek, Dik Mevius, Engeline van Duijkeren, and Dick Heederik. 2010. Methicillin Resistant Staphylococcus aureus ST398 in Veal Calf Farming: Human MRSA Carriage Related with Animal Antimicrobial Usage and Farm Hygiene. *PLOS ONE*, 5(6):e10990.

Luca Guardabassi, Stefan Schwarz, and David H. Lloyd. 2004. Pet animals as reservoirs of antimicrobial-resistant bacteria. *Journal of Antimicrobial Chemotherapy*, 54(2):321–332.

Jeremy Howard and Sebastian Ruder. 2018. Universal language model fine-tuning for text classification. In *Proceedings of the 56th Annual Meeting of the Association for Computational Linguistics (Volume 1: Long Papers)*, pages 328–339, Melbourne, Australia.

B. Hur, L. Y. Hardefeldt, K. Verspoor, T. Baldwin, and J. R. Gilkerson. 2019. Using natural language processing and VetCompass to understand antimicrobial usage patterns in Australia. *Australian Veterinary Journal*, 97(8):298–300.

Brian A. Hur, Laura Y. Hardefeldt, Karin M. Verspoor, Timothy Baldwin, and James R. Gilkerson. 2020. Describing the antimicrobial usage patterns of companion animal veterinary practices; free text analysis of more than 4.4 million consultation records. *PLOS ONE*, 15(3):1–15.

Jing Jiang and ChengXiang Zhai. 2007. Instance weighting for domain adaptation in NLP. In *Proceedings of the 45th Annual Meeting of the Association of Computational Linguistics*, pages 264–271.

GCA Johnston and JM Lumsden. Antimicrobial susceptiblity of bacterial isolates from 27 thoroughbreds with arytenoid chondropathy. In *Proceedings of the 39th Bain Fallon Memorial Lectures*, volume 2017:11. Equine Veterinarians Australia.

Anirudh Joshi, Timothy Baldwin, Richard O Sinnott, and Cecile Paris. 2018. UniMelb at SemEval-2018 task 12: Generative implication using LSTMs, Siamese networks and semantic representations with synonym fuzzing. In *Proceedings of The 12th International Workshop on Semantic Evaluation*, pages 1124–1128.

Noel Kennedy, Dave C Brodbelt, David B Church, and Dan G O'Neill. 2019. Detecting false-positive disease references in veterinary clinical notes without manual annotations. *NPJ Digital Medicine*, 2(1):1–7.

Mahnoosh Kholghi, Laurianne Sitbon, Guido Zuccon, and Anthony Nguyen. 2016. Active learning: a step towards automating medical concept extraction. *Journal of the American Medical Informatics Association*, 23(2):289–296.

Svetlana Kiritchenko and Colin Cherry. 2011. Lexically-triggered hidden markov models for clinical document coding. In *Proceedings of the 49th Annual Meeting of the Association for Computational Linguistics: Human Language Technologies-Volume 1*, pages 742–751.

Claudia Leacock and Martin Chodorow. 1998. Combining local context and WordNet similarity for word sense identification. *WordNet: An electronic lexical database*, 49(2):265–283.

Jinhyuk Lee, Wonjin Yoon, Sungdong Kim, Donghyeon Kim, Sunkyu Kim, Chan Ho So, and Jaewoo Kang. 2019. BioBERT: a pre-trained biomedical language representation model for biomedical text mining. *arXiv:1901.08746 [cs]*. ArXiv: 1901.08746.

Min Li, Zhihui Fei, Min Zeng, Fang-Xiang Wu, Yaohang Li, Yi Pan, and Jianxin Wang. 2018a. Automated icd-9 coding via a deep learning approach. *IEEE/ACM transactions on computational biology and bioinformatics*, 16(4):1193–1202.

Yitong Li, Timothy Baldwin, and Trevor Cohn. 2018b. What's in a domain? learning domain-robust text representations using adversarial training. In *Proceedings of the 2018 Conference of the North American Chapter of the Association for Computational Linguistics: Human Language Technologies, Volume 2 (Short Papers)*, pages 474–479, New Orleans, Louisiana.

Yitong Li, Trevor Cohn, and Timothy Baldwin. 2017. Robust training under linguistic adversity. In *Proceedings of the 15th Conference of the EACL (EACL 2017)*, pages 21–27, Valencia, Spain.

David H. Lloyd. 2007. Reservoirs of antimicrobial resistance in pet animals. *Clinical Infectious Diseases: An Official Publication of the Infectious Diseases Society of America*, 45 Suppl 2:S148–152.

Lajanugen Logeswaran, Ming-Wei Chang, Kenton Lee, Kristina Toutanova, Jacob Devlin, and Honglak Lee. 2019. Zero-Shot Entity Linking by Reading Entity Descriptions. *arXiv:1906.07348 [cs]*. ArXiv: 1906.07348.

Xiaofei Ma, Peng Xu, Zhiguo Wang, Ramesh Nallapati, and Bing Xiang. 2019. Domain Adaptation with BERT-based Domain Classification and Data Selection. In *Proceedings of the 2nd Workshop on Deep Learning Approaches for Low-Resource NLP (DeepLo 2019)*, pages 76–83, Hong Kong, China.

Paul McGreevy, Peter Thomson, Navneet K Dhand, David Raubenheimer, Sophie Masters, Caroline S Mansfield, Timothy Baldwin, Ricardo J Soares Magalhaes, Jacquie Rand, Peter Hill, Anne Peaston, James Gilkerson, Martin Combs, Shane Raidal, Peter Irwin, Peter Irons, Richard Squires, David Brodbelt, and Jeremy Hammond. 2017. VetCompass Australia: A National Big Data Collection System for Veterinary Science. page 15.

Allen Nie, Ashley Zehnder, Rodney L. Page, Arturo L. Pineda, Manuel A. Rivas, Carlos D. Bustamante, and James Zou. 2018. DeepTag: inferring all-cause diagnoses from clinical notes in under-resourced medical domain. *arXiv:1806.10722 [cs]*. ArXiv: 1806.10722.

NIH. 2019. SNOMED CT.

Dan G. O'Neill, Alison M. Skipper, Jade Kadhim, David B. Church, Dave C. Brodbelt, and Rowena M. A. Packer. 2019. Disorders of Bulldogs under primary veterinary care in the UK in 2013. *PLOS ONE*, 14(6):e0217928.

C. Pulcini, E. Botelho-Nevers, O. J. Dyar, and S. Harbarth. 2014. The impact of infectious disease specialists on antibiotic prescribing in hospitals. *Clinical Microbiology and Infection*, 20(10):963–972.

Alec Radford, Jeffrey Wu, Rewon Child, David Luan, Dario Amodei, and Ilya Sutskever. 2019. Language Models are Unsupervised Multitask Learners.

I. Roca, M. Akova, F. Baquero, J. Carlet, M. Cavaleri, S. Coenen, J. Cohen, D. Findlay, I. Gyssens, O. E. Heure, G. Kahlmeter, H. Kruse, R. Laxminarayan, E. Liébana, L. López-Cerero, A. MacGowan, M. Martins, J. Rodríguez-Baño, J. M. Rolain, C. Segovia, B. Sigauque, E. Tacconelli, E. Wellington, and J. Vila. 2015. The global threat of antimicrobial resistance: science for intervention. *New Microbes and New Infections*, 6:22–29.

I. M. Rollo, J. Williamson, and R. L. Plackett. 1952. Acquired Resistance To Penicillin And To Neoarsphenamine In Spirochaeta Recurrentis. *British Journal of Pharmacology and Chemotherapy*, 7(1):33–41.

Özlem Uzuner, Imre Solti, and Eithon Cadag. 2010. Extracting medication information from clinical text. *Journal of the American Medical Informatics Association*, 17(5):514–518.

Jason Wei and Kai Zou. 2019. EDA: Easy data augmentation techniques for boosting performance on text classification tasks. In *Proceedings of the 2019 Conference on Empirical Methods in Natural Language Processing and the 9th International Joint Conference on Natural Language Processing (EMNLP-IJCNLP)*, pages 6382–6388, Hong Kong, China.

Ledell Wu, Fabio Petroni, Martin Josifoski, Sebastian Riedel, and Luke Zettlemoyer. 2019. Zero-shot Entity Linking with Dense Entity Retrieval. *arXiv:1911.03814 [cs]*. ArXiv: 1911.03814.

Yuhui Zhang, Allen Nie, Ashley Zehnder, Rodney L. Page, and James Zou. 2019. VetTag: improving automated veterinary diagnosis coding via large-scale language modeling. *NPJ Digital Medicine*, 2(1).

Benchmark and Best Practices for Biomedical Knowledge Graph Embeddings

David Chang[1], Ivana Balažević[2], Carl Allen[2], Daniel Chawla[1]
Cynthia Brandt[1], Richard Andrew Taylor[1]
[1]Yale Center for Medical Informatics, Yale University
[2]School of Informatics, University of Edinburgh, UK
{david.chang, richard.taylor}@yale.edu
{ivana.balazevic, carl.allen}@ed.ac.uk

Abstract

Much of biomedical and healthcare data is encoded in discrete, symbolic form such as text and medical codes. There is a wealth of expert-curated biomedical domain knowledge stored in knowledge bases and ontologies, but the lack of reliable methods for learning knowledge representation has limited their usefulness in machine learning applications. While text-based representation learning has significantly improved in recent years through advances in natural language processing, attempts to learn biomedical concept embeddings so far have been lacking. A recent family of models called knowledge graph embeddings have shown promising results on general domain knowledge graphs, and we explore their capabilities in the biomedical domain. We train several state-of-the-art knowledge graph embedding models on the SNOMED-CT knowledge graph, provide a benchmark with comparison to existing methods and in-depth discussion on best practices, and make a case for the importance of leveraging the multi-relational nature of knowledge graphs for learning biomedical knowledge representation. The embeddings, code, and materials will be made available to the community[1].

1 Introduction

A vast amount of biomedical domain knowledge is stored in knowledge bases and ontologies. For example, SNOMED Clinical Terms (SNOMED-CT)[2] is the most widely used clinical terminology in the world for documentation and reporting in healthcare, containing hundreds of thousands of medical terms and their relations, organized in a polyhierarchical structure. SNOMED-CT can be thought of as a knowledge graph: a collection of triples consisting of a head entity, a relation, and a tail entity, denoted (h, r, t). SNOMED-CT is one of over a

hundred terminologies under the Unified Medical Language System (UMLS) (Bodenreider, 2004), which provides a metathesaurus that combines millions of biomedical concepts and relations under a common ontological framework. The unique identifiers assigned to the concepts as well as the Resource Release Format (RRF) standard enable interoperability and reliable access to information. The UMLS and the terminologies it encompasses are a crucial resource for biomedical and healthcare research.

One of the main obstacles in clinical and biomedical natural language processing (NLP) is the ability to effectively represent and incorporate domain knowledge. A wide range of downstream applications such as entity linking, summarization, patient-level modeling, and knowledge-grounded language models could all benefit from improvements in our ability to represent domain knowledge. While recent advances in NLP have dramatically improved textual representation (Alsentzer et al., 2019), attempts to learn analogous dense vector representations for biomedical concepts in a terminology or knowledge graph (*concept embeddings*) so far have several drawbacks that limit their usability and wide-spread adoption. Further, there is currently no established best practice or benchmark for training and comparing such embeddings. In this paper, we explore knowledge graph embedding (KGE) models as alternatives to existing methods and make the following contributions:

- We train five recent KGE models on SNOMED-CT and demonstrate their advantages over previous methods, making a case for the importance of leveraging the multi-relational nature of knowledge graphs for biomedical knowledge representation.

- We establish a suite of benchmark tasks to enable fair comparison across methods and include much-needed discussion on best prac-

---

[1] https://github.com/dchang56/snomed_kge
[2] https://www.nlm.nih.gov/healthit/snomedct

Proceedings of the BioNLP 2020 workshop, pages 167–176
Online, July 9, 2020 ©2020 Association for Computational Linguistics

tices for working with biomedical knowledge graphs.

- We also serve the general KGE community by providing benchmarks on a new dataset with real-world relevance.

- We make the embeddings, code, and other materials publicly available and outline several avenues of future work to facilitate progress in the field.

2 Related Work and Background

2.1 Biomedical concept embeddings

Early attempts to learn biomedical concept embeddings have applied variants of the skip-gram model (Mikolov et al., 2013) on large biomedical or clinical corpora. Med2Vec (Choi et al., 2016) learned embeddings for 27k ICD-9 codes by incorporating temporal and co-occurrence information from patient visits. Cui2Vec (Beam et al., 2019) used an extremely large collection of multimodal medical data to train embeddings for nearly 109k concepts under the UMLS.

These corpus-based methods have several drawbacks. First, the corpora are inaccessible due to data use agreements, rendering them irreproducible. Second, these methods tend to be data-hungry and extremely data inefficient for capturing domain knowledge. In fact, one of the main limitations of language models in general is their reliance on the distributional hypothesis, essentially making use of mostly co-occurrence level information in the training corpus (Peters et al., 2019). Third, they do a poor job of achieving sufficient concept coverage: Cui2Vec, despite its enormous training data, was only able to capture 109k concepts out of over 3 million concepts in the UMLS, drastically limiting its downstream usability.

A more recent trend has been to apply network embedding (NE) methods directly on a knowledge graph that represents structured domain knowledge. NE methods such as Node2Vec (Grover and Leskovec, 2016) learn embeddings for nodes in a network (graph) by applying a variant of the skip-gram model on samples generated using random walks, and they have shown impressive results on node classification and link prediction tasks on a wide range of network datasets. In the biomedical domain, CANode2Vec (Kotitsas et al., 2019) applied several NE methods on single-relation subsets of the SNOMED-CT graph, but the lack of

comparison to existing methods and the disregard for the heterogeneous structure of the knowledge graph substantially limit its significance.

Notably, Snomed2Vec (Agarwal et al., 2019) applied NE methods on a clinically relevant multi-relational subset of the SNOMED-CT graph and provided comparisons to previous methods to demonstrate that applying NE methods directly on the graph is more data efficient, yields better embeddings, and gives explicit control over the subset of concepts to train on. However, one major limitation of NE approaches is that they relegate relationships to mere indicators of connectivity, discarding the semantically rich information encoded in multi-relational, heterogeneous knowledge graphs.

We posit that applying KGE methods on a knowledge graph is more principled and should therefore yield better results. We now provide a brief overview of the KGE literature and describe our experiments in Section 3.

2.2 Knowledge Graph Embeddings

Knowledge graphs are collections of facts in the form of ordered triples ($\mathbf{h}$, $\mathbf{r}$, $\mathbf{t}$), where entity $\mathbf{h}$ is related to entity $\mathbf{t}$ by relation $\mathbf{r}$. Because knowledge graphs are often incomplete, an ability to infer unknown facts is a fundamental task (link prediction). A series of recent KGE models approach link prediction by learning embeddings of entities and relations based on a scoring function that predicts a probability that a given triple is a fact.

RESCAL (Nickel et al., 2011) represents relations as a bilinear product between subject and object entity vectors. Although a very expressive model, RESCAL is prone to overfitting due to the large number of parameters in the full rank relation matrix, increasing quadratically with the number of relations in the graph.

DistMult (Yang et al., 2015) is a special case of RESCAL with a diagonal matrix per relation, reducing overfitting. However, by limiting linear transformations on entity embeddings to a stretch, DistMult cannot model asymmetric relations.

ComplEx (Trouillon et al., 2016) extends DistMult to the complex domain, enabling it to model asymmetric relations by introducing complex conjugate operations into the scoring function.

SimplE (Kazemi and Poole, 2018) modifies Canonical Polyadic (CP) decomposition (Hitchcock, 1927) to allow two embeddings for each entity (head and tail) to be learned dependently.

A recent model TuckER (Balažević et al., 2019) is shown to be a fully expressive, linear model that subsumes several tensor factorization based approaches including all models described above.

TransE (Bordes et al., 2013) is an example of an alternative *translational* family of KGE models, which regard a relation as a translation (vector offset) from the subject to the object entity vectors. Translational models have an additive component in the scoring function, in contrast to the multiplicative scoring functions of bilinear models.

RotatE (Sun et al., 2019) extends the notion of translation to rotation in the complex plane, enabling the modeling of symmetry/antisymmetry, inversion, and composition patterns in knowledge graph relations.

We restrict our experiments to five models due to their available implementation under a common, scalable platform (Zhu et al., 2019): TransE, ComplEx, DistMult, SimplE, and RotatE.

3 Experimental Setup

3.1 Data

Given the complexity of the UMLS, we detail our preprocessing steps to generate the final dataset. We subset the 2019AB version of the UMLS to SNOMED_CT_US terminology, taking all active concepts and relations in the MRCONSO.RRF and MRREL.RRF files. We extract semantic type information from MRSTY.RRF and semantic group information from the Semantic Network website[3] to filter concepts and relations to 8 broad semantic groups of interest: Anatomy (**ANAT**), Chemicals & Drugs (**CHEM**), Concepts & Ideas (**CONC**), Devices (**DEVI**), Disorders (**DISO**), Phenomena (**PHEN**), Physiology (**PHYS**), and Procedures (**PROC**). We also exclude specific semantic types deemed unnecessary. A full list of the semantic types included in the dataset and their broader semantic groups can be found in the Supplements.

The resulting list of triples comprises our final knowledge graph dataset. Note that the UMLS includes reciprocal relations (ISA and INVERSE_ISA), making the graph bidirectional. A random split results in train-to-test leakage, which can inflate the performance of weaker models (Dettmers et al., 2018). We fix this by ensuring reciprocal relations are in the same split, not across splits. Descriptive statistics of the final dataset are shown in Table 1. After splitting, we also ensure

there are no unseen entities or relations in the validation and test sets by simply moving them to the train set. More details and the code used for data preparation are included in the Supplements.

Descriptions	Statistics
Entities	293,884
Relation types	170
Facts	2,073,848
- Train	1,965,032
- Valid / Test	48,936 / 49,788

Table 1: Statistics of the final SNOMED dataset.

3.2 Implementation

Considering the non-trivial size of SNOMED-CT and the importance of scalability and consistent implementation for running experiments, we use GraphVite (Zhu et al., 2019) for the KGE models. GraphVite is a graph embedding framework that emphasizes scalability, and its speedup relative to existing implementations is well-documented[4]. While the backend is written largely in C++, a Python interface allows customization. We make our customized Python code available. We use the five models available in GraphVite in our experiments: TransE, ComplEx, DistMult, SimplE, and RotatE. While we restrict our current work to these models, future work should also consider other state-of-the-art models such as TuckER (Balažević et al., 2019) and MuRP (Balažević et al., 2019), especially since MuRP is shown to be particularly effective for graphs with hierarchical structure. Pretrained embeddings for Cui2Vec and Snomed2Vec were used as provided by the authors, with dimensionality 500 and 200, respectively.

All experiments were run on 3 GTX-1080ti GPUs, and final runs took ∼6 hours on a single GPU. Hyperparameters were either tuned on the validation set for each model: margin (4, 6, 8, 10) and learning_rate (5e-4, 1e-4, 5e-5, 1e-5); set: num_negative (60), dim (512), num_epoch (2000); or took default values from GraphVite. The final hyperparameter configuration can be found in the Appendix.

3.3 Evaluation and Benchmark

3.3.1 KGE Link Prediction

A standard evaluation task in the KGE literature is link prediction. However, NE methods also use

---

[3]https://semanticnetwork.nlm.nih.gov

[4]https://github.com/DeepGraphLearning/graphVite

link prediction as a standard evaluation task. While both predict whether two nodes are connected, NE link prediction performs binary classification on a balanced set of positive and negative edges based on the assumption that the graph is complete. In contrast, knowledge graphs are typically assumed incomplete, making link prediction for KGE a ranking-based task in which the model's scoring function is used to rank candidate samples without relying on ground truth negatives. In this paper, link prediction refers to the latter ranking-based KGE method.

Candidate samples are generated for each triple in the test set using all possible entities as the target entity, where the target can be set to `head`, `tail`, or `both`. For example, if the target is `tail`, the model predicts scores for all possible candidates for the tail entity in (h, r, ?). For a test set with 50k triples and 300k possible unique entities, the model calculates scores for fifteen billion candidate triples. The candidates are filtered to exclude triples seen in the train, validation, and test sets, so that known triples do not affect the ranking and cause false negatives. Several ranking-based metrics are computed based on the sorted scores. Note that SNOMED-CT contains a *transitive closure* file, which lists explicit transitive closures for the hierarchical relations `ISA` and `INVERSE_ISA` (if A `ISA` B, and B `ISA` C, then the transitive closure includes A `ISA` C). This file should be included in the file list used to filter candidates to best enable the model to learn hierarchical structure.

Typical link prediction metrics include Mean Rank (MR), Mean Reciprocal Rank (**MRR**), and Hits@k (**H@k**). MR is considered to be sensitive to outliers and unreliable as a metric. Guu et al. proposed using Mean Quantile (**MQ**) as a more robust alternative to MR and MRR. We use MQ_{100} as a more challenging version of MQ that introduces a cut-off at the top 100th ranking, appropriate for the large numbers of possible entities. Link prediction results are reported in Table 2.

3.3.2 Embedding Evaluation

For fair comparison with existing methods, we perform some of the benchmark tasks for assessing medical concept embeddings proposed by Beam et al.. However, we discuss their methodological flaws in Section 5 and suggest more appropriate evaluation methods.

Since non-KGE methods are not directly comparable on tasks that require both relation and concept embeddings, to compare embeddings across methods we perform entity semantic classification, which requires only concept embeddings.

We generate a dataset for entity classification by taking the intersection of the concepts covered in all (7) models, comprising 39k concepts with 32 unique semantic types and 4 semantic groups. We split the data into train and test sets with 9:1 ratio, and train a simple linear layer with 0.1 dropout and no further hyperparameter tuning. The single linear layer for classification assesses the linear separability of semantic information in the entity embedding space for each model. Results for semantic type and group classification are reported in Table 3.

4   Visualization

We first discuss the embedding visualizations obtained through LargeVis (Tang et al., 2016), an efficient large-scale dimensionality reduction technique available as an application in GraphVite.

Figure 1 shows concept embeddings for RotatE, ComplEx, Snomed2Vec, and Cui2Vec, with colors corresponding to broad semantic groups. Cui2Vec embeddings show structure but not coherent semantic clusters. Snomed2Vec shows tighter groupings of entities, though the clusters are patchy and scattered across the embedding space. ComplEx produces globular clusters centered around the origin, with clearer boundaries between groups. RotatE gives visibly distinct clusters with clear group separation that appear intuitive: entities of the Physiology semantic group (black) overlap heavily with those of Disorders (magenta); also entities under the Concepts semantic group (red) are relatively scattered, perhaps due to their abstract nature, compared to more concrete entities like Devices (cyan), Anatomy (blue), and Chemicals (green), which form tighter clusters.

Interestingly, the embedding visualizations for the 5 KGE models fall into 2 types: RotatE and TransE produce well-separated clusters while ComplEx, DistMult and SimplE produce globular clusters around the origin. Since the plots for each type appear almost indistinguishable we show one from each (RotatE and ComplEx). We attribute the characteristic difference between the two model types to the nature of their scoring functions: RotatE and TransE have an additive component while ComplEx, DistMult and SimplE are multiplicative.

Figure 2 shows more fine-grained semantic structure by coloring 5 selected semantic types under

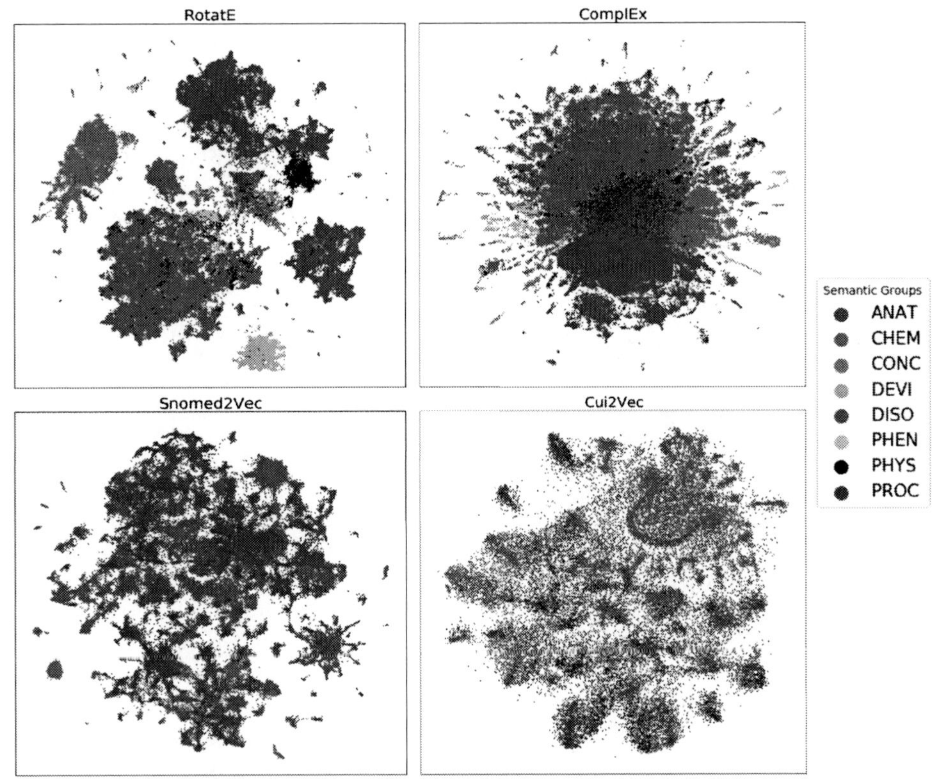

Figure 1: Concept embedding visualization (RotatE, ComplEx, Snomed2Vec, Cui2Vec) by semantic group.

the Procedures semantic group and greying out the rest. We see that RotatE produces subclusters that are also intuitive. *Laboratory procedures* are well-separated on their own, *health care activity* and *educational activity* overlap significantly, and *diagnostic procedures* and *therapeutic or preventative procedures* overlap significantly. ComplEx also reveals subclusters with globular shape, and Snomed2Vec captures *laboratory procedures* well but leaves other types scattered. These observations are consistent across other semantic groups. We include similar visualizations for the Chemicals & Drugs semantic group in the Supplements.

While semantic class information is not the only significant aspect of SNOMED-CT, since the SNOMED-CT graph is largely organized around semantic group and type information, it is promising that embeddings learned (without supervision) preserve it.

5 Results

5.1 Link Prediction

Table 2 shows results for the link prediction task for the 5 KGE models on SNOMED-CT. Having

Model	MRR	MQ_{100}	H@1	H@10
TransE	.346	.739	.212	.597
ComplEx	.461	.761	.360	.652
DistMult	.420	.752	.309	.626
SimplE	.432	.735	.337	.615
RotatE	.317	.742	.162	.599
$TransE_{FB}$	.294	-	-	.465
$TransE_{WN}$	.226	-	-	.501
$RotatE_{FB}$	.338	-	.241	.533
$RotatE_{WN}$	.476	-	.428	.571

Table 2: Link prediction results: for the 5 KGE models on SNOMED-CT (top); and for TransE and RotatE on two standard KGE datasets (Sun et al., 2019) (bottom).

no previous results to compare to, we include performance of TransE and RotatE on two standard KGE benchmark datasets for reference: FB15k-237 (14,541 entities, 237 relations, and 310,116 triples) and WN18RR (40,943 entities, 11 relations, and 93,003 triples). Given that SNOMED-CT is larger and arguably a more complex knowledge graph than the two datasets, the link prediction results suggest that the KGE models learn a reasonable

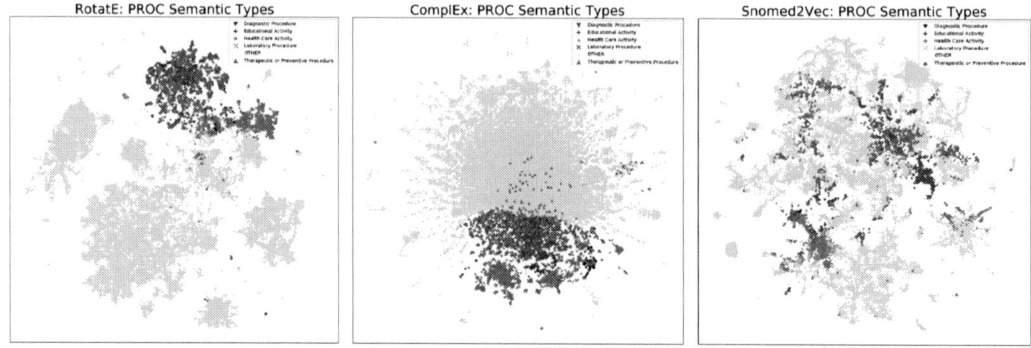

Figure 2: Visualization of selected semantic types under the Procedures semantic group for RotatE, ComplEx, and Snomed2Vec. Semantic types with more than 2,000 entities were subsampled to 1,200 for visibility. Cui2Vec (not shown) was similar to Snomed2Vec but more dispersed.

representation of SNOMED-CT. We include sample model outputs for the top 10 entity scores for link prediction in the Supplements.

5.2 Embedding Evaluation and Relation Prediction

Test set accuracy for entity semantic type (**STY**) and semantic group (**SG**) classification are reported in Table 3. In accordance with the visualizations of semantic clusters (Figures 1 and 2), the KGE and NE methods perform significantly better than the corpus-based method (Cui2Vec). Notably, TransE and RotatE attain near-perfect accuracy for the broader semantic group classification (4 classes). ComplEx, DistMult, and SimplE perform slighty worse, Snomed2Vec slightly below them, and Cui2Vec falls behind by a significant margin. We see a greater discrepancy in relative performance by model type in semantic type classification (32 classes), in which more fine-grained semantic information is required.

Two advantages of the semantic type and group entity classification tasks are: (i) information is provided by the UMLS, making the task non-proprietary and standardized; (ii) it readily shows whether a model preserves the semantic structure of the ontology, an important aspect of the data. The tasks can also easily be modified for custom data and specific domains, e.g. class labels for genes and proteins relevant to a particular biomedical application can be used in classification to assess how well the model captures relevant domain-specific information.

For comparison to related work, we also examine the benchmark tasks to assess medical concept embeddings based on *statistical power* and *cosine similarity bootstrapping*, proposed by Beam et al.. For a given known relationship pair (e.g. x cause_of y), a null distribution of pairwise cosine similarity scores is computed by bootstrapping 10,000 samples of the same semantic category as x and y respectively. The cosine similarity of the observed sample is compared to the 95th percentile of the bootstrap distribution (statistical significance at the 0.05 level). The authors claim that, when applied to a collection of known relationships (causative, comorbidity, etc), the procedure estimates the fraction of true relationships discovered given a tolerance for some false positive rate. Following this, we report the statistical power of all 7 models for two of the tasks: *semantic type* and *causative relationships*. The former (**ST**) aims to assess a model's ability to determine if two concepts share the same semantic type. The latter consists of two relation types: cause_of (**Co**) and causative_agent_of (**CA**). Results are reported in Table 3. The cosine similarity bootstrap results, particularly for the causative relationship tasks, illustrate a major flaw in the protocol. While Snomed2Vec and Cui2Vec attain similar statistical powers for **CA** and **Co**, we see large discrepancies between the two tasks for the KGE models, especially for ComplEx, DistMult, and SimplE, which produce globular embedding clusters. Examining the dataset, we observe that the cause_of relations occur mostly between concepts *within* the same semantic group/cluster (e.g. Disorder), whereas the causative_agent_of relations occur between concepts in *different* semantic groups/clusters (e.g. Chemicals to Disorders). The large discrepancy in **CA** task results

	Entity Classification		Cosine-Sim Bootstrap			Relation Prediction		
Model	**SG (4)**	**STY (32)**	**ST**	**CA**	**Co**	**MRR**	**H@1**	**H@10**
Snomed2Vec	.944	.769	.387	**.903**	.894	-	-	-
Cui2Vec	.891	.673	.416	.584	.559	-	-	-
TransE	**.993**	**.827**	**.579**	.765	**.978**	.800	.727	**.965**
ComplEx	.956	.786	.249	.001	.921	.731	.606	.914
DistMult	.971	.794	.275	.014	.971	.734	.569	.946
SimplE	.953	.768	.242	.011	.791	**.854**	**.803**	.946
RotatE	**.995**	**.829**	.544	.242	.943	**.849**	**.799**	.957

Table 3: Results for (i) entity classification of semantic type and group (test accuracy); (ii) selected tasks from (Beam et al., 2019); and (iii) relation prediction. Best results in bold.

for the KGE models is because using cosine similarity embeds the assumption that all related entities are close, regardless of the relation type. The assumption that cosine similarity in the concept embedding space is an appropriate measure of a diverse range of relatedness (a much broader abstraction that subsumes semantic similarity and causality), renders this evaluation protocol unsuitable for assessing a model's ability to capture specific types of relational information in the embeddings. Essentially, all that can be said about the cosine similarity-based procedure is that it assesses how close entities are in that space as measured by cosine distance. It does not reveal the nature of their relationship or what kind of relational information is encoded in the space to begin with.

In contrast, KGE methods explicitly model relations and are better equipped to make inferences about the relational structure of the knowledge graph embeddings. Thus, we propose *relation prediction* as a standard evaluation task for assessing a model's ability to capture information about relations in the knowledge graph. We simply modify the link prediction task described above to accommodate `relation` as a target (formulated as (h, ?, t), generating ranking-based metrics for the model's ability to prioritize the correct relation type given a pair of concepts. This provides a more principled and interpretable way to evaluate the models' relation representations directly based on the model prediction. The last 3 columns of Table 3 report relation prediction metrics for the 5 KGE models. In particular, RotatE and SimplE perform well, attaining around 0.8 Hits@1 and around 0.85 MRR.

We conduct error analysis to gain further insight by categorizing relation types into 6 groups based on the *cardinality* and *homogeneity* of their source and target semantic groups. If the set of unique head or tail entities for a relation type in the dataset belongs to only one semantic group, then it has a cardinality of 1, and a cardinality of `many` otherwise. If the mapping of the source semantic groups to the target semantic groups are one-to-one (e.g. DISO to DISO and CHEM to CHEM), then it is considered homogeneous. We report relation prediction metrics for each of the 6 groups of relation types for RotatE and ComplEx in Table 4.

We see that RotatE gives impressive relation prediction performance for all groups except for many-to-many-homogeneous, a seemingly challenging group of relations containing ambiguous and synonymous relation types, e.g. `possibly_equivalent_to`, `same_as`, `refers_to`, `isa`. The full list of M-M-hom relations are shown in the Appendix. In contrast, ComplEx struggles with a wider array of relation types, suggesting that it is generally less able to model different types than RotatE. The last two rows under each model show per-relation results for the causative relationships mentioned previously: `cause_of` and `causative_agent_of`. RotatE again shows significantly better results compared to ComplEx, in line with its theoretically superior representation capacity (Sun et al., 2019).

6   Discussion

Based on our findings, we recommend the use of KGE models to leverage the multi-relational nature of knowledge graphs for learning biomedical concept and relation embeddings; and of appropriate evaluation tasks such as link prediction, entity classification and relation prediction for fair comparison across models. We also encourage analysis beyond standard validation metrics, e.g. visualization, examining model predictions, reporting metrics for different relation groupings and devis-

Relation	MRR	H@1	H@10	Count
ComplEx				
1-1-hom	.600	.319	.944	72
M-M-hom	.605	.417	.877	29,028
M-1	.683	.557	.884	2,509
1-M	.738	.640	.916	2,497
1-1	.889	.817	.995	420
M-M	.867	.819	.941	15,044
Co	.706	.662	.779	145
CA	.857	.822	.908	303
RotatE				
M-M-hom	.784	.718	.934	29,028
M-M	.973	.944	.992	15,044
M-1	.971	.945	.998	2,509
1-M	.975	.953	.998	2,497
1-1	.985	.959	1.	420
1-1-hom	.972	.976	1.	72
Co	.803	.738	.890	145
CA	.996	.993	1.	303

Table 4: Relation prediction results for RotatE and ComplEx by category of relation type (last two rows relate to causative relation types).

ing problem or domain-specific validation tasks. A further promising evaluation task is the triple prediction proposed in (Allen et al., 2019), which we leave for future work. A more ideal way to assess concept embeddings in biomedical NLP applications and patient-level modeling would be to design a suite of benchmark downstream tasks that incorporate the embeddings, but that warrants a rigorous paper of its own and is left for future work.

We believe this paper serves the biomedical NLP community as an introduction to KGEs and their evaluation and analyses, and also the KGE community by providing a potential standard benchmark dataset with real-world relevance.

7 Conclusion and Future Work

We present results from applying 5 leading KGE models to the SNOMED-CT knowledge graph and compare them to related work through visualizations and evaluation tasks, making a case for the importance of using models that leverage the multi-relation nature of knowledge graphs for learning biomedical knowledge representation. We discuss best practices for working with biomedical knowledge graphs and evaluating the embeddings learned from them, proposing link prediction, entity classification, and relation prediction as standard evaluation tasks. We encourage researchers to engage in further validation through visualizations, error analyses based on model predictions, examining stratified metrics, and devising domain-specific tasks that can assess the usefulness of the embeddings for a given application domain.

There are several immediate avenues of future work. While we focus on the SNOMED-CT dataset and the KGE models implemented in GraphVite, other biomedical terminologies such as the Gene Ontology (The Gene Ontology Consortium, 2018) and RxNorm (Nelson et al., 2011) could be explored and more recent KGE models, e.g. TuckER (Balažević et al., 2019) and MuRP (Balažević et al., 2019), applied. Additional sources of information could also potentially be incorporated, such as textual descriptions of entities and relations. In preliminary experiments, we initialized entity and relation embeddings with the embeddings of their textual descriptors extracted using Clinical Bert (Alsentzer et al., 2019), but it did not yield gains. This may suggest that the concept and language spaces are substantially different and strategies to jointly train with linguistic and knowledge graph information require further study. Other sources of information include entity types (e.g. UMLS semantic type) and paths, or multi-hop generalizations of the 1-hop relations (triples) typically used in KGE models (Guu et al., 2015). Notably, CoKE trains contextual knowledge graph embeddings using path-level information under an adapted version of the BERT training paradigm (Wang et al., 2019).

Lastly, the usefulness of biomedical knowledge graph embeddings should be investigated in downstream applications in biomedical NLP such as information extraction, concept normalization and entity linking, computational fact checking, question answering, summarization, and patient trajectory modeling. In particular, entity linkers act as a bottleneck between text and concept spaces, and leveraging KGEs could help develop sophisticated tools to parse existing biomedical and clinical text datasets for concept-level annotations and additional insights. Well performing entity linkers may then enable training knowledge-grounded large-scale language models like KnowBert (Peters et al., 2019). Overall, methods for learning and incorporating domain-specific knowledge representation are still at an early stage and further discussions are needed.

References

Khushbu Agarwal, Tome Eftimov, Raghavendra Addanki, Sutanay Choudhury, Suzanne Tamang, and Robert Rallo. 2019. Snomed2Vec: Random Walk and Poincare Embeddings of a Clinical Knowledge Base for Healthcare Analytics. *arXiv:1907.08650 [cs, stat].* ArXiv: 1907.08650.

Carl Allen, Ivana Balazevic, and Timothy M. Hospedales. 2019. On Understanding Knowledge Graph Representation. *arXiv:1909.11611 [cs, stat].*

Emily Alsentzer, John Murphy, William Boag, Wei-Hung Weng, Di Jin, Tristan Naumann, and Matthew McDermott. 2019. Publicly available clinical BERT embeddings. In *Proceedings of the 2nd Clinical Natural Language Processing Workshop*, pages 72–78, Minneapolis, Minnesota, USA. Association for Computational Linguistics.

Ivana Balažević, Carl Allen, and Timothy M Hospedales. 2019. Tucker: Tensor factorization for knowledge graph completion. In *Empirical Methods in Natural Language Processing*.

Ivana Balažević, Carl Allen, and Timothy Hospedales. 2019. Multi-relational poincar\'e graph embeddings. In *Advances in Neural Information Processing Systems*.

Andrew L. Beam, Benjamin Kompa, Allen Schmaltz, Inbar Fried, Griffin Weber, Nathan P. Palmer, Xu Shi, Tianxi Cai, and Isaac S. Kohane. 2019. Clinical Concept Embeddings Learned from Massive Sources of Multimodal Medical Data. *arXiv:1804.01486 [cs, stat].* ArXiv: 1804.01486.

Olivier Bodenreider. 2004. The Unified Medical Language System (UMLS): integrating biomedical terminology. *Nucleic Acids Research*, 32(90001):267D–270.

Antoine Bordes, Nicolas Usunier, Alberto Garcia-Duran, Jason Weston, and Oksana Yakhnenko. 2013. Translating embeddings for modeling multi-relational data. In C. J. C. Burges, L. Bottou, M. Welling, Z. Ghahramani, and K. Q. Weinberger, editors, *Advances in Neural Information Processing Systems 26*, pages 2787–2795. Curran Associates, Inc.

Edward Choi, Mohammad Taha Bahadori, Elizabeth Searles, Catherine Coffey, Michael Thompson, James Bost, Javier Tejedor-Sojo, and Jimeng Sun. 2016. Multi-layer Representation Learning for Medical Concepts. In *Proceedings of the 22nd ACM SIGKDD International Conference on Knowledge Discovery and Data Mining - KDD '16*, pages 1495–1504, San Francisco, California, USA. ACM Press.

Tim Dettmers, Minervini Pasquale, Stenetorp Pontus, and Sebastian Riedel. 2018. Convolutional 2d knowledge graph embeddings. In *Proceedings of the 32th AAAI Conference on Artificial Intelligence*, pages 1811–1818.

Aditya Grover and Jure Leskovec. 2016. node2vec: Scalable feature learning for networks. In *Proceedings of the 22nd ACM SIGKDD International Conference on Knowledge Discovery and Data Mining*.

Kelvin Guu, John Miller, and Percy Liang. 2015. Traversing knowledge graphs in vector space. In *Proceedings of the 2015 Conference on Empirical Methods in Natural Language Processing*, pages 318–327, Lisbon, Portugal. Association for Computational Linguistics.

Frank L. Hitchcock. 1927. The expression of a tensor or a polyadic as a sum of products. *Journal of Mathematics and Physics*, 6(1-4):164–189.

Seyed Kazemi and David Poole. 2018. Simple embedding for link prediction in knowledge graphs. In *Advances in Neural Information Processing Systems 32*.

Sotiris Kotitsas, Dimitris Pappas, Ion Androutsopoulos, Ryan McDonald, and Marianna Apidianaki. 2019. Embedding Biomedical Ontologies by Jointly Encoding Network Structure and Textual Node Descriptors. In *Proceedings of the 18th BioNLP Workshop and Shared Task*, pages 298–308, Florence, Italy. Association for Computational Linguistics.

Tomas Mikolov, Ilya Sutskever, Kai Chen, Greg S Corrado, and Jeff Dean. 2013. Distributed representations of words and phrases and their compositionality. In C. J. C. Burges, L. Bottou, M. Welling, Z. Ghahramani, and K. Q. Weinberger, editors, *Advances in Neural Information Processing Systems 26*, pages 3111–3119. Curran Associates, Inc.

Stuart Nelson, Kelly Zeng, John Kilbourne, Tammy Powell, and Robin Moore. 2011. Normalized names for clinical drugs: RxNorm at 6 years. *JAMIA*, 18(4):441–448.

Maximilian Nickel, Volker Tresp, and Hans-Peter Kriegal. 2011. A three-way model for collective learning on multi-relational data. In *Proceedings of the 28th International Conference on Machine Learning*, pages 809–816.

Matthew E. Peters, Mark Neumann, Robert L Logan, Roy Schwartz, Vidur Joshi, Sameer Singh, and Noah A. Smith. 2019. Knowledge enhanced contextual word representations. In *EMNLP*.

Zhiqing Sun, Zhi-Hong Deng, Jian-Yun Nie, and Jian Tang. 2019. Rotate: Knowledge graph embedding by relational rotation in complex space. In *International Conference on Learning Representations*.

Jian Tang, Jingzhou Liu, Ming Zhang, and Qiaozhu Mei. 2016. Visualizing large-scale and high-dimensional data. In *Proceedings of the 25th International Conference on World Wide Web*, pages 287–297. International World Wide Web Conferences Steering Committee.

The Gene Ontology Consortium. 2018. The Gene Ontology Resource: 20 years and still GOing strong. *Nucleic Acids Research*, 47(D1):D330–D338.

Théo Trouillon, Johannes Welbl, Sebastian Riedel, Éric Gaussier, and Guillaume Bouchard. 2016. Complex embeddings for simple link prediction. In *International Conference on Machine Learning (ICML)*, volume 48, pages 2071–2080.

Quan Wang, Pingping Huang, Haifeng Wang, Songtai Dai, Wenbin Jiang, Jing Liu, Yajuan Lyu, and Hua Wu. 2019. Coke: Contextualized knowledge graph embedding. *arXiv:1911.02168*.

Bishan Yang, Scott Wen-tau Yih, Xiaodong He, Jianfeng Gao, and Li Deng. 2015. Embedding Entities and Relations for Learning and Inference in Knowledge Bases. In *Proceedings of the International Conference on Learning Representations (ICLR) 2015*.

Zhaocheng Zhu, Shizhen Xu, Meng Qu, and Jian Tang. 2019. Graphvite: A high-performance cpu-gpu hybrid system for node embedding. In *The World Wide Web Conference*, pages 2494–2504. ACM.

Extensive Error Analysis and a Learning-Based Evaluation of Medical Entity Recognition Systems to Approximate User Experience

Isar Nejadgholi, Kathleen C. Fraser and Berry De Bruijn
National Research Council Canada
{isar.nejadgholi, kathleen.fraser,berry.debruijn}@nrc-cnrc.gc.ca

Abstract

When comparing entities extracted by a medical entity recognition system with gold standard annotations over a test set, two types of mismatches might occur, label mismatch or span mismatch. Here we focus on span mismatch and show that its severity can vary from a serious error to a fully acceptable entity extraction due to the subjectivity of span annotations. For a domain-specific BERT-based NER system, we showed that 25% of the errors have the same labels and overlapping span with gold standard entities. We collected expert judgement which shows more than 90% of these mismatches are accepted or partially accepted by the user. Using the training set of the NER system, we built a fast and lightweight entity classifier to approximate the user experience of such mismatches through accepting or rejecting them. The decisions made by this classifier are used to calculate a learning-based F-score which is shown to be a better approximation of a forgiving user's experience than the relaxed F-score. We demonstrated the results of applying the proposed evaluation metric for a variety of deep learning medical entity recognition models trained with two datasets.

1 Introduction

Named entity recognition (NER) in medical texts involves the automated recognition and classification of relevant medical/clinical entities, and has numerous applications including information extraction from clinical narratives (Meystre et al., 2008), identifying potential drug interactions and adverse affects (Harpaz et al., 2014; Liu et al., 2016), and de-identification of personal health data (Dernoncourt et al., 2017).

In recent years, medical NER systems have improved over previous baseline performance by incorporating developments such as deep learning models (Yadav and Bethard, 2018), contextual word embeddings (Zhu et al., 2018; Si et al., 2019), and domain-specific word embeddings (Alsentzer et al., 2019; Lee et al., 2019; Peng et al., 2019). Typically, research groups report their results using common evaluation metrics (most often precision, recall, and F-score) on standardized data sets. While this facilitates exact comparison, it is difficult to know whether modest gains in F-score are associated with significant qualitative differences in the system performance, and how the benefits and drawbacks of different embedding types are reflected in the output of the NER system.

This work aims to investigate the types of errors and their proportion in the output of modern deep learning models for medical NER. We suggest that an evaluation metric should be a close reflection of what users experience when using the model. We investigate different types of errors that are penalized by exact F-score and identify a specific error type where there is high degrees of disagreement between the human user experience and what exact F-score measures: namely, errors where the extracted entity is correctly labeled, but the span only overlaps with the annotated entity rather than matching perfectly. We obtain expert human judgement for 5296 such errors, ranking the severity of the error in terms of end user experience. We then compare the commonly used F-score metrics with human perception, and investigate if there is a way to automatically analyze such errors as part of the system evaluation. The code that calculates the number of different types of errors given the predictions of an NER model and the corresponding annotations is available upon request and will be released at `https://github.com/nrc-cnrc/NRC-MedNER-Eval` after publication. We will also release the collected expert judgements so that other researchers can use it as a benchmark for further investigation about this type of errors.

177

Proceedings of the BioNLP 2020 workshop, pages 177–186
Online, July 9, 2020 ©2020 Association for Computational Linguistics

2 What do NER Evaluation Metrics Measure?

An output entity from an NER system can be incorrect for two reasons: either the span is wrong, or the label is wrong (or both). Although entity-level exact F-score (also called strict F-score) is established as the most common metric for comparing NER models, exact F-score is the least forgiving metric in that it only credits a prediction when both the span and the label exactly match the annotation.

Other evaluation metrics have been proposed. The Message Understanding Conference (MUC) used an evaluation which took into account different types of errors made by the system (Chinchor and Sundheim, 1993). Building on that work, the SemEval 2013 Task 9.1 (recognizing and labelling pharmacological substances in biomedical text) employed four different evaluations: *strict match*, in which label and span match the gold standard exactly, *exact boundary match*, in which the span boundaries match exactly regardless of label, *partial boundary match*, in which the span boundaries partially match regardless of label, and *type match*, in which the label is correct and the span overlaps with the gold standard (Segura Bedmar et al., 2013). The latter metric, also commonly known as *inexact match*, has been used to compute inexact or relaxed F-score in the i2b2 2010 clinical NER challenge (Uzuner et al., 2011). Relaxed F-score and exact F-score are the most frequently used evaluation metrics for measuring the performance of medical NER systems (Yadav and Bethard, 2018). Other biomedical NER evaluations have accepted a span as a match as long as either the right or left boundary is correct (Tsai et al., 2006). In BioNLP shared task 2013, the accuracy of the boundaries is relaxed or measured based on similarity of entities (Bossy et al., 2013). Another strategy is to annotate all possible spans for an entity and accept any matches as correct (Yeh et al., 2005), although this detailed level of annotation is rare.

Here, we focus on the differences between what F-score measures and the user experience. In the case of a correct label with a span mismatch, it is not always obvious that the user is experiencing an error, due to the subjectivity of span annotations (Tsai et al., 2006; Kipper-Schuler et al., 2008). Existing evaluation metrics treat all such span mismatches equally, either penalizing them all (exact F-score), rewarding them all (relaxed F-score), or based on oversimplified rules that do not general-

ize across applications and data sets. We use both human judgement and a learning-based approach to evaluate span mismatch errors and the resulting gap between what F-score measures and what a human user experiences. We only consider the information extraction task and not any specific downstream task.

3 Types of Errors in NER systems

While the SemEval 2013 Task 9.1 categorized different types of matches for the purpose of evaluation, we further categorize mismatches for the sake of error analysis. We consider five types of mismatches between annotation and prediction of the NER system. Reporting and comparing the number of these mismatches alongside an averaged score such as F-score can shed light on the differences of NER systems.

- **Mismatch Type-1, Complete False Positive**: An entity is predicted by the NER model, but is not annotated in the hand-labelled text.

- **Mismatch Type-2, Complete False Negative**: A hand labelled entity is not predicted by the model.

- **Mismatch Type-3, Wrong label, Right span**: A hand-labelled entity and a predicted one have the same spans but different tags.

- **Mismatch Type-4, Wrong label, Overlapping span**: A hand-labelled entity and a predicted one have overlapping spans but different tags.

- **Mismatch Type-5, Right label, Overlapping span**: A hand-labelled entity and a predicted one have overlapping spans and the same tags.

We focus on Type-5 errors and show that treating these mismatches is not a trivial task. Previous works have shown that some Type-5 mismatches are completely wrong predictions while others are fully acceptable predictions resulting from the subjectivity and inconsistency of span annotations (Tsai et al., 2006).

Figure 1 shows several examples of error Type-5. In the first example, *an adenosine - thallium stress test* is annotated as a *test*, while the NER system extracts *thallium stress test* as a *test*. Here, what NER extracted is partially correct but misses an

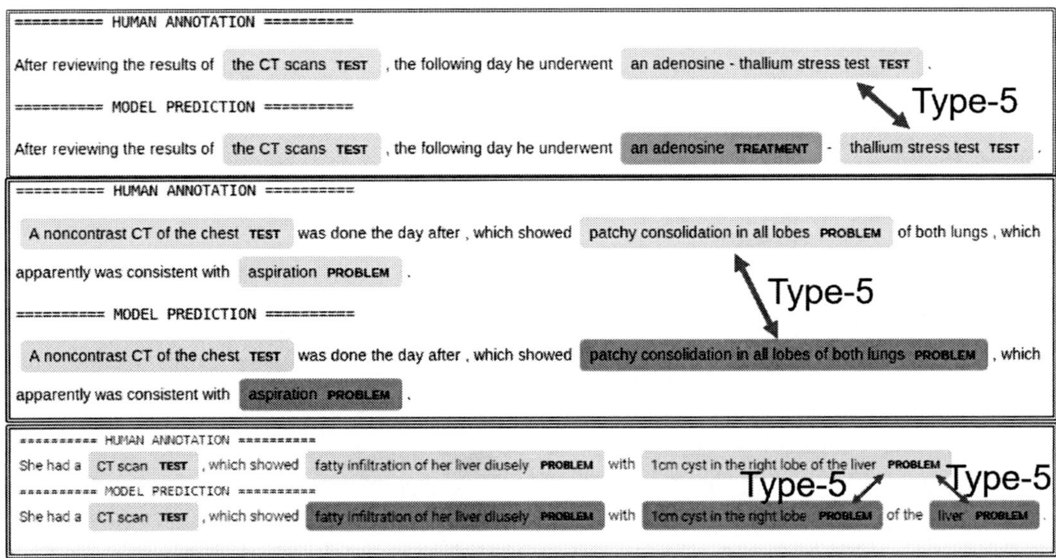

Figure 1: Examples of Type-5 error. We used the visualisation tool developed in (Zhu et al., 2018)

important part of the entity. Whether the extracted entity is acceptable may depend on the downstream task. In the next sentence, *patchy consolidation in all lobes* is annotated as a *problem*, but the NER system extracted *patchy consolidation in all lobes of both longs* as the *problem*. Here, the system's prediction is more complete than the annotated entity, and so it appears to be a fully acceptable prediction. In the last example, according to human annotation, *1cm cyst in the right lobe of the liver* is a *problem*, but the NER system extracts two entities from the same phrase, 1) *1cm cyst in the right lobe* as a *problem* and 2) *liver* as another *problem*. While the first extracted entity is correct and may be acceptable the second one is completely wrong.

4 Datasets and Models

We consider two medical text datasets, one clinical and the other biomedical. We analyse the errors of three models for each dataset to cover a variety of deep learning models.

i2b2 dataset: The i2b2 dataset of annotated clinical notes was introduced by (Uzuner et al., 2011) in a shared task on entity recognition and relation extraction. The texts, consisting of de-identified discharge summaries, have been annotated for three entity types: problems, tests, and treatments. There are two versions of this dataset, as the version that was released to the wider NLP community contains fewer texts than in the original shared task. We use the second version, which has become an important benchmark in the literature on clinical NER (Bhatia et al., 2019; Zhu et al., 2018). There are 170 documents (16520 entities) in the i2b2 train set and 256 documents (31161 entities) in its test set.

The i2b2 dataset was annotated by community annotators with carefully crafted guidelines. The ground truth generated by the community obtained F-measures above 0.90 against the ground truth of the experts (Uzuner et al., 2011).

MedMentions dataset: The MedMentions dataset was released in 2019 and contains 4,392 abstracts from biomedical articles on PubMed (Mohan and Li, 2019). The abstracts are annotated for UMLS concepts and semantic types. The fully annotated dataset contains 127 semantic types and these classes are highly-imbalanced. The creators of the dataset also provide a version which has been annotated with only a subset of the most relevant concepts, called 'st21pv' (*21 semantic types from preferred vocabularies*); we consider this version in the current work. While fewer papers have been published on MedMentions to date, it represents an interesting challenge to NLP systems due to its imbalanced and high number of classes, and some observed inconsistencies in the annotations (Fraser et al., 2019). There are 3513 documents (162,908 entities) in the st21pv train set and 879 documents (40,101 entities) in the test set.

MedMentions was annotated by a team of professional annotators with rich experience in biomedical content curation. The precision of the annotation in MedMention is estimated as 97.3% (Mohan and Li, 2019).

Model Structures: We explore a variety of NER deep learning models. For all the models we follow the commonly used deep learning structure consisting of a pretrained embedding model and supervised prediction layers. For embedding, we explore three different models: a non-contextualized embedding model (Glove), general domain contextualized embedding model (BERT pretrained on general domain text) and a domain-specific contextualized embedding model (BERT pretrained on domain-specific text corpora). For the i2b2 dataset, we consider *Glove+bi-LSTM+CRF* (Pennington et al., 2014), *BERT+linear* (Devlin et al., 2018) and *ClinicalBERT+linear* (Alsentzer et al., 2019) models. For the st21pv MedMentions dataset, we consider *Glove+bi-LSTM+CRF*, *BERT+linear* and *BioBERT+linear* models (Lee et al., 2019). Clinical BERT is pretrained on clinical notes (similar to i2b2) and BioBERT is pretrained on biomedical articles from PubMed (similar to st21pv).

5 Analysis of Error Types Across Models and Datasets

Further investigation of Type-5 errors is only worthwhile if a significant proportion of the errors belong to this group. We looked at the distribution of error types across datasets and NER models, described in Section 4, and visualized the results in Figure 2. By calculating the distribution of error types, we observed that for all assessed models at least 20% of the errors are recognized as Type-5 mismatches.

Moreover, for both datasets, we observed that better NER models generate more Type-5 errors. Models based on general BERT outperform glove-based models in terms of both exact and relaxed f-score and they also generate relatively more Type-5 errors. Same pattern is observed when comparing domain-specific BERT models with general BERT models. This observation may be explained with the fact that contextualized embeddings combine the meaning of words through attention mechanism and the span information might be more vague in the resulting representation. Figure 3 shows exact F-score, relaxed F-score and the proportion of Type-5 mismatches to the total number of errors, for all the models and datasets. This analysis implies that proper handling of Type-5 errors becomes more

important for comparison of modern strong NER systems.

6 Expert Judgement on Type-5 Errors

We considered an information extraction task and asked a medical doctor to assess the Type-5 errors made by the BioBERT NER model on the st21pv dataset and either confirm or reject the extracted entity with granular scores. Our goal is to: 1) investigate the proportion of Type-5 extracted entities that are acceptable, 2) set a benchmark of human experience from Type-5 errors.

Human Judgement Scheme: The following scoring scheme is used by the expert for scoring the acceptability of Type-5 mismatches for the BioBERT-based model trained with the st21vp dataset. The Type-5 mismatches are identified and the expert is given the original sentence in the test set, the annotated (gold-standard) entity, and the entity predicted by the NER model for all 5296 Type-5 mismatches.

SCORE = 1: The predicted entity is wrong and gets rejected. For example, while *gene transfer* is annotated as a *research_activity* in the test set, the NER extracted *gene* as *research_activity*.

SCORE = 2: The predicted entity is correct but an important piece of information is missing when seen in the full sentence. The prediction is partially accepted by the expert. For example, *injury of lung* is labeled as *injury_or_poisoning* in the test set, but the NER extracts only the word *injury* as *injury_or_poisoning*.

SCORE = 3: The predicted entity is correct but could be more complete. The prediction is accepted by the expert. The entity *normal HaCaT lines* is annotated as *anatomical_structure* in the test set but the NER extracts only *HaCaT lines* with the same label.

SCORE = 4: The predicted entity is equally correct and is accepted by the expert. As an example the annotated entity in test set is *196b-5p*, as an *anatomical_structure* but the NER extracts *-196b-5p*, as an entity with the same tag.

SCORE = 5: The predicted entity is more complete than the annotated entity and is accepted by the expert. The annotated entity in the test set is *drugs* with the tag *chemical* and the NER extracts *Alzheimer's drugs* with the same tag.

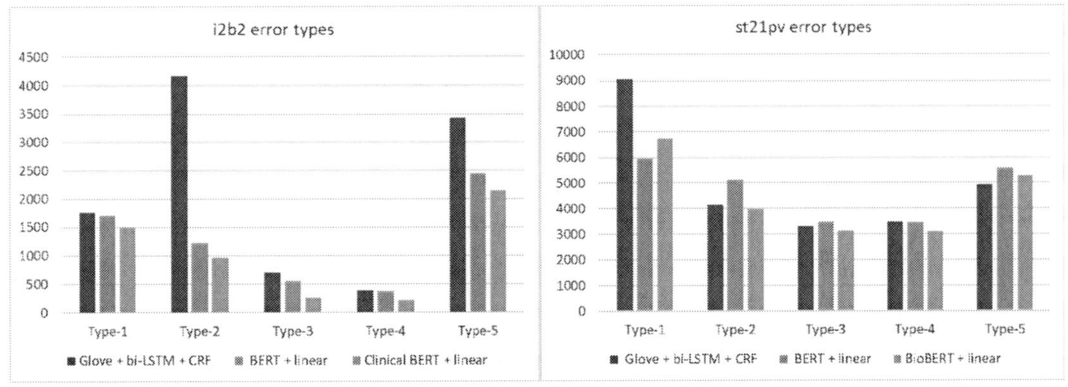

Figure 2: Types of errors made on the i2b2 and MedMentions-st21pv datasets

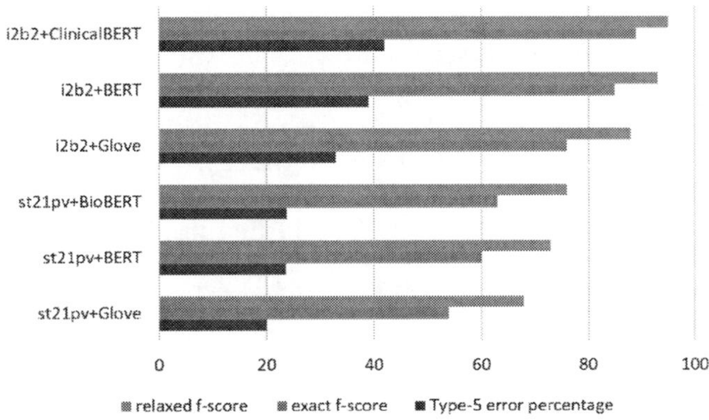

Figure 3: The change of relative proportion of Type-5 errors across dataset and models as the f-scores change

Results of Human Judgement Analysis: The results of the expert judgement are summarized in Figure 4.

- Almost 40% of the Type-5 errors are scored as 5. This means that in 40% of the cases the prediction of the NER is more complete than the entity labeled in its test set.

- 70% of the extracted entities scored 3 or above and are fully accepted by the expert.

- 21% of the Type-5 mismatches are scored as 2. These are accepted as a correct entity extraction when seen out of the context, but in the context of a given sentence they lack an important piece of information. Depending on the downstream tasks, they might be an acceptable prediction or not.

- Only 9% of the extracted entities are totally rejected by the expert.

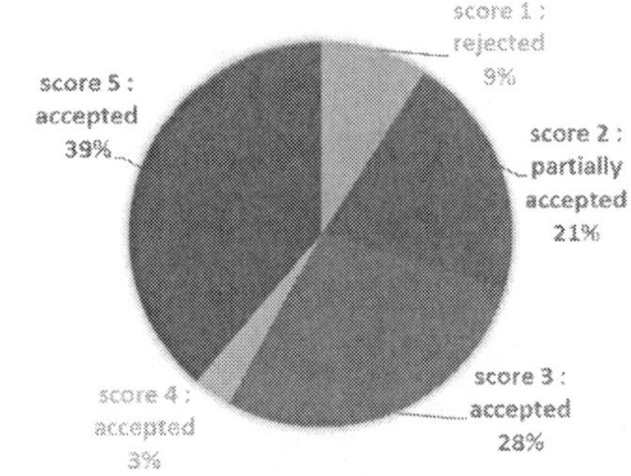

Figure 4: Results of expert judgement for Type-5 mismatches of the BioBERT-based NER model trained with MedMentions-st21pv dataset.

7 Entity Classifier for Automatic Refining of Type-5 Mismatches

We propose that an entity classifier can be trained to predict the tag of entities extracted by the NER model and the predicted tag can be used to distin-

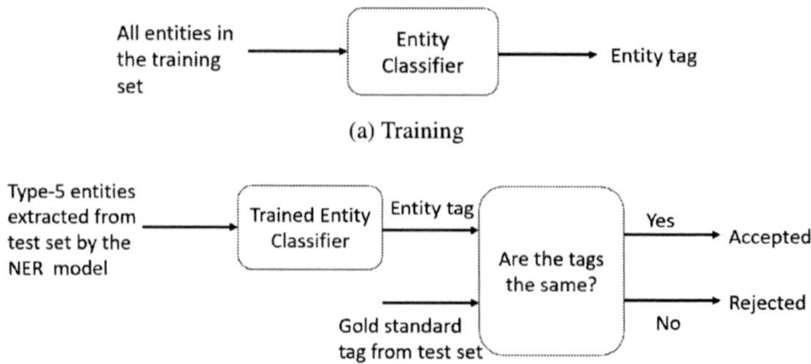

(a) Training

(b) Using for evaluation of the NER model

Figure 5: Workflow of the proposed entity classifier

guish between acceptable and unacceptable Type-5 errors. Figure 5 shows the workflow of the proposed method. Using the training dataset of the NER model, we train an entity classifier with gold standard entities as inputs and their assigned tags as outputs. For this classifier, the span is given and the tag is the only information that has to be learned. Although the full context of the sentence helps the NER model to learn a better representation of the entity, many entities can be classified without seeing the full sentence and this is what the entity classifier learns.

For Type-5 entities, the human annotators and the NER already agree on the tag and it is only the span that is in disagreement. So, the intuition here is that the entity classifier can confirm or reject the tag predicted by NER, given the identified span. This classifier is meant to play a third party role that has seen the variety of span annotations in the training dataset and performs the task that the human expert did in Section 6. This classifier is trained once for each dataset and is not dependent on the type of the NER model.

7.1 Building the Training Data for the Entity Classifier

In order to build a training dataset for the entity classifier, we extracted pairs of *(entity, tag)* from the IOB annotated dataset. The entity classifier should also be able to identify cases where the extracted entity does not belong to any of the pre-defined tags. For this reason we add the label *other* to the list of tags of the classifier. To find examples of the *other* class, we used the spaCy library (Honnibal and Montani, 2017) to extract all the noun chunks that are out of the boundaries of tagged entities and randomly chose a number of them. We limited the

size of the *other* class to the average size of classes related to the existing tags.

7.2 Classifier Structure

For the classifier structure, we chose to use a DistilBERT model (Sanh et al., 2019) with a linear prediction layer. DistilBERT is a distilled version of BERT that is an optimum choice when fast inference is required. Since this classifier is going to be used for evaluation and error analysis and is not the main focus of building an NER model, the lightweight and fast inference is an important practical criterion. We train the classifier only one epoch for both datasets. When trained on the train set and tested on the test set, we achieved 89% F-score for i2b2 and 77% F-score for st21pv dataset.

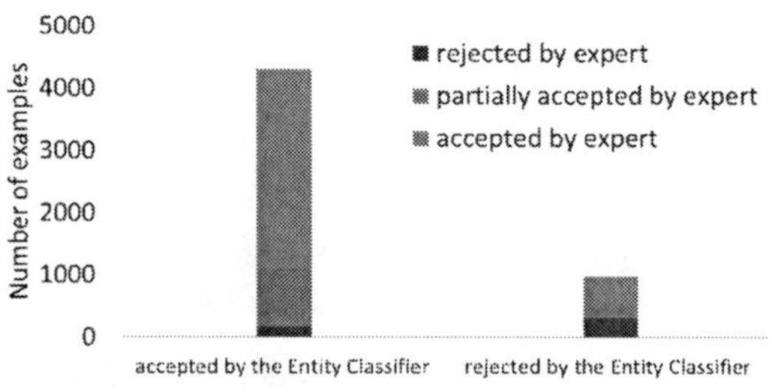

Figure 6: Comparison of decisions made by the human expert and the entity classifier for the Type-5 mismatches of BioBERT NER and st21pv dataset.

7.3 Using the Entity Classifier for Refining Type-5 Mismatches

By building the entity classifier, our goal is to refine the Type-5 errors and separate the acceptable predictions of the NER from the unacceptable. For instance, in the last example shown in Figure 1

Annotated in test set	Tag in test set	Extracted by NER	Tag from Entity classifier	Decision
Central pathology	biomedical discipline	Central	Spatial concept	Reject
Therapies	healthcare activity	Agonist Therapies	healthcare activity	Accept

Table 1: Examples of accepted and rejected Type-5 mismatches using the entity classifier (st21pv dataset).

there are two Type-5 errors. We feed the two extracted entities *'1 cm cyst in the right lobe'* and *'liver'* to the entity classifier trained for i2b2 dataset. The classifier predicts the tag *'problem'* for the extracted entity *'1 cm cyst in the right lobe'* and *'Other'* for the extracted entity *'liver'*. Using these predictions we decide that the first entity is acceptable, since although the span of the extracted entity does not match the annotation, the classifier still recognizes it as a member of the correct class. We reject the extracted entity *'liver'* as a *'problem'* since the classifier recognizes it as not being a *'problem'*. Table 1 shows examples of rejected and accepted Type-5 mismatches from the st21pv dataset.

7.4 Comparing the Classifier and the Expert

Figure 6 shows the comparison between the expert's judgment and the classifier's judgement about Type-5 mismatches for the BioBERT NER model on st21 pv dataset.

Our analysis shows that 96% of the entities accepted by the classifier are also accepted or partially accepted by the expert, and 86% of the entities accepted or partially accepted by the expert are accepted by the classifier as well. The classifier and the expert disagree about 17% of the entities. In 24% of the disagreements, the probabilities assigned to the tags generated by the entity classifier are low (less than 0.5) and our manual investigation shows that the classifier's prediction is mostly wrong in these cases. These mistakes mostly occurs in 5 classes namely *anatomical_structure, biologic_function, chemical, finding* and *health_care_activity*.

We also observed that this classifier is not able to distinguish between accepted and partially accepted entities extracted by the NER model, which is one of the limitations of this method. The probabilities assigned to the tags is 0.89 ± 0.17 for accepted entities, 0.88 ± 0.17 for partially accepted entities, and 0.78 ± 0.23 for rejected entities.

8 Refining Type-5 Mismatches Across Datasets and Models

Figure 7 shows how the entity classifier refines Type-5 errors across models and datasets. Consistently, a significant proportion of Type-5 errors are accepted by the entity classifier. For example, for the Glove-based model trained on i2b2 dataset, the entity classifier accepts 90% of Type-5 errors which is 26.6% of the total number of the errors penalized by the exact f-score. The proportion of accepted Type-5 mismatches to the total number of errors is 31.11% for *i2b2+BERT*, 33.23% for *i2b2+ClinicalBERT*, 17.95% for *st21pv+Glove*, 19.55% for *st21pv+BERT* and 19.36% for *st21pv+BioBERT*. To sum up, about 20% to 30% of the mismatches penalized by exact f-score are accepted by the entity classifier.

9 Learning-Based F-score

The trained entity classifier can be leveraged for F-score calculation. Here, instead of penalizing all the type-5 mismatches as in exact F-score or rewarding all of them in relaxed F-score, we penalize the type-5 mismatches that are rejected by the classifier and reward the rest of them. In other words, this F-score penalizes errors of Type-1, Type-2, Type-3, Type-4 and the Rejected Type-5 mismatches. Accepted Type-5 mismatches and exact matches are rewarded.

9.1 Evaluation of the Learning-Based F-score

We use the expert judgement collected in Section 6 to quantify human experience for the BioBERT-based NER model on st21pv dataset and then use that as a benchmark to evaluate the proposed learning-based F-score. We consider two scenarios based on the scores described in Section 6, 1) a strict user that only accepts scores equal to or above 3, 2) a forgiving user that accepts scores equal to or above 2. We calculated the F-score for each scenario and investigated the error of exact F-score, relaxed F-score and the proposed F-score to each of these scenarios. Table 2 shows that in applications where strict evaluation of the NER

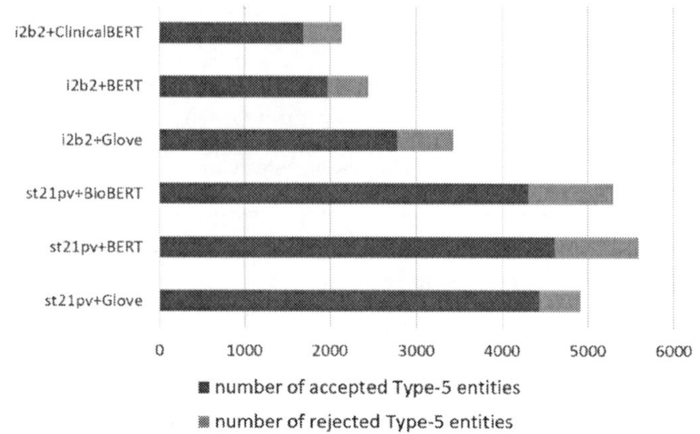

Figure 7: Number of accepted/rejected Type-5 mismatches by the entity classifier

F-score	err. wtr strict user	err. wtr forgiving user
Exact	**-4.3%**	-5.5%
Proposed	5.9%	**4.7%**
Relaxed	8.2%	7.1%

Table 2: Comparing F-scores with human experience.

is needed, exact F-score is better than both proposed and relaxed f-score and results in the least error with respect to the human experience. However, in cases that partially accepted entities can be considered as useful predictions, the proposed method results in the least disagreement with human experience. A better classifier would be able to model human preferences better, and thus make the learning-based F-score a stronger alternative to exact or relaxed F-scores. Another important finding from Table 2 that when choosing between exact and relaxed F-score, exact is the better metric to choose.

Figure 8 shows how the proposed F-score can be compared with exact and relaxed F-score. We only have annotations for the BioBERT+stpv dataset and for the rest of the models we cannot evaluate the F-score with respect to human experience. As expected, from this figure we observe that for all the models, the proposed F-score is a forgiving one and is much closer to the relaxed F-score than the exact F-score.

10 Discussion

We highlighted the fact that when we evaluate NER systems by comparing extracted and annotated entities across a test set, for a significant part of the errors that are penalized by the exact F-score, the la-bel is recognized correctly and the span has overlap with the annotated entity. We referred to this type of error as Type-5 mismatch and for six NER models (3 model structures and 2 datasets) showed that at least 20% of the errors belong to this category. The previous literature has raised the issue that in the case of medical NER, many such predictions are valid and useful entity extractions and penalizing them is a flaw of evaluation metrics. However, distinguishing between acceptable and unacceptable predictions when the label is correct and the span overlaps is not trivial.

We argue that the best evaluation metric is the one that reflects the human experience of the system best. We collected human judgement about a all Type-5 errors made by a NER model based on BioBERT embeddings, trained with st21pv dataset and showed that almost 70% of such errors are completely acceptable and only 10% of them are rejected by the user. The rest of the predictions are acceptable entities for the associated tags but lack important information when seen in the context.

Setting human experience as a benchmark, we suggested that expert judgement can be approximated by a decision made by an entity classifier. The entity classifier can be trained using the training set of an NER. While the NER model looks at the context and identifies the type of the entity and a partially correct span, this classifier looks at the extracted entities out of context and decides whether with the partially correct span, the extracted entity can still belong to the predicted class or not. The entity classifier trained on st21pv dataset accepts more than 80% of Type-5 errors made by BioBERT-based NER model trained with the same

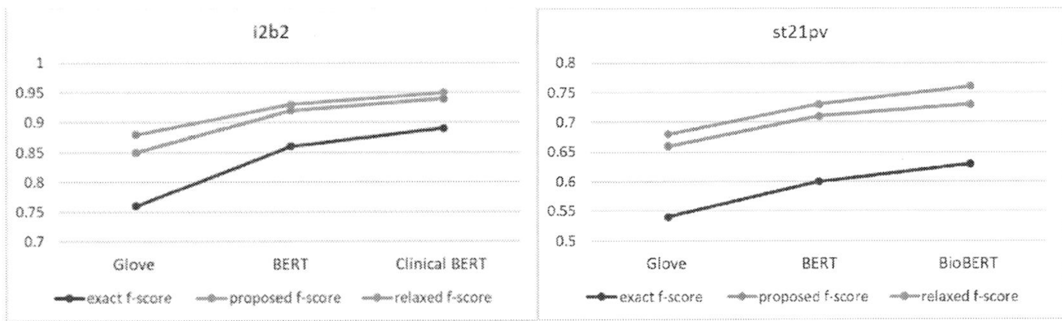

Figure 8: Comparison of f-scores.

dataset, 96% of which is also accepted by the expert user. The proposed entity classifier is trained for each NER training set once and can be used to evaluate any NER model trained on that dataset, regardless of the structure of the NER model being evaluated. We used a computationally inexpensive model structure and encourage researchers to use this model in order to automatically evaluate Type-5 mismatches. Reporting the distribution of errors across all error types and also accepted and rejected Type-5 errors, will allow us to compare our models in a variety of dimensions and sheds light on how these models behave differently for detecting labels and spans.

Accepting some Type-5 errors as useful predictions can be translated to F-score calculation by not penalizing the accepted entity extractions. We did this calculation separately for the cases that were accepted by human expert or the classifier, and showed that the F-score resulting from the classifier is closer to the judgement of a forgiving user than both the exact and the relaxed F-score. In cases where a strict evaluation of the system is desired, exact F-score is a better approximation of human experience, due to the fact that the entity classifier is a forgiving one and accepts most of the cases that are partially accepted by the expert.

We only collected human judgement on the decisions made by NER model for one model and one dataset. Further investigation is needed to confirm or reject our observations and to investigate the limitations and potential capabilities of training an entity classifier alongside a NER model and using that for error analysis. Also, further research is needed to find a way of distinguishing between partially accepted and accepted entity extractions, which is a necessary tool for measuring the experi-

ence of a strict user. Using extra sources of training data other than the NER training dataset may be a way to improve the judgements of the entity classifier. We used this classifier for error analysis and refining of Type-5 errors. In future work, we can look at the possibility of using this classifier as a refining tool for all types of mismatches or a post-processing tool without the need for annotation to identify the types of mismatches.

11 Conclusion

Medical NER systems that are based on most recent deep learning structures generate a high amount of outputs that match with the hand-labelled entities in terms of tag but only overlap in the span. While the exact f-score penalizes all of these predictions and relaxed f-score credits all of them, a human user accepts a significant proportion of them as valid entities and rejects the rest.
A reformatted version of the NER training dataset can be used to train an entity classifier for evaluation of extracted entities with right label and overlapping span. We showed that there is a high degree of agreement between human expert and this entity classifier in accepting or rejecting span mismatches. This classifier is used to calculate a learning-based evaluation metric that outperforms relaxed F-score in approximating the experience of a forgiving user.

References

Emily Alsentzer, John Murphy, William Boag, Wei-Hung Weng, Di Jin, Tristan Naumann, and Matthew McDermott. 2019. Publicly available clinical BERT embeddings. In *Proceedings of the 2nd Clinical Natural Language Processing Workshop*, pages 72–

78, Minneapolis, Minnesota, USA. Association for Computational Linguistics.

Parminder Bhatia, Busra Celikkaya, and Mohammed Khalilia. 2019. Joint entity extraction and assertion detection for clinical text. In *Proceedings of the 57th Annual Meeting of the Association for Computational Linguistics*, pages 954–959.

Robert Bossy, Wiktoria Golik, Zorana Ratkovic, Philippe Bessières, and Claire Nédellec. 2013. BioNLP shared task 2013–an overview of the bacteria biotope task. In *Proceedings of the BioNLP Shared Task 2013 Workshop*, pages 161–169.

Nancy Chinchor and Beth Sundheim. 1993. MUC-5 evaluation metrics. In *Fifth Message Understanding Conference (MUC-5): Proceedings of a Conference Held in Baltimore, Maryland, August 25-27, 1993*.

Franck Dernoncourt, Ji Young Lee, Ozlem Uzuner, and Peter Szolovits. 2017. De-identification of patient notes with recurrent neural networks. *Journal of the American Medical Informatics Association*, 24(3):596–606.

Jacob Devlin, Ming-Wei Chang, Kenton Lee, and Kristina Toutanova. 2018. Bert: Pre-training of deep bidirectional transformers for language understanding. *arXiv preprint arXiv:1810.04805*.

Kathleen C Fraser, Isar Nejadgholi, Berry De Bruijn, Muqun Li, Astha LaPlante, and Khaldoun Zine El Abidine. 2019. Extracting UMLS concepts from medical text using general and domain-specific deep learning models. *EMNLP-IJCNLP 2019*, page 157.

Rave Harpaz, Alison Callahan, Suzanne Tamang, Yen Low, David Odgers, Sam Finlayson, Kenneth Jung, Paea LePendu, and Nigam H Shah. 2014. Text mining for adverse drug events: the promise, challenges, and state of the art. *Drug Safety*, 37(10):777–790.

Matthew Honnibal and Ines Montani. 2017. spaCy 2: Natural language understanding with Bloom embeddings, convolutional neural networks and incremental parsing. To appear.

Karin Kipper-Schuler, Vinod Kaggal, James Masanz, Philip Ogren, and Guergana Savova. 2008. System evaluation on a named entity corpus from clinical notes. In *Language resources and evaluation conference, LREC*, pages 3001–3007.

Jinhyuk Lee, Wonjin Yoon, Sungdong Kim, Donghyeon Kim, Sunkyu Kim, Chan Ho So, and Jaewoo Kang. 2019. BioBERT: a pre-trained biomedical language representation model for biomedical text mining. *Bioinformatics*.

Shengyu Liu, Buzhou Tang, Qingcai Chen, and Xiaolong Wang. 2016. Drug-drug interaction extraction via convolutional neural networks. *Computational and Mathematical Methods in Medicine*.

Stéphane M Meystre, Guergana K Savova, Karin C Kipper-Schuler, and John F Hurdle. 2008. Extracting information from textual documents in the electronic health record: a review of recent research. *Yearbook of Medical Informatics*, 17(01):128–144.

Sunil Mohan and Donghui Li. 2019. MedMentions: A large biomedical corpus annotated with UMLS concepts. *arXiv preprint arXiv:1902.09476*.

Yifan Peng, Shankai Yan, and Zhiyong Lu. 2019. Transfer learning in biomedical natural language processing: An evaluation of bert and elmo on ten benchmarking datasets. In *Proceedings of the 2019 Workshop on Biomedical Natural Language Processing (BioNLP 2019)*, pages 58–65.

Jeffrey Pennington, Richard Socher, and Christopher D Manning. 2014. Glove: Global vectors for word representation. In *Proceedings of the 2014 conference on empirical methods in natural language processing (EMNLP)*, pages 1532–1543.

Victor Sanh, Lysandre Debut, Julien Chaumond, and Thomas Wolf. 2019. Distilbert, a distilled version of bert: smaller, faster, cheaper and lighter. *arXiv preprint arXiv:1910.01108*.

Isabel Segura Bedmar, Paloma Martínez, and María Herrero Zazo. 2013. Semeval-2013 task 9: Extraction of drug-drug interactions from biomedical texts (ddiextraction 2013). volume 2, pages 341–350. Association for Computational Linguistics.

Yuqi Si, Jingqi Wang, Hua Xu, and Kirk Roberts. 2019. Enhancing clinical concept extraction with contextual embeddings. *Journal of the American Medical Informatics Association*, 26(11):1297–1304.

Richard Tzong-Han Tsai, Shih-Hung Wu, Wen-Chi Chou, Yu-Chun Lin, Ding He, Jieh Hsiang, Ting-Yi Sung, and Wen-Lian Hsu. 2006. Various criteria in the evaluation of biomedical named entity recognition. *BMC Bioinformatics*, 7(1):92.

Özlem Uzuner, Brett R South, Shuying Shen, and Scott L DuVall. 2011. 2010 i2b2/VA challenge on concepts, assertions, and relations in clinical text. *Journal of the American Medical Informatics Association*, 18(5):552–556.

Vikas Yadav and Steven Bethard. 2018. A survey on recent advances in named entity recognition from deep learning models. In *Proceedings of the 27th International Conference on Computational Linguistics*, pages 2145–2158.

Alexander Yeh, Alexander Morgan, Marc Colosimo, and Lynette Hirschman. 2005. BioCreAtIvE task 1a: gene mention finding evaluation. *BMC Bioinformatics*, 6(S1):S2.

Henghui Zhu, Ioannis Ch Paschalidis, and Amir Tahmasebi. 2018. Clinical concept extraction with contextual word embedding. *arXiv preprint arXiv:1810.10566*.

A Data-driven Approach for Noise Reduction in Distantly Supervised Biomedical Relation Extraction

Saadullah Amin* Katherine Ann Dunfield* Anna Vechkaeva Günter Neumann

German Research Center for Artificial Intelligence (DFKI)
Multilinguality and Language Technology Lab
{saadullah.amin, katherine.dunfield, anna.vechkaeva, guenter.neumann}@dfki.de

Abstract

Fact triples are a common form of structured knowledge used within the biomedical domain. As the amount of unstructured scientific texts continues to grow, manual annotation of these texts for the task of relation extraction becomes increasingly expensive. Distant supervision offers a viable approach to combat this by quickly producing large amounts of labeled, but considerably noisy, data. We aim to reduce such noise by extending an entity-enriched relation classification BERT model to the problem of multiple instance learning, and defining a simple data encoding scheme that *significantly* reduces noise, reaching state-of-the-art performance for distantly-supervised biomedical relation extraction. Our approach further encodes knowledge about the direction of relation triples, allowing for increased focus on relation learning by reducing noise and alleviating the need for joint learning with knowledge graph completion.

1 Introduction

Relation extraction (RE) remains an important natural language processing task for understanding the interaction between entities that appear in texts. In supervised settings (GuoDong et al., 2005; Zeng et al., 2014; Wang et al., 2016), obtaining fine-grained relations for the biomedical domain is challenging due to not only the annotation costs, but the added requirement of domain expertise. Distant supervision (DS), however, provides a meaningful way to obtain large-scale data for RE (Mintz et al., 2009; Hoffmann et al., 2011), but this form of data collection also tends to result in an increased amount of noise, as the target relation may not always be expressed (Takamatsu et al., 2012; Ritter et al., 2013). Exemplified in Figure 1, the last two

---

* Equal contribution
On behalf of the PRECISE4Q consortium

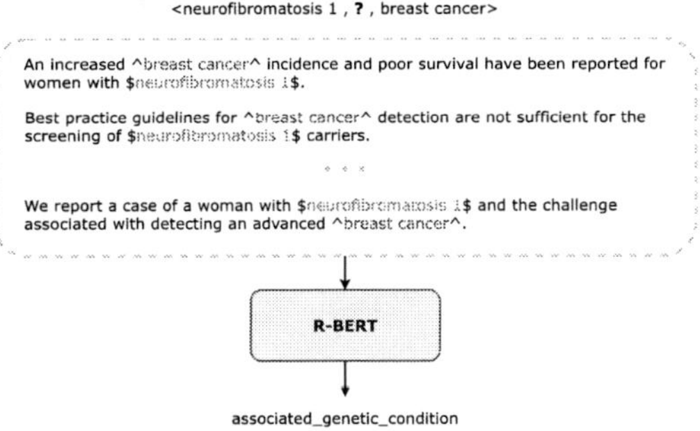

Figure 1: Example of a distantly supervised bag of sentences for a knowledge base tuple (*neurofibromatosis 1, breast cancer*) with special order sensitive entity markers to capture the position and the latent relation direction with BERT for predicting the missing relation.

sentences can be seen as potentially *noisy* evidence, as they do not explicitly express the given relation.

Since individual instance labels may be unknown (Wang et al., 2018), we instead build on the recent findings of Wu and He (2019) and Soares et al. (2019) in using positional markings and latent relation direction (Figure 1), as a signal to mitigate noise in *bag-level* multiple instance learning (MIL) for distantly supervised biomedical RE. Our approach greatly simplifies previous work by Dai et al. (2019) with following contributions:

- We extend *sentence-level* relation enriched BERT (Wu and He, 2019) to *bag-level* MIL.

- We demonstrate that the simple applications of this model under-perform and require knowledge base order-sensitive markings, *k-tag*, to achieve state-of-the-art performance. This data encoding scheme captures the latent relation direction and provides a simple way to reduce noise in distant supervision.

- We make our code and data creation pipeline publicly available: https://github.com/suamin/umls-medline-distant-re

Proceedings of the BioNLP 2020 workshop, pages 187–194
Online, July 9, 2020 ©2020 Association for Computational Linguistics

2 Related Work

In MIL-based distant supervision for *corpus-level* RE, earlier works rely on the assumption that at least one of the evidence samples represent the target relation in a triple (Riedel et al., 2010; Hoffmann et al., 2011; Surdeanu et al., 2012). Recently, piecewise convolutional neural networks (PCNN) (Zeng et al., 2014) have been applied to DS (Zeng et al., 2015), with notable extensions in selective attention (Lin et al., 2016) and the modelling of noise dynamics (Luo et al., 2017). Han et al. (2018a) proposed a joint learning framework for knowledge graph completion (KGC) and RE with mutual attention, showing that DS improves downstream KGC performance, while KGC acts as an indirect signal to filter textual noise. Dai et al. (2019) extended this framework to biomedical RE, using improved KGC models, ComplEx (Trouillon et al., 2017) and SimplE (Kazemi and Poole, 2018), as well as additional auxiliary tasks of entity-type classification and named entity recognition to mitigate noise.

Pre-trained language models, such as BERT (Devlin et al., 2019), have been shown to improve the downstream performance of many NLP tasks. Relevant to distant RE, Alt et al. (2019) extended the OpenAI Generative Pre-trained Transformer (GPT) model (Radford et al., 2019) for *bag-level* MIL with selective attention (Lin et al., 2016). Sun et al. (2019) enriched pre-training stage with KB entity information, resulting in improved performance. For *sentence-level* RE, Wu and He (2019) proposed an entity marking strategy for BERT (referred to here as R-BERT) to perform relation classification. Specifically, they mark the entity boundaries with special tokens following the order they appear in the sentence. Likewise, Soares et al. (2019) studied several data encoding schemes and found marking entity boundaries important for *sentence-level* RE. With such encoding, they further proposed a novel pre-training scheme for distributed relational learning, suited to few-shot relation classification (Han et al., 2018b).

Our work builds on these findings, in particular, we extend the BERT model (Devlin et al., 2019) for *bag-level* MIL, similar to Alt et al. (2019). More importantly, noting the significance of *sentence-ordered* entity marking in *sentence-level* RE (Wu and He, 2019; Soares et al., 2019), we introduce the *knowledge-based* entity marking strategy suited to *bag-level* DS. This naturally encodes the information stored in KB, reducing the inherent noise.

3 Bag-level MIL for Distant RE

3.1 Problem Definition

Let $\mathcal{E}$ and $\mathcal{R}$ represent the set of entities and relations from a knowledge base $\mathcal{KB}$, respectively. For $h, t \in \mathcal{E}$ and $r \in \mathcal{R}$, let $(h, r, t) \in \mathcal{KB}$ be a fact triple for an ordered tuple (h, t). We denote all such (h, t) tuples by a set $\mathcal{G}^+$, i.e., there exists some $r \in \mathcal{R}$ for which the triple (h, r, t) belongs to the KB, called *positive groups*. Similarly, we denote by $\mathcal{G}^-$ the set of *negative groups*, i.e., for all $r \in \mathcal{R}$, the triple (h, r, t) does not belong to KB. The union of these groups is represented by $\mathcal{G} = \mathcal{G}^+ \cup \mathcal{G}^-$ [1]. We denote by $\mathcal{B}_g = [s_g^{(1)}, ..., s_g^{(m)}]$ an unordered sequence of sentences, called *bag*, for $g \in \mathcal{G}$ such that the sentences contain the group $g = (h, t)$, where the bag size m can vary. Let f be a function that maps each element in the bag to a low-dimensional relation representation $[\mathbf{r}_g^{(1)}, ..., \mathbf{r}_g^{(m)}]$. With o, we represent the bag aggregation function, that maps instance level relation representation to a final bag representation $\mathbf{b}_g = o(f(\mathcal{B}_g))$. The goal of distantly supervised *bag-level* MIL for *corpus-level* RE is then to predict the missing relation r given the bag.

3.2 Entity Markers

Wu and He (2019) and Soares et al. (2019) showed that using special markers for entities with BERT in the order they appear in a sentence encodes the positional information that improves the performance of *sentence-level* RE. It allows the model to focus on target entities when, possibly, other entities are also present in the sentence, implicitly doing entity disambiguation and reducing noise. In contrast, for *bag-level* distant supervision, the noisy channel be attributed to several factors for a given triple (h, r, t) and bag $\mathcal{B}_g$:

1. Evidence sentences may not express the relation.

2. Multiple entities appearing in the sentence, requiring the model to disambiguate target entities among other.

3. The direction of missing relation.

4. Discrepancy between the order of the target entities in the sentence and knowledge base.

To address (1), common approaches are to learn a negative relation class NA and use better bag aggregation strategies (Lin et al., 2016; Luo et al.,

---

[1]The sets are disjoint, $\mathcal{G}^+ \cap \mathcal{G}^- = \varnothing$

2017; Alt et al., 2019). For (2), encoding positional information is important, such as, in PCNN (Zeng et al., 2014), that takes into account the relative positions of *head* and *tail* entities (Zeng et al., 2015), and in (Wu and He, 2019; Soares et al., 2019) for *sentence-level* RE. To account for (3) and (4), multi-task learning with KGC and mutual attention has proved effective (Han et al., 2018a; Dai et al., 2019). Simply extending sentence sensitive marking to *bag-level* can be adverse, as it enhances (4) and even if the composition is uniform, it distributes the evidence sentence across several bags. On the other hand, expanding relations to multiple sub-classes based on direction (Wu and He, 2019), enhances class imbalance and also distributes supporting sentences. To jointly address (2), (3) and (4), we introduce KB sensitive encoding suitable for *bag-level* distant RE.

Formally, for a group $g = (h, t)$ and a matching sentence $s_g^{(i)}$ with tokens $(x_0, ..., x_L)^2$, we add special tokens $ and ˆ to mark the entity spans as:

Sentence ordered: Called *s-tag*, entities are marked in the order they appear in the sentence. Following Soares et al. (2019), let $s_1 = (i, j)$ and $s_2 = (k, l)$ be the index pairs with $0 < i < j - 1, j < k, k \leq l - 1$ and $l \leq L$ delimiting the entity mentions $e_1 = (x_i, ..., x_j)$ and $e_2 = (x_k, ..., x_l)$ respectively. We mark the boundary of s_1 with $ and s_2 with ˆ. Note, e_1 and e_2 can be either h or t.

KB ordered: Called *k-tag*, entities are marked in the order they appear in the KB. Let $s_h = (i, j)$ and $s_t = (k, l)$ be the index pairs delimiting head (h) and tail (t) entities, irrespective of the order they appear in the sentence. We mark the boundary of s_h with $ and s_t with ˆ.

The *s-tag* annotation scheme is followed by Soares et al. (2019) and Wu and He (2019) for span identification. In Wu and He (2019), each relation type $r \in \mathcal{R}$ is further expanded to two sub-classes as $r(e_1, e_2)$ and $r(e_2, e_1)$ to capture direction, while holding the *s-tag* annotation as fixed. For DS-based RE, since the ordered tuple (h, t) is given, the task is reduced to relation classification without direction. This side information is encoded in data with *k-tag*, covering (2) but also (3) and (4). To account for (1), we also experiment with selective attention (Lin et al., 2016) which has been widely used in other works (Luo et al., 2017; Han et al., 2018a; Alt et al., 2019).

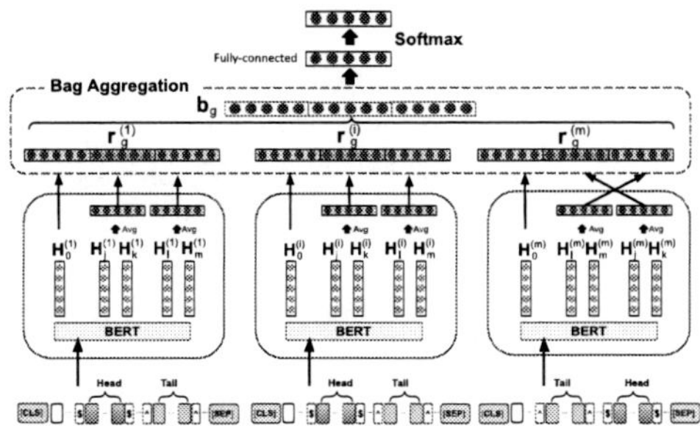

Figure 2: Multiple instance learning (MIL) based *bag-level* relation classification BERT with KB ordered entity marking (Section 3.2). Special markers $ and ˆ always delimit the span of *head* (h_s, h_e) and *tail* (t_s, t_e) entities regardless of their order in the sentence. The markers captures the *positions* of entities and latent relation *direction*.

3.3 Model Architecture

BERT (Devlin et al., 2019) is used as our base sentence encoder, specifically, BioBERT (Lee et al., 2020), and we extend R-BERT (Wu and He, 2019) to *bag-level* MIL. Figure 2 shows the model's architecture with *k-tag*. Consider a bag $\mathcal{B}_g$ of size m for a group $g \in \mathcal{G}$ representing the ordered tuple (h, t), with corresponding spans $S_g = [(s_h^{(1)}, s_t^{(1)}), ..., (s_h^{(m)}, s_t^{(m)})]$ obtained with *k-tag*, then for a pair of sentences in the bag and spans, $(s^{(i)}, (s_h^{(i)}, s_t^{(i)}))$, we can represent the model in three steps, such that the first two steps represent the map f and the final step o, as follows:

1. SENTENCE ENCODING: BERT is applied to the sentence and the final hidden state $\mathbf{H}_0^{(i)} \in \mathbb{R}^d$, corresponding to the [CLS] token, is passed through a linear layer[3] $\mathbf{W}^{(1)} \in \mathbb{R}^{d \times d}$ with $tanh(.)$ activation to obtain the global sentence information in $\mathbf{h}_0^{(i)}$.

2. RELATION REPRESENTATION: For the head entity, represented by the span $s_h^{(i)} = (j, k)$ for $k > j$, we apply average pooling $\frac{1}{k-j+1} \sum_{n=j}^{k} \mathbf{H}_n^{(i)}$, and similarly for the tail entity with span $s_t^{(i)} = (l, m)$ for $m > l$, we get $\frac{1}{m-l+1} \sum_{n=l}^{m} \mathbf{H}_n^{(i)}$. The pooled representations are then passed through a shared linear layer $\mathbf{W}^{(2)} \in \mathbb{R}^{d \times d}$ with $tanh(.)$ activation to get $\mathbf{h}_h^{(i)}$ and $\mathbf{h}_t^{(i)}$. To get the final latent relation representation, we concatenate the pooled entities representation with [CLS] as $\mathbf{r}_g^{(i)} = [\mathbf{h}_0^{(i)}; \mathbf{h}_h^{(i)}; \mathbf{h}_t^{(i)}] \in \mathbb{R}^{3d}$.

$^2 x_0 =$ [CLS] and $x_L =$ [SEP]

^{3}Each linear layer is implicitly assumed with a bias vector

189

3. BAG AGGREGATION: After applying the first
two steps to each sentence in the bag, we obtain
$[\mathbf{r}_g^{(1)}, ..., \mathbf{r}_g^{(m)}]$. With a final linear layer consisting
of a relation matrix $\mathbf{M}_r \in \mathbb{R}^{|\mathcal{R}| \times 3d}$ and a bias vec-
tor $\mathbf{b}_r \in \mathbb{R}^{|\mathcal{R}|}$, we aggregate the bag information
with o in two ways:

Average: The bag elements are averaged as:

$$\mathbf{b}_g = \frac{1}{m} \sum_{i=1}^{m} \mathbf{r}_g^{(i)}$$

Selective attention (Lin et al., 2016): For a row $\mathbf{r}$
in $\mathbf{M}_r$ representing the relation $r \in \mathcal{R}$, we get the
attention weights as:

$$\alpha_i = \frac{\exp(\mathbf{r}^T \mathbf{r}_g^{(i)})}{\sum_{j=1}^{m} \exp(\mathbf{r}^T \mathbf{r}_g^{(j)})}$$

$$\mathbf{b}_g = \sum_{i=1}^{m} \alpha_i \mathbf{r}_g^{(i)}$$

Following $\mathbf{b}_g$, a *softmax* classifier is applied to pre-
dict the probability $p(r|\mathbf{b}_g; \theta)$ of relation r being a
true relation with θ representing the model parameters, where we minimize the cross-entropy loss
during training.

4 Experiments

4.1 Data

Similar to (Dai et al., 2019), UMLS[4] (Bodenreider,
2004) is used as our KB and MEDLINE abstracts[5]
as our text source. A data summary is shown in
Table 1 (see Appendix A for details on the data creation pipeline). We approximate the same statistics
as reported in Dai et al. (2019) for relations and
entities, but it is important to note that the data does
not contain the same samples. We divided triples
into train, validation and test sets, and following
(Weston et al., 2013; Dai et al., 2019), we make
sure that there is no overlapping facts across the
splits. Additionally, we add another constraint, i.e.,
there is no sentence-level overlap between the training and held-out sets. To perform groups negative
sampling, for the collection of evidence sentences
supporting NA relation type bags, we extend KGC
open-world assumption to *bag-level* MIL (see A.3).
20% of the data is reserved for testing, and of the
remaining 80%, we use 10% for validation and the
rest for training.

---

[4]We use 2019 release: `umls-2019AB-full`

[5]`https://www.nlm.nih.gov/bsd/medline.`
`html`

Table 1: Overall statistics of the data.

Triples	Entities	Relations	Pos. Groups	Neg. Groups
169,438	27,403	355	92,070	64,448

4.2 Models and Evaluation

We compare each tagging scheme, *s-tag* and *k-tag*,
with average (*avg*) and selective attention (*attn*) bag
aggregation functions. To test the setup of Wu and
He (2019), which follows *s-tag*, we expand each
relation type (*exprels*) $r \in \mathcal{R}$ to two sub-classes
$r(e_1, e_2)$ and $r(e_2, e_1)$ indicating relation direction
from first entity to second and vice versa. For all
experiments, we used batch size 2, bag size 16 with
sampling (see A.4 for details on bag composition),
learning rate $2e^{-5}$ with linear decay, and 3 epochs.
As the standard practice (Weston et al., 2013), evaluation is performed through constructing candidate
triples by combining the entity pairs in the test set
with all relations (except NA) and ranking the resulting triples. The extracted triples are matched
against the test triples and the precision-recall (PR)
curve, area under the PR curve (AUC), F1 measure,
and Precision@k, for k in $\{100, 200, 300, 2000,
4000, 6000\}$ are reported.

4.3 Results

Performance metrics are shown in Table 2 and plots
of the resulting PR curves in Figure 3. Since our
data differs from Dai et al. (2019), the AUC cannot
be directly compared. However, Precision@k indicates the general performance of extracting the true
triples, and can therefore be compared. Generally,
models annotated with *k-tag* perform significantly
better than other models, with *k-tag+avg* achieving
state-of-the-art Precision@$\{2k,4k,6k\}$ compared
to the previous best (Dai et al., 2019). The best
model of Dai et al. (2019) uses PCNN sentence
encoder, with additional tasks of SimplE (Kazemi
and Poole, 2018) based KGC and KG-attention,
entity-type classification and named entity recognition. In contrast our data-driven method, *k-tag*,
greatly simplifies this by directly encoding the KB
information, i.e., order of the *head* and *tail* entities and therefore, the latent relation direction.
Consider again the example in Figure 1 where our
source triple (h, r, t) is (*neurofibromatosis 1, asso-
ciated_genetic_condition, breast cancer*), and only
last sentence has the same order of entities as KB.
This discrepancy is conveniently resolved (note in
Figure 2, for last sentence the extracted entities

Table 2: Relation extraction results for different model configurations and data splits.

Model	Bag Agg.	AUC	F1	P@100	P@200	P@300	P@2k	P@4k	P@6k
Dai et al. (2019)	-	-	-	-	-	-	.913	.829	.753
s-tag	*avg*	.359	.468	.791	.704	.649	.504	.487	.481
	attn	.122	.225	.587	.563	.547	.476	.441	.418
s-tag+exprels	*avg*	.383	.494	.508	.519	.521	.507	.508	.511
	attn	.114	.216	.459	.476	.482	.504	.496	.484
k-tag	*avg*	**.684**	**.649**	**.974**	**.983**	**.986**	**.983**	**.977**	**.969**
	attn	.314	.376	.967	.941	.925	.857	.814	.772

sentence order is flipped to KG order when concatenating, unlike *s-tag*) with *k-tag*. We remark that such knowledge can be seen as learned, when jointly modeling with KGC, however, considering the task of *bag-level* distant RE only, the KG triples are *known* information and we utilize this information explicitly with *k-tag* encoding.

As PCNN (Zeng et al., 2015) can account for the relative positions of head and tail entities, it also performs better than the models tagged with *s-tag* using sentence order. Similar to Alt et al. (2019)[6], we also note that the pre-trained contextualized models result in sustained long tail performance. *s-tag+exprels* reflects the direct application of Wu and He (2019) to *bag-level* MIL for distant RE. In this case, the relations are explicitly extended to model entity direction appearing first to second in the sentence, and vice versa. This implicitly introduces independence between the two sub-classes of the same relation, limiting the gain from shared knowledge. Likewise, with such expanded relations, class imbalance is further enhanced to more fine-grained classes.

Though selective attention (Lin et al., 2016) has been shown to improve the performance of distant RE (Luo et al., 2017; Han et al., 2018a; Alt et al., 2019), models in our experiments with such an attention mechanism significantly underperformed, in each case bumping the area under the PR curve and making it flatter. We note that more than 50% of bags are under-sized, in many cases, with only 1-2 sentences, requiring repeated over-sampling to match fixed bag size, therefore, making it difficult for attention to learn a distribution over the bag with repetitions, and further adding noise. For such cases, the distribution should ideally be close to uniform, as is the case with averaging, resulting in better performance.

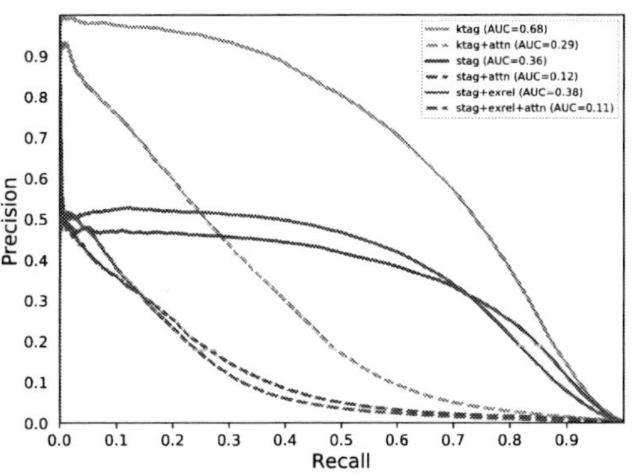

Figure 3: Precision-Recall (PR) curve for different models. We see that the models with *k-tag* perform better than the *s-tag* with average aggregation showing consistent performance for long-tail relations.

5 Conclusion

This work extends BERT to *bag-level* MIL and introduces a simple data-driven strategy to reduce the noise in distantly supervised biomedical RE. We note that the *position* of entities in sentence and the *order* in KB encodes the latent *direction* of relation, which plays an important role for learning under such noise. With a relatively simple methodology, we show that this can sufficiently be achieved by reducing the need for additional tasks and highlighting the importance of data quality.

Acknowledgements

The authors would like to thank the anonymous reviewers for helpful feedback. The work was partially funded by the European Union's Horizon 2020 research and innovation programme under grant agreement No. 777107 through the project Precise4Q and by the German Federal Ministry of Education and Research (BMBF) through the project DEEPLEE (01IW17001).

---

[6]Their model does not use any entity marking strategy.

References

Christoph Alt, Marc Hübner, and Leonhard Hennig. 2019. Fine-tuning pre-trained transformer language models to distantly supervised relation extraction. In *Proceedings of the 57th Annual Meeting of the Association for Computational Linguistics*, pages 1388–1398.

Olivier Bodenreider. 2004. The unified medical language system (UMLS): integrating biomedical terminology. *Nucleic acids research*, 32(suppl_1):D267–D270.

Qin Dai, Naoya Inoue, Paul Reisert, Ryo Takahashi, and Kentaro Inui. 2019. Distantly supervised biomedical knowledge acquisition via knowledge graph based attention. In *Proceedings of the Workshop on Extracting Structured Knowledge from Scientific Publications*, pages 1–10.

Jacob Devlin, Ming-Wei Chang, Kenton Lee, and Kristina Toutanova. 2019. BERT: Pre-training of deep bidirectional transformers for language understanding. In *Proceedings of the 2019 Conference of the North American Chapter of the Association for Computational Linguistics: Human Language Technologies, Volume 1 (Long and Short Papers)*, pages 4171–4186.

Zhou GuoDong, Su Jian, Zhang Jie, and Zhang Min. 2005. Exploring various knowledge in relation extraction. In *Proceedings of the 43rd annual meeting on association for computational linguistics*, pages 427–434. Association for Computational Linguistics.

Xu Han, Tianyu Gao, Yuan Yao, Deming Ye, Zhiyuan Liu, and Maosong Sun. 2019. OpenNRE: An open and extensible toolkit for neural relation extraction. In *Proceedings of EMNLP-IJCNLP: System Demonstrations*, pages 169–174.

Xu Han, Zhiyuan Liu, and Maosong Sun. 2018a. Neural knowledge acquisition via mutual attention between knowledge graph and text. In *Thirty-Second AAAI Conference on Artificial Intelligence*.

Xu Han, Hao Zhu, Pengfei Yu, Ziyun Wang, Yuan Yao, Zhiyuan Liu, and Maosong Sun. 2018b. FewRel: A large-scale supervised few-shot relation classification dataset with state-of-the-art evaluation. In *Proceedings of the 2018 Conference on Empirical Methods in Natural Language Processing*, pages 4803–4809.

Raphael Hoffmann, Congle Zhang, Xiao Ling, Luke Zettlemoyer, and Daniel S Weld. 2011. Knowledge-based weak supervision for information extraction of overlapping relations. In *Proceedings of the 49th Annual Meeting of the Association for Computational Linguistics: Human Language Technologies-Volume 1*, pages 541–550. Association for Computational Linguistics.

Seyed Mehran Kazemi and David Poole. 2018. Simple embedding for link prediction in knowledge graphs. In *Advances in neural information processing systems*, pages 4284–4295.

Jinhyuk Lee, Wonjin Yoon, Sungdong Kim, Donghyeon Kim, Sunkyu Kim, Chan Ho So, and Jaewoo Kang. 2020. BioBERT: a pre-trained biomedical language representation model for biomedical text mining. *Bioinformatics*, 36(4):1234–1240.

Adam Lerer, Ledell Wu, Jiajun Shen, Timothee Lacroix, Luca Wehrstedt, Abhijit Bose, and Alex Peysakhovich. 2019. PyTorch-BigGraph: A large-scale graph embedding system. In *Proceedings of the 2nd SysML Conference*, Palo Alto, CA, USA.

Yankai Lin, Shiqi Shen, Zhiyuan Liu, Huanbo Luan, and Maosong Sun. 2016. Neural relation extraction with selective attention over instances. In *Proceedings of the 54th Annual Meeting of the Association for Computational Linguistics (Volume 1: Long Papers)*, pages 2124–2133.

Bingfeng Luo, Yansong Feng, Zheng Wang, Zhanxing Zhu, Songfang Huang, Rui Yan, and Dongyan Zhao. 2017. Learning with noise: Enhance distantly supervised relation extraction with dynamic transition matrix. *arXiv preprint arXiv:1705.03995*.

Mike Mintz, Steven Bills, Rion Snow, and Dan Jurafsky. 2009. Distant supervision for relation extraction without labeled data. In *Proceedings of the Joint Conference of the 47th Annual Meeting of the ACL and the 4th International Joint Conference on Natural Language Processing of the AFNLP: Volume 2-Volume 2*, pages 1003–1011. Association for Computational Linguistics.

Alec Radford, Jeffrey Wu, Rewon Child, David Luan, Dario Amodei, and Ilya Sutskever. 2019. Language models are unsupervised multitask learners. *OpenAI Blog*, 1(8):9.

Sebastian Riedel, Limin Yao, and Andrew McCallum. 2010. Modeling relations and their mentions without labeled text. In *Joint European Conference on Machine Learning and Knowledge Discovery in Databases*, pages 148–163. Springer.

Alan Ritter, Luke Zettlemoyer, Mausam, and Oren Etzioni. 2013. Modeling missing data in distant supervision for information extraction. *Transactions of the Association for Computational Linguistics*, 1:367–378.

Livio Baldini Soares, Nicholas FitzGerald, Jeffrey Ling, and Tom Kwiatkowski. 2019. Matching the blanks: Distributional similarity for relation learning. In *Proceedings of the 57th Annual Meeting of the Association for Computational Linguistics*, pages 2895–2905.

Yu Sun, Shuohuan Wang, Yukun Li, Shikun Feng, Xuyi Chen, Han Zhang, Xin Tian, Danxiang Zhu, Hao Tian, and Hua Wu. 2019. ERNIE: Enhanced representation through knowledge integration. *arXiv preprint arXiv:1904.09223*.

Mihai Surdeanu, Julie Tibshirani, Ramesh Nallapati, and Christopher D Manning. 2012. Multi-instance multi-label learning for relation extraction. In *Proceedings of the 2012 joint conference on empirical methods in natural language processing and computational natural language learning*, pages 455–465. Association for Computational Linguistics.

Shingo Takamatsu, Issei Sato, and Hiroshi Nakagawa. 2012. Reducing wrong labels in distant supervision for relation extraction. In *Proceedings of the 50th Annual Meeting of the Association for Computational Linguistics: Long Papers-Volume 1*, pages 721–729. Association for Computational Linguistics.

Théo Trouillon, Christopher R Dance, Éric Gaussier, Johannes Welbl, Sebastian Riedel, and Guillaume Bouchard. 2017. Knowledge graph completion via complex tensor factorization. *The Journal of Machine Learning Research*, 18(1):4735–4772.

Linlin Wang, Zhu Cao, Gerard De Melo, and Zhiyuan Liu. 2016. Relation classification via multi-level attention cnns. In *Proceedings of the 54th Annual Meeting of the Association for Computational Linguistics (Volume 1: Long Papers)*, pages 1298–1307.

Xinggang Wang, Yongluan Yan, Peng Tang, Xiang Bai, and Wenyu Liu. 2018. Revisiting multiple instance neural networks. *Pattern Recognition*, 74:15–24.

Jason Weston, Antoine Bordes, Oksana Yakhnenko, and Nicolas Usunier. 2013. Connecting language and knowledge bases with embedding models for relation extraction. In *Conference on Empirical Methods in Natural Language Processing*, pages 1366–1371.

Shanchan Wu and Yifan He. 2019. Enriching pre-trained language model with entity information for relation classification. In *Proceedings of the 28th ACM International Conference on Information and Knowledge Management*, pages 2361–2364.

Daojian Zeng, Kang Liu, Yubo Chen, and Jun Zhao. 2015. Distant supervision for relation extraction via piecewise convolutional neural networks. In *Proceedings of the 2015 conference on empirical methods in natural language processing*, pages 1753–1762.

Daojian Zeng, Kang Liu, Siwei Lai, Guangyou Zhou, and Jun Zhao. 2014. Relation classification via convolutional deep neural network. In *Proceedings of COLING 2014, the 25th International Conference on Computational Linguistics: Technical Papers*, pages 2335–2344.

A Data Pipeline

In this section, we explain the steps taken to create the data for distantly-supervised (DS) biomedical relation extraction (RE). We highlight the importance of a data creation pipeline as the quality of data plays a key role in the downstream performance of our model. We note that a pipeline is likewise important for generating reproducible results, and contributes toward the possibility of having either a benchmark dataset or a repeatable set of rules.

A.1 UMLS processing

The fact triples were obtained for English concepts, filtering for RO relation types only (Dai et al., 2019). We collected 9.9M (CUI_head, relation_text, CUI_tail) triples, where CUI represents the concept unique identifier in UMLS.

A.2 MEDLINE processing

From 34.4M abstracts, we extracted 160.4M unique sentences. To perform fast and scalable search, we use the Trie data structure[7] to index all the textual descriptions of UMLS entities. In obtaining a clean set of sentences, we set the minimum and maximum sentence character length to 32 and 256 respectively, and further considered only those sentences where matching entities are mentioned only once. The latter decision is to lower the noise that may come when only one instance of multiple occurrences is marked for a matched entity. With these constraints, the data was reduced to 118.7M matching sentences.

A.3 Groups linking and negative sampling

Recall the entity groups $\mathcal{G} = \mathcal{G}^+ \cup \mathcal{G}^-$ (Section 3.1). For training with NA relation class, we generate hard negative samples with an open-world assumption (Soares et al., 2019; Lerer et al., 2019) suited to *bag-level* multiple instance learning (MIL). From 9.9M triples, we removed the relation type and collected 9M CUI groups in the form of (h, t). Since each CUI is linked to more than one textual form, all of the text combinations for two entities must be considered for a given pair, resulting in 531M textual groups $\mathcal{T}$ for the 586 relation types.

Next, for each matched sentence, let $\mathcal{P}_s^2$ denote the size 2 permutations of entities present in the sentence, then $\mathcal{T} \cap \mathcal{P}_s^2$ return groups which are *present in KB* and *have matching evidence* (positive

---

[7]`https://github.com/vi3k6i5/flashtext`

groups, $\mathcal{G}^+$). Simultaneously, with a probability of $\frac{1}{2}$, we remove the h or t entity from this group and replace it with a novel entity e in the sentence, such that the resulting group (e, t) or (h, e) belongs to $\mathcal{G}^-$. This method results in sentences that are seen both for the true triple, as well as for the invalid ones. Further using the constraints that the relation group sizes must be between 10 to 1500, we find 354^8 relation types (approximately the same as Dai et al. (2019)) with 92K positive groups and 2.1M negative groups, which were reduced to 64K by considering a random subset of 70% of the positive groups. Table 1 provides these summary statistics.

A.4 Bag composition and data splits

For bag composition, we created bags of constant size by randomly under- or over-sampling the sentences in the bag (Han et al., 2019) to avoid larger bias towards common entities (Soares et al., 2019). The true distribution had a long tail, with more than 50% of the bags having 1 or 2 sentences. We defined a bag to be *uniform*, if the special markers represent the same entity in each sentence, either h or t. If the special markers can take on both h or t, we consider that bag to have a *mix* composition. The *k-tag* scheme, on the other hand, naturally generates uniform bags. Further, to support the setting of Wu and He (2019), we followed the *s-tag* scheme and expanded the relations by adding a suffix to denote the directions as $r(e_1, e_2)$ or $r(e_2, e_1)$, with the exception of the NA class, resulting in 709 classes. For fair comparisons with *k-tag*, we generated uniform bags with *s-tag* as well, by keeping e_1 and e_2 the same per bag. Due to these bag composition and class expansion (in one setting, *exprels*) differences, we generated three different splits, supporting each scheme, with the same test sets in cases where the classes are not expanded and a different test set when the classes are expanded. Table A.1 shows the statistics for these splits.

Table A.1: Different data splits.

Model	Set Type	Triples	Triples (w/o NA)	Groups	Sentences (Sampled)
k-tag	train	92,972	48,563	92,972	1,487,552
	valid	13,555	8,399	15,963	255,408
	test	33,888	20,988	38,860	621,760
s-tag	train	91,555	47,588	125,852	2,013,632
	valid	13,555	8,399	22,497	359,952
	test	33,888	20,988	55,080	881,280
s-tag+exprels	train	125,155	71,402	125,439	2,007,024
	valid	22,604	16,298	22,607	361,712
	test	55,083	39,282	55,094	881,504

---

[8]355 including NA relation

Global Locality in Biomedical Relation and Event Extraction

Elaheh ShafieiBavani, Antonio Jimeno Yepes, Xu Zhong, David Martinez Iraola
IBM Research Australia
{elaheh.shafieibavani, david.martinez.iraola1}@ibm.com
{antonio.jimeno, peter.zhong}@au1.ibm.com

Abstract

Due to the exponential growth of biomedical literature, event and relation extraction are important tasks in biomedical text mining. Most work only focus on relation extraction, and detect a single entity pair mention on a short span of text, which is not ideal due to long sentences that appear in biomedical contexts. We propose an approach to both relation and event extraction, for simultaneously predicting relationships between all mention pairs in a text. We also perform an empirical study to discuss different network setups for this purpose. The best performing model includes a set of multi-head attentions and convolutions, an adaptation of the transformer architecture, which offers self-attention the ability to strengthen dependencies among related elements, and models the interaction between features extracted by multiple attention heads. Experiment results demonstrate that our approach outperforms the state of the art on a set of benchmark biomedical corpora including BioNLP 2009, 2011, 2013 and BioCreative 2017 shared tasks.

1 Introduction

Event and relation extraction has become a key research topic in natural language processing with a variety of practical applications especially in the biomedical domain, where these methods are widely used to extract information from massive document sets, such as scientific literature and patient records. This information contains the interactions between named entities such as protein-protein, drug-drug, chemical-disease, and more complex events.

Relations are usually described as typed, sometimes directed, pairwise links between defined named entities (Björne et al., 2009). Event extraction differs from relation extraction in the sense that an event has an annotated trigger word (e.g., a verb), and could be an argument of other events to connect more than two entities. Event extraction is a more complicated task compared to relation extraction due to the tendency of events to capture the semantics of texts. For clarity, Figure 1 shows an example from the GE11 shared task corpus that includes two nested events.

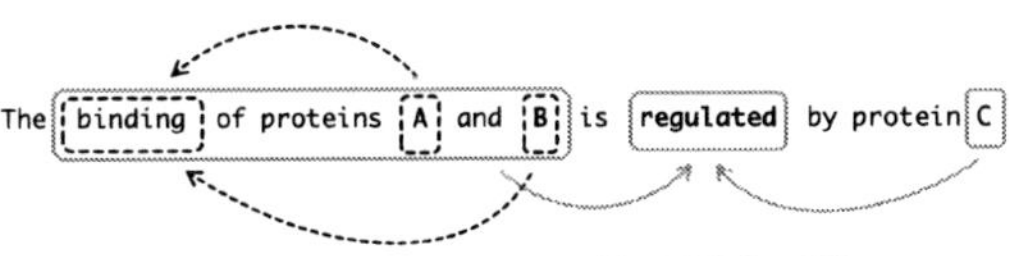

Figure 1: Example of nested events from GE11 shared task

Recently, deep neural network models obtain state-of-the-art performance for event and relation extraction. Two major neural network architectures for this purpose include Convolutional Neural Networks (CNNs) (Santos et al., 2015; Zeng et al., 2015) and Recurrent Neural Networks (RNNs) (Mallory et al., 2015; Verga et al., 2015; Zhou et al., 2016). While CNNs can capture the local features based on the convolution operations and are more suitable for addressing short sentence sequences, RNNs are good at learning long-term dependency features, which are considered more suitable for dealing with long sentences. Therefore, combining the advantages of both models is the key point for improving biomedical event and relation extraction performance (Zhang et al., 2018).

However, encoding long sequences to incorporate long-distance context is very expensive in RNNs (Verga et al., 2018) due to their computational dependence on the length of the sequence. In addition, computations could not be parallelized since each token's representation requires as input the representation of its previous token. In contrast, CNNs can be executed entirely in parallel

Proceedings of the BioNLP 2020 workshop, pages 195–204
Online, July 9, 2020 ©2020 Association for Computational Linguistics

across the sequence, and have shown good performance in event and relation extraction (Björne and Salakoski, 2018). However, the amount of context incorporated into a single token's representation is limited by the depth of the network, and very deep networks can be difficult to learn (Hochreiter, 1998).

To address these problems, self-attention networks (Parikh et al., 2016; Lin et al., 2017) come into play. They have shown promising empirical results in various natural language processing tasks, such as information extraction (Verga et al., 2018), machine translation (Vaswani et al., 2017) and natural language inference (Shen et al., 2018). One of their strengths lies in their high parallelization in computation and flexibility in modeling dependencies regardless of distance by explicitly attending to all the elements. In addition, their performance can be improved by multi-head attention (Vaswani et al., 2017), which projects the input sequence into multiple subspaces and applies attention to the representation in each subspace.

In this paper, we propose a new neural network model that combines multi-head attention mechanisms with a set of convolutions to provide global locality in biomedical event and relation extraction. Convolutions capture the local structure of text, while self-attention learns the global interaction between each pair of words. Hence, our approach models locality for self-attention while the interactions between features are learned by multi-head attentions. The experiment results over the biomedical benchmark corpora show that providing global locality outperforms the existing state of the art for biomedical event and relation extraction. The proposed architecture is shown in Figure 2.

Conducting a set of experiments over the corpora of the shared tasks for BioNLP 2009, 2011 and 2013, and BioCreative 2017, we compare the performance of our model with the best-performing system (TEES) (Björne and Salakoski, 2018) in the shared tasks. The results we achieve via precision, recall, and F-score demonstrate that our model obtains state-of-the-art performance. We also empirically assess three variants of our model and elaborate on the results further in the experiments.

The rest of the paper is organized as follows. Section 2 summarizes the background. The data, and the proposed approach are explained in Sections 3 and 4 respectively. Section 5 explains the experiments and discusses the achieved results. Finally, Section 6 summarizes the findings of the paper and presents future work.

2 Background

Biomedical event and relation extraction have been developed thanks to the contribution of corpora generated for community shared tasks (Kim et al., 2009, 2011; Nédellec et al., 2013; Segura Bedmar et al., 2011, 2013; Krallinger et al., 2017). In these tasks, relevant biomedical entities such as genes, proteins and chemicals are given and the information extraction methods aim to identify relations alone or relations and events together within a sentence span.

A variety of methods have been evaluated on these tasks, which range from rule based methods to more complex machine learning methods, either supported by shallow or deep learning approaches. Some of the deep learning based methods include CNNs (Björne and Salakoski, 2018; Santos et al., 2015; Zeng et al., 2015) and RNNs (Li et al., 2019; Mallory et al., 2015; Verga et al., 2015; Zhou et al., 2016). CNNs will identify local context relations while their performance may suffer when entities need to be identified in a broader context. On the other hand, RNNs are difficult to parallelize while they do not fully solve the long dependency problem (Verga et al., 2018). Moreover, such approaches are proposed for relation extraction, but not to extract nested events. In this work, we intend to improve over existing methods. We combine a set of parallel multi-head attentions with a set of 1D convolutions to provide global locality in biomedical event and relation extraction. Our approach models locality for self-attention while the interactions between features are learned by multi-head attentions. We evaluate our model on data from the shared tasks for BioNLP 2009, 2011 and 2013, and BioCreative 2017.

The BioNLP Event Extraction tasks provide the most complex corpora with often large sets of event types and at times relatively small corpus sizes. Our proposed approach achieves higher performance on the GE09, GE11, EPI11, ID11, REL11, GE13, CG13 and PC13 BioNLP Shared Task corpora, compared to the top performing system (TEES) (Björne and Salakoski, 2018) for both relation and event extraction in these tasks. Since the annotations for the test sets of the BioNLP Shared Task corpora are not provided, we uploaded our predictions to the task organizers' servers for evaluation.

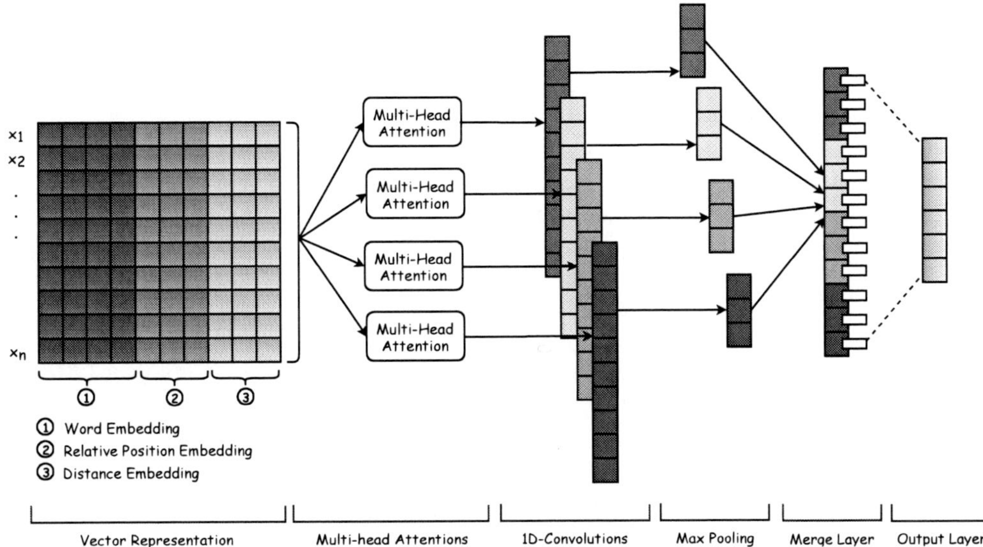

Figure 2: Our model architecture for biomedical event and relation extraction: The embedding vectors are merged together before the multi-head attention and convolution layers. The global max pooling is then applied to the results of these operations. Finally, the output layer shows the predicted labels.

The CHEMPROT corpus in the BioCreative VI Chemical–Protein relation extraction task (CP17) also provides a standard comparison with current methods in relation extraction. The CHEMPROT corpus is relatively large compared to its low number of five relation types. Our model outperforms the best-performing system (TEES) (Björne and Salakoski, 2018) for relation extraction in this task.

3 Data

We develop and evaluate our approach on a number of event and relation extraction corpora. These corpora originate from three BioNLP Shared Tasks (Kim et al., 2009; Björne and Salakoski, 2011; Nédellec et al., 2013) and the BioCreative VI Chemical–Protein relation extraction task (Krallinger et al., 2017). The BioNLP corpora cover various domains of molecular biology and provide the most complex event annotations. The BioCreative corpora use pairwise relation annotations. Table 1 shows information about these corpora.

For further analysis and experiments, we also used the AMIA gene-mutation corpus available in (Jimeno Yepes et al., 2018). The training/testing sets contain 2656/385 mentions of mutations, and 2799/280 of genes or proteins, and 1617/130 rela-

Corpus	Domain	E	I	S
GE09	Molecular Biology	10	6	11380
GE11	Molecular Biology	10	6	14958
EPI11	Epigenetics and PTM:s	16	6	11772
ID11	Infection Diseases	11	7	5118
REL11	Entity Relations	1	2	11351
GE13	Molecular Biology	15	6	8369
CG13	Cancer Genetics	42	9	5938
PC13	Pathway Curation	24	9	5040
CP17	Chemical-Protein Int.	-	5	24594

Table 1: Information about the domain, number of event and entity types (E), number of event argument and relation types (I), and number of sentences (S), related to the corpora of the biomedical shared tasks

tions between genes and mutations. We extracted about 30% of the training set as the validation set.

4 Model

We propose a new biomedical event extraction model that is mainly built upon multi-head attentions to learn the global interactions between each pair of tokens, and convolutions to provide locality. The proposed neural network architecture consists of 4 parallel multi-head attentions followed by a set of 1D convolutions with window sizes 1, 3, 5 and 7. Our model attends to the most important tokens in the input features[1], and enhances the feature extraction of dependent elements across multiple heads, irrespective of their distance. Moreover, we model locality for multi-head attentions by restricting the attended tokens to local regions via convolutions.

The relation and event extraction task is modelled as a graph representation of events and relations (Björne and Salakoski, 2018). Entities and event triggers are nodes, and relations and event arguments are the edges that connect them. An event is modelled as a trigger node and its set of outgoing edges. Relation and event extraction are performed through the following classification tasks: (i) Entity and Trigger Detection, which is a named-entity recognition task where entities and event triggers in a sentence span are detected to generate the graph nodes; (ii) Relation and Event Detection, where relations and event arguments are predicted for all valid pairs of entity and trigger nodes to create the graph edges; (iii) Event Duplication, where each event is classified as an event or a negative which causes unmerging in the graph[2]; (iv) Modifier Detection, in which event modality (speculation or negation) is detected. In relation extraction tasks where entities are given, only the second classification task is partially used.

The same network architecture is used for all four classification tasks, with the number of predicted labels changing between tasks.

4.1 Inputs

The input is modelled in the context of a sentence window, centered around the target entity, relation or event. The sentence is modelled as a linear sequence of word tokens. Following the work in (Björne and Salakoski, 2018), we use a set of embedding vectors as the input features, where each unique word token is mapped to the relevant vector space embeddings. We use the pre-trained 200-dimensional word2vec vectors (Mikolov et al., 2013) induced on a combination of the English Wikipedia and the millions of biomedical research articles from PubMed and PubMed Central (Moen and Ananiadou, 2013), along with the 8-dimensional embeddings of relative positions, and distances learned from the input corpus. Following the work in (Zeng et al., 2014), we use Distance features, where the relative distances to tokens of interest are mapped to their own vectors. We also consider Relative Position features to identify the locations and roles (i.e., entities, event triggers, and arguments) of tokens in the classified structure. Finally, these embeddings with their learned weights[3] are concatenated together to shape an n-dimensional vector e_i for each word token. This merged input sequence is then processed by a set of parallel multi-head attentions followed by convolutional layers.

4.2 Multi-head Attention

Self-attention networks produce representations by applying attention to each pair of tokens from the input sequence, regardless of their distance. According to the previous work (Vaswani et al., 2017), multi-head attention applies self-attention multiple times over the same inputs using separately normalized parameters (attention heads) and combines the results, as an alternative to applying one pass of attention with more parameters. The intuition behind this modeling decision is that dividing the attention into multiple heads makes it easier for the model to learn to attend to different types of relevant information with each head. The self-attention updates input embeddings e_i by performing a weighted sum over all tokens in the sequence, weighted by their importance for modeling token i. Given an input sequence $E = \{e_1, ..., e_I\} \in \mathbb{R}^{I \times d}$, the model first projects each input to a key k, value v, and query q, using separate affine transformations with ReLU activations (Glorot et al., 2011). Here, k, v, and q are each in $\mathbb{R}^{\frac{d}{H}}$, where d indicates the hidden size, and H is the number of heads. The attention weights a_{ij}^h for head h between tokens i and j are computed using scaled dot-product attention:

$$a_{ij}^h = \sigma(\frac{q_i^{h^T} k_j^h}{\sqrt{d}}) \tag{1}$$

$$o_i^h = \sum_j v_j^h \odot s_{ij}^h$$

where o_i^h is the output of the attention head h. $\odot$ denotes element-wise multiplication and σ indicates a softmax along the jth dimension. The scaled attention is meant to aid optimization by flattening the softmax and better distributing the gradients (Vaswani et al., 2017). The outputs of the individual attention heads are concatenated into o_i as: $o_i = [o_i^1; ...; o_i^H]$. Herein, all layers use residual

---

[1] We choose different embeddings for each task/dataset to be in line with TEES.

[2] Since events are n-ary relations, event nodes may overlap.

[3] The only exception is for the word vectors, where the original weights are used to provide generalization to words outside the task's training corpus.

connections between the output of the multi-headed attention and its input. Layer normalization (Lei Ba et al., 2016), $LN(.)$, is then applied to the output: $m_i = LN(e_i + o_i)$. The multi-head attention layer uses a softmax activation function.

4.3 Convolutions

The multi-head attentions are then followed by a set of parallel 1D convolutions with window sizes 1, 3, 5 and 7. Adding these explicit n-gram modelings helps the model to learn to attend to local features. Our convolutions use the ReLU activation function. We use $C(.)$ to denote a convolutional operator. The convolutional portion of the model is given by:

$$c_i = ReLU(C(m_i)) \qquad (2)$$

Global max pooling is then applied to each 1D convolution and the resulting features are merged together into an output vector.

4.4 Classification

Finally, the output layer performs the classification, where each label is represented by one neuron. The classification layer uses the sigmoid activation function. Classification is performed as multilabel classification where each example may have zero, one or multiple positive labels.

We use the *adam optimizer* with *binary crossentropy* and a learning rate of 0.001. Dropout of 0.1 is also applied at two steps of merging input features and global max pooling to provide generalization.

5 Experiments and Results

We have conducted a set of experiments to evaluate our proposed approach over the benchmark biomedical corpora. In addition to evaluating our main model (4MHA-4CNN), we have evaluated the performance of three variants of our proposed approach: (i) 4MHA: 4 parallel multi-head attentions apply self-attention multiple times over the input features; (ii) 1MHA: only 1 multi-head attention applies self-attention to the input features; (iii) 4CNN-4MHA: multiple self-attentions are applied to the input features via a set of 1D convolutions[4]. The 4CNN architecture matches the best performing configuration (4CNN - mixed 5 X ensemble)[5] used by TEES (Björne and Salakoski, 2018), which

is composed of four 1D convolutions with window sizes 1, 3, 5 and 7. In our models and TEES, we set the number of filters for the convolutions to 64. The number of heads for multi-head attentions is also set to 8. The reported results of TEES are achieved by running their out-of-the-box system for different tasks.

Since training a single model can be prone to overfitting if the validation set is too small (Björne and Salakoski, 2018), we use mixed 5 model ensemble, which takes 5-best models (out of 20), ranked with micro-averaged F-score on randomized train/validation set split, and considers their averaged predictions. These ensemble predictions are calculated for each label as the average of all the models' predicted confidence scores. Precision, recall, and F-score of the proposed approach and its variants are compared to TEES in Table 2. Our model (4MHA-4CNN) obtains the state-of-the-art results compared to those of the top performing system (TEES) in different shared tasks: BioNLP (GE09, GE11, EPI11, ID11, REL11, GE13, CG13, PC13), BioCreative (CP17), and the AMIA dataset.

Analyzing the results, we observe that the proposed 4MHA-4CNN model has the best F-score in the majority of datasets except for EPI11, ID11 and CG13, where the proposed MHA models (i.e., 1MHA and 4MHA) have the best F-score and recall. These tasks are related to epigenetics and post-translational modifications (EPI11), infection diseases (ID11) and cancer genetics (CG13), where events typically require long dependencies in most of the cases. It explains why the MHA-alone models are better than when combined with convolutions. The F-scores achieved by 4MHA-4CNN and 4MHA models on GE09 dataset are also very close. In many cases, when using the configurations in which MHA is applied to the input features, both precision and recall are better compared to other configurations. Moreover, having four parallel MHAs applied to the input features outperforms 1MHA and the other potential variants[6].

In terms of precision, the advantage of applying 4CNN versus 4MHA to the merged input features depends on the dataset. On PC13, the precision when using 4CNN on the merged input features is much higher compared to other configurations, but the recall is significantly lower.

The proposed 4MHA-4CNN model has also

---

[4]We also conducted experiments with 1CNN-1MHA and 1MHA-1CNN, which are excluded due to the poor performance.

[5]We use 4CNN to represent this configuration.

[6]The experiment with 8MHA, and multiple MHAs one after the other on the whole sequence are excluded from the paper due to the poor perfromance.

Task	Precision	Recall	F-score	Approach
GE09	<u>65.73</u>	44.72	53.23	TEES 4CNN
	65.01	**46.83**	**54.44**	Proposed 4MHA
	64.37	45.19	53.10	Proposed 1MHA
	61.99	45.51	52.48	Proposed 4CNN-4MHA
	65.98	<u>45.60</u>	<u>53.93</u>	Proposed 4MHA-4CNN
GE11	66.09	46.62	54.68	TEES 4CNN
	66.19	<u>48.67</u>	<u>56.09</u>	Proposed 4MHA
	<u>66.26</u>	48.60	56.07	Proposed 1MHA
	67.07	47.61	55.69	Proposed 4CNN-4MHA
	66.12	**49.34**	**56.51**	Proposed 4MHA-4CNN
EPI11	63.31	46.73	53.78	TEES 4CNN
	63.71	**50.73**	<u>56.48</u>	Proposed 4MHA
	66.38	<u>49.85</u>	**56.94**	Proposed 1MHA
	63.60	45.72	53.20	Proposed 4CNN-4MHA
	<u>65.43</u>	48.55	55.74	Proposed 4MHA-4CNN
ID11	<u>70.14</u>	44.36	54.35	TEES 4CNN
	66.63	**48.65**	<u>56.24</u>	Proposed 4MHA
	71.64	<u>46.99</u>	**56.75**	Proposed 1MHA
	68.92	41.04	51.44	Proposed 4CNN-4MHA
	69.05	44.91	54.43	Proposed 4MHA-4CNN
REL11	71.26	62.37	66.52	TEES 4CNN
	<u>71.56</u>	63.78	<u>67.45</u>	Proposed 4MHA
	68.55	<u>64.39</u>	66.40	Proposed 1MHA
	71.02	55.53	62.33	Proposed 4CNN-4MHA
	71.91	**65.39**	**68.50**	Proposed 4MHA-4CNN
GE13	**62.22**	39.96	<u>48.66</u>	TEES 4CNN
	<u>60.68</u>	40.35	48.47	Proposed 4MHA
	60.21	<u>40.75</u>	48.60	Proposed 1MHA
	58.14	37.66	45.71	Proposed 4CNN-4MHA
	59.76	**41.65**	**49.09**	Proposed 4MHA-4CNN
CG13	66.08	49.05	56.30	TEES 4CNN
	<u>65.92</u>	**53.50**	**59.06**	Proposed 4MHA
	67.02	<u>52.49</u>	<u>58.87</u>	Proposed 1MHA
	61.91	48.02	54.09	Proposed 4CNN-4MHA
	65.47	51.71	57.78	Proposed 4MHA-4CNN
PC13	**63.49**	43.37	51.54	TEES 4CNN
	59.45	**49.90**	<u>54.26</u>	Proposed 4MHA
	<u>60.64</u>	47.25	53.11	Proposed 1MHA
	57.61	43.23	49.39	Proposed 4CNN-4MHA
	60.51	<u>49.43</u>	**54.41**	Proposed 4MHA-4CNN
CP17	73.00	45.00	56.00	TEES 4CNN
	70.00	**58.00**	**63.00**	Proposed 4MHA
	77.00	48.00	58.00	Proposed 1MHA
	77.00	44.00	56.00	Proposed 4CNN-4MHA
	75.00	<u>50.00</u>	<u>60.00</u>	Proposed 4MHA-4CNN
AMIA	84.41	87.52	85.90	TEES 4CNN
	83.73	88.51	86.01	Proposed 4MHA
	<u>85.12</u>	<u>89.50</u>	<u>87.31</u>	Proposed 1MHA
	85.02	89.01	87.00	Proposed 4CNN-4MHA
	85.21	**90.11**	**87.53**	Proposed 4MHA-4CNN

Table 2: Precision, Recall and F-score, measured on the corpora of various shared tasks for our models, and the state of the art. The best scores (the first and the second highest scores) for each task are bolded and highlighted, respectively. All the results (except those of CP17 and AMIA) are evaluated using the official evaluation program/server of each task.

good recall, except for EPI11, ID11, and CG13, where 4MHA is better. As mentioned before, the addition of convolutions after the multi-head attentions might be less useful in these three sets, since sentences in these topics describe interactions for which long context dependencies are present.

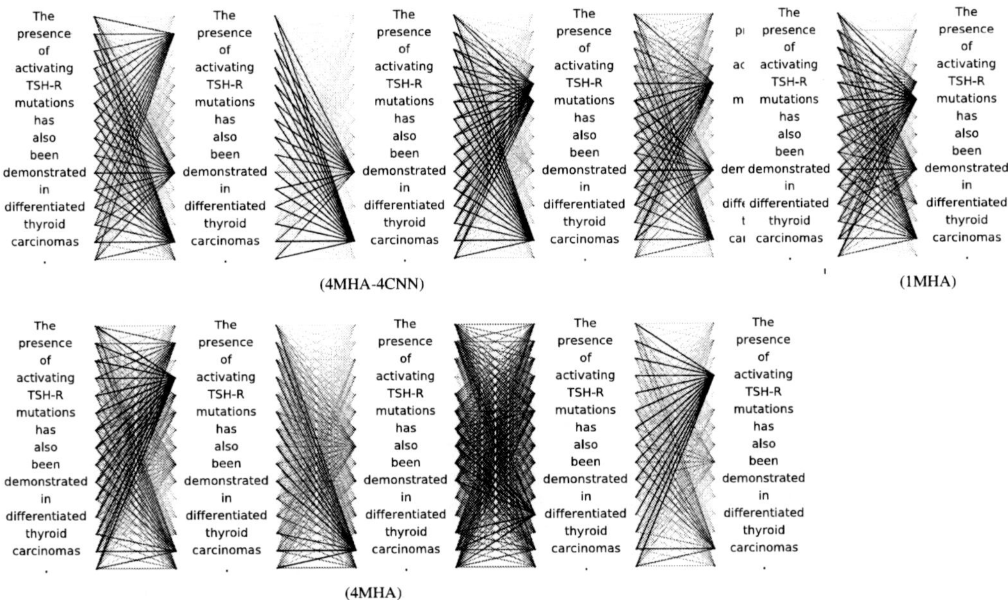

Figure 3: Visualization of multi-head attention in different architectures

Overall, our observations support the hypothesis that higher recall/F-score is obtained in configurations in which 4MHA is applied first to the merged input features, where CNNs are not as convenient as MHAs to deal with long dependencies.

5.1 Discussion

Besides improving the previous state of the art, the results indicate that combining multi-head attention with convolution provides an effective performance compared to individual components. Among the variants of our model, 4MHA also outperforms TEES over all the shared tasks reported in Table 2. Even though convolutions are quite effective (Björne and Salakoski, 2018) on their own, multi-head attentions improve their performance being able to deal with longer dependencies.

Figure 3 shows the multi-head attention (sum of the attention of all heads) of the "relation and event detection" classification task for different proposed network architectures (4MHA-4CNN, 1MHA, and 4MHA) on a sample sentence *"The presence of activating TSH-R mutations has also been demonstrated in differentiated thyroid carcinomas."*. In the 4MHA and 4MHA-4CNN models, the four multi-head attention layers contribute distinctively different attentions from each other. This allows the 4MHA and 4MHA-4CNN models to independently exploit more relationships between the to-kens than the 1MHA model. In addition, the convolutions make the 4MHA-4CNN model have more focused attentions on certain important tokens than the 4MHA model.

Considering the computational complexity, according to the work in (Vaswani et al., 2017), self-attention has a cost that is quadratic with the length of the sequence, while the convolution cost is quadratic with respect to the representation dimension of the data. The representation dimension of the data is typically higher compared to the length of individual sentences. Outperforming convolutions in terms of computational complexity and F-score, multi-head attention mechanisms seem to be better suited. Although the addition of convolutions after the multi-head makes the model more expensive, the lower representation dimension of the filters reduces the cost.

5.2 Error Analysis

We have performed error analysis on the baseline system (TEES), and our approach[7] over the gene-mutation AMIA and CP17 datasets[8], and observed the following sources of error.

---

[7] We consider the same configuration for the convolutions in both TEES and our approach.

[8] We only use these datasets for error analysis due to the limited access to the gold set of other datasets. Hence, this error analysis only covers relation extraction.

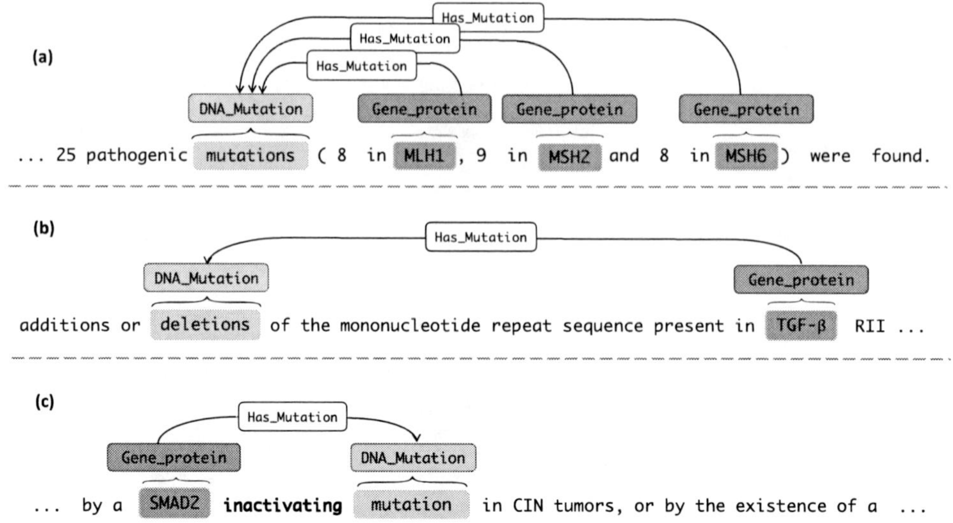

Figure 4: Error analysis of TEES and our approach over the gene-mutation AMIA dataset

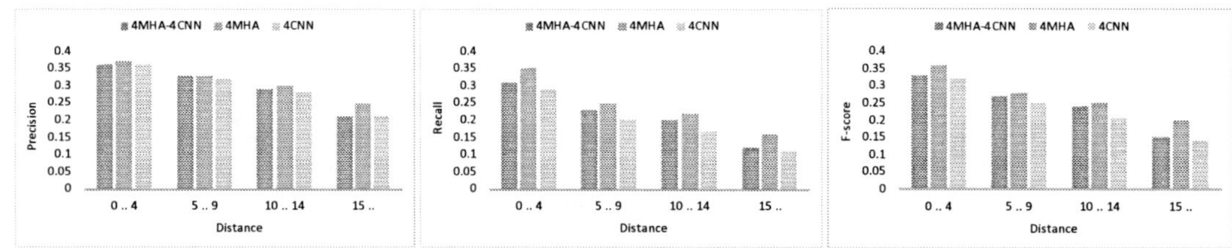

Table 3: Empirical evaluation of long-distance dependencies on CP17

Relations involving multiple entities: This is a major source of false negatives for TEES, while our approach exhibits a more robust behavior and achieves full recall. The reason would be the ability of multi-head attention to jointly attend to information from different representation subspaces at different positions (Vaswani et al., 2017). In an example from the AMIA dataset (Figure 4 (a)), there is a "has_mutation" relationship between the term "mutations" and the three gene-protein entities of "MLH1", "MSH2", and "MSH6". While the state-of-the-art approach only finds the relation between the mutation and the first gene-protein (MLH1) and ignores the other two relations, our approach captures the relations between the mutation and all three entities (MLH1, MSH2, and MSH6).

Long-distance dependencies: TEES also seems to have difficulty in annotating long-distance relations, as in the missed relation between "deletions" and "TGF-β" in an example from the AMIA

dataset (Figure 4 (b)), which is captured by our approach. We explored this issue further by plotting the performance of different proposed architectures and that of TEES over different distances. We relied on the CP17 dataset, since the test set is considerably larger than AMIA. We performed this analysis for the best performing network architecture proposed (4MHA-4CNN) along with 4MHA and 4CNN architectures separately as the individual components, to study how these architectures behave in capturing distant relations. We measure the distance as the number of tokens between the farthest entities involved in a relation, by employing the tokenization carried out by the TEES pre-processing tool. The results are provided in Figure 3. Regardless of the evaluation metric used, we observe that the scores decrease at longer distances, and 4MHA outperforms the other two architectures, which lies in the ability of multi-head attention to capture long distance dependencies. This experiment shows how 4MHA provides globality in 4MHA-4CNN, which slightly outperforms 4CNN in longer distances.

Negative or speculative contexts: Regarding the false positives for TEES that are generally well handled by our system, the annotation of speculative or negative language seems to be problematic. For instance, as depicted in Figure 4 (c), TEES incorrectly captures the relation between "mutation" and "SMAD2", despite the negative cue, "inactivating". Even though our approach correctly ignores this false positive in the short distance, it still captures speculative long dependencies, which motivates a natural extension of our work in future.

6 Conclusion

We have proposed a novel architecture based on multi-head attention and convolutions, which deals with the long dependencies typical of biomedical literature. The results show that this architecture outperforms the state of the art on existing biomedical information extraction corpora. While multi-head attention identifies long dependencies in extracting relations and events, convolutions provide the additional benefit of capturing more local relations, which improves the performance of existing approaches. The finding that CNN-before-MHA is outperformed by MHA-before-CNN is very interesting and could be used as a competitive baseline for future work.

Our ongoing work includes generalizing our findings to other non-biomedical information extraction tasks. Current work is focused on event and relation extraction from a single short/long sentence; we would like to experiment with additional contents to study the behaviour of these models across sentence boundaries (Verga et al., 2018). Finally, we intend to extend our approach to deal with speculative contexts by considering more semantic linguistic features, e.g., sense embeddings (Rothe and Schütze, 2015) on biomedical literature.

References

Jari Björne, Juho Heimonen, Filip Ginter, Antti Airola, Tapio Pahikkala, and Tapio Salakoski. 2009. Extracting complex biological events with rich graph-based feature sets. In *Proceedings of the Workshop on Current Trends in Biomedical Natural Language Processing: Shared Task*, pages 10–18. Association for Computational Linguistics.

Jari Björne and Tapio Salakoski. 2011. Generalizing biomedical event extraction. In *Proceedings of the BioNLP Shared Task 2011 Workshop*, pages 183–191. Association for Computational Linguistics.

Jari Björne and Tapio Salakoski. 2018. Biomedical event extraction using convolutional neural networks and dependency parsing. In *Proceedings of the BioNLP 2018 workshop*, pages 98–108.

Xavier Glorot, Antoine Bordes, and Yoshua Bengio. 2011. Deep sparse rectifier neural networks. In *Proceedings of the fourteenth international conference on artificial intelligence and statistics*, pages 315–323.

Sepp Hochreiter. 1998. The vanishing gradient problem during learning recurrent neural nets and problem solutions. *International Journal of Uncertainty, Fuzziness and Knowledge-Based Systems*, 6(02):107–116.

Antonio Jimeno Yepes, Andrew MacKinlay, Natalie Gunn, Christine Schieber, Noel Faux, Matthew Downton, Benjamin Goudey, and Richard L Martin. 2018. A hybrid approach for automated mutation annotation of the extended human mutation landscape in scientific literature. In *AMIA Annual Symposium Proceedings*, volume 2018, page 616. American Medical Informatics Association.

Jin-Dong Kim, Tomoko Ohta, Sampo Pyysalo, Yoshinobu Kano, and Jun'ichi Tsujii. 2009. Overview of bionlp'09 shared task on event extraction. In *Proceedings of the Workshop on Current Trends in Biomedical Natural Language Processing: Shared Task*, pages 1–9. Association for Computational Linguistics.

Jin-Dong Kim, Yue Wang, Toshihisa Takagi, and Akinori Yonezawa. 2011. Overview of genia event task in bionlp shared task 2011. In *Proceedings of the BioNLP Shared Task 2011 Workshop*, pages 7–15. Association for Computational Linguistics.

Martin Krallinger, Obdulia Rabal, Saber A Akhondi, et al. 2017. Overview of the biocreative vi chemical-protein interaction track. In *Proceedings of the sixth BioCreative challenge evaluation workshop*, volume 1, pages 141–146.

Jimmy Lei Ba, Jamie Ryan Kiros, and Geoffrey E Hinton. 2016. Layer normalization. *arXiv preprint arXiv:1607.06450*.

Diya Li, Lifu Huang, Heng Ji, and Jiawei Han. 2019. Biomedical event extraction based on knowledge-driven tree-lstm. In *NAACL-HLT*.

Zhouhan Lin, Minwei Feng, Cicero Nogueira dos Santos, Mo Yu, Bing Xiang, Bowen Zhou, and Yoshua Bengio. 2017. A structured self-attentive sentence embedding. *arXiv preprint arXiv:1703.03130*.

Emily K Mallory, Ce Zhang, Christopher Re, and Russ B Altman. 2015. Large-scale extraction of gene interactions from full-text literature using deepdive. *Bioinformatics*, 32(1):106–113.

Tomas Mikolov, Ilya Sutskever, Kai Chen, Greg S Corrado, and Jeff Dean. 2013. Distributed representations of words and phrases and their compositionality. In *Advances in neural information processing systems*, pages 3111–3119.

SPFGH Moen and Tapio Salakoski2 Sophia Ananiadou. 2013. Distributional semantics resources for biomedical text processing. *Proceedings of LBM*, pages 39–44.

Claire Nédellec, Robert Bossy, Jin-Dong Kim, Jung-Jae Kim, Tomoko Ohta, Sampo Pyysalo, and Pierre Zweigenbaum. 2013. Overview of bionlp shared task 2013. In *Proceedings of the BioNLP Shared Task 2013 Workshop*, pages 1–7.

Ankur P Parikh, Oscar Täckström, Dipanjan Das, and Jakob Uszkoreit. 2016. A decomposable attention model for natural language inference. *arXiv preprint arXiv:1606.01933*.

Sascha Rothe and Hinrich Schütze. 2015. Autoextend: Extending word embeddings to embeddings for synsets and lexemes. *arXiv preprint arXiv:1507.01127*.

Cicero Nogueira dos Santos, Bing Xiang, and Bowen Zhou. 2015. Classifying relations by ranking with convolutional neural networks. *arXiv preprint arXiv:1504.06580*.

Isabel Segura Bedmar, Paloma Martínez, and María Herrero Zazo. 2013. Semeval-2013 task 9: Extraction of drug-drug interactions from biomedical texts (ddiextraction 2013). Association for Computational Linguistics.

Isabel Segura Bedmar, Paloma Martinez, and Daniel Sánchez Cisneros. 2011. The 1st ddiextraction-2011 challenge task: Extraction of drug-drug interactions from biomedical texts.

Tao Shen, Tianyi Zhou, Guodong Long, Jing Jiang, Shirui Pan, and Chengqi Zhang. 2018. Disan: Directional self-attention network for rnn/cnn-free language understanding. In *Thirty-Second AAAI Conference on Artificial Intelligence*.

Ashish Vaswani, Noam Shazeer, Niki Parmar, Jakob Uszkoreit, Llion Jones, Aidan N Gomez, Łukasz Kaiser, and Illia Polosukhin. 2017. Attention is all you need. In *Advances in neural information processing systems*, pages 5998–6008.

Patrick Verga, David Belanger, Emma Strubell, Benjamin Roth, and Andrew McCallum. 2015. Multilingual relation extraction using compositional universal schema. *arXiv preprint arXiv:1511.06396*.

Patrick Verga, Emma Strubell, and Andrew McCallum. 2018. Simultaneously self-attending to all mentions for full-abstract biological relation extraction. *arXiv preprint arXiv:1802.10569*.

Daojian Zeng, Kang Liu, Yubo Chen, and Jun Zhao. 2015. Distant supervision for relation extraction via piecewise convolutional neural networks. In *Proceedings of the 2015 Conference on Empirical Methods in Natural Language Processing*, pages 1753–1762.

Daojian Zeng, Kang Liu, Siwei Lai, Guangyou Zhou, Jun Zhao, et al. 2014. Relation classification via convolutional deep neural network.

Yijia Zhang, Hongfei Lin, Zhihao Yang, Jian Wang, Shaowu Zhang, Yuanyuan Sun, and Liang Yang. 2018. A hybrid model based on neural networks for biomedical relation extraction. *Journal of biomedical informatics*, 81:83–92.

Peng Zhou, Wei Shi, Jun Tian, Zhenyu Qi, Bingchen Li, Hongwei Hao, and Bo Xu. 2016. Attention-based bidirectional long short-term memory networks for relation classification. In *Proceedings of the 54th Annual Meeting of the Association for Computational Linguistics (Volume 2: Short Papers)*, volume 2, pages 207–212.

An Empirical Study of Multi-Task Learning on BERT
for Biomedical Text Mining

Yifan Peng   **Qingyu Chen**   **Zhiyong Lu**
National Center for Biotechnology Information
National Library of Medicine, National Institutes of Health
Bethesda, MD, USA
{yifan.peng, qingyu.chen, zhiyong.lu}@nih.gov

Abstract

Multi-task learning (MTL) has achieved remarkable success in natural language processing applications. In this work, we study a multi-task learning model with multiple decoders on varieties of biomedical and clinical natural language processing tasks such as text similarity, relation extraction, named entity recognition, and text inference. Our empirical results demonstrate that the MTL fine-tuned models outperform state-of-the-art transformer models (e.g., BERT and its variants) by 2.0% and 1.3% in biomedical and clinical domains, respectively. Pairwise MTL further demonstrates more details about which tasks can improve or decrease others. This is particularly helpful in the context that researchers are in the hassle of choosing a suitable model for new problems. The code and models are publicly available at https://github.com/ncbi-nlp/bluebert.

1   Introduction

Multi-task learning (MTL) is a field of machine learning where multiple tasks are learned in parallel while using a shared representation (Caruana, 1997). Compared with learning multiple tasks individually, this joint learning effectively increases the sample size for training the model, thus leads to performance improvement by increasing the generalization of the model (Zhang and Yang, 2017). This is particularly helpful in some applications such as medical informatics where (labeled) datasets are hard to collect to fulfill the data-hungry needs of deep learning.

MTL has long been studied in machine learning (Ruder, 2017) and has been used successfully across different applications, from natural language processing (Collobert and Weston, 2008; Luong et al., 2016; Liu et al., 2019c), computer vision (Wang et al., 2009; Liu et al., 2019a; Chen et al., 2019), to health informatics (Zhou et al., 2011; He et al., 2016; Harutyunyan et al., 2019). MTL has also been studied in biomedical and clinical natural language processing (NLP) such as named entity recognition and normalization and the relation extraction. However, most of these studies focus on either one task with multi corpora (Khan et al., 2020; Wang et al., 2019b) or multi-tasks on a single corpus (Xue et al., 2019; Li et al., 2017; Zhao et al., 2019).

To bridge this gap, we investigate the use of MTL with transformer-based models (BERT) on multiple biomedical and clinical NLP tasks. We hypothesize the performance of the models on individual tasks (especially in the same domain) can be improved via joint learning. Specifically, we compare three models: the independent single-task model (BERT), the model refined via MTL (called MT-BERT-Refinement), and the model fine-tuned for each task using MT-BERT-Refinement (called MT-BERT-Fine-Tune). We conduct extensive empirical studies on the Biomedical Language Understanding Evaluation (BLUE) benchmark (Peng et al., 2019), which offers a diverse range of text genres (biomedical and clinical text) and NLP tasks (such as text similarity, relation extraction, and named entity recognition). When learned and fine-tuned on biomedical and clinical domains separately, we find that MTL achieved over 2% performance on average, created new state-of-the-art results on four BLUE benchmark tasks. We also demonstrate the use of multi-task learning to obtain a single model that still produces state-of-the-art performance on all tasks. This positive answer will be very helpful in the context that researchers are in the hassle of choosing a suitable model for new problems where training resources are limited.

Our contribution in this work is three-fold: (1) We conduct extensive empirical studies on 8 tasks from a diverse range of text genres. (2) We

205

Proceedings of the BioNLP 2020 workshop, pages 205–214
Online, July 9, 2020 ©2020 Association for Computational Linguistics

demonstrate that the MTL fine-tuned model (MT-BERT-Fine-Tune) achieved state-of-the-art performance on average and there is still a benefit to utilizing the MTL refinement model (MT-BERT-Refinement). Pairwise MTL, where two tasks were trained jointly, further demonstrates which tasks can improve or decrease other tasks. (3) We make codes and pre-trained MT models publicly available.

The rest of the paper is organized as follows. We first present related work in Section 2. Then, we describe the multi-task learning in Section 3, followed by our experimental setup, results, and discussion in Section 4. We conclude with future work in the last section.

2   Related work

Multi-tasking learning (MTL) aims to improve the learning of a model for task t by using the knowledge contained in the tasks where all or a subset of tasks are related (Zhang and Yang, 2017). It has long been studied and has applications on neural networks in the natural language processing domain (Caruana, 1997). Collobert and Weston (2008) proposed to jointly learn six tasks such as part-of-speech tagging and language modeling in a time-decay neural network. Changpinyo et al. (2018) summarized recent studies on applying MTL in sequence tagging tasks. Bingel and Søgaard (2017) and Martínez Alonso and Plank (2017) focused on conditions under which MTL leads to gain in NLP, and suggest that certain data features such as learning curve and entropy distribution are probably better predictors of MTL gains.

In the biomedical and clinical domains, MTL has been studied mostly in two directions. One is to apply MTL on a single task with multiple corpora. For example, many studies focused on named entity recognition (NER) tasks (Crichton et al., 2017; Wang et al., 2019a,b). Zhang et al. (2018), Khan et al. (2020), and Mehmood et al. (2019) integrated MTL in the transformer-based networks (BERT), which is the state-of-the-art language representation model and demonstrated promising results to extract biomedical entities from literature. Yang et al. (2019) extracted clinical named entity from Electronic Medical Records using LSTM-CRF based model. Besides NER, Li et al. (2018) and Li and Ji (2019) proposed to use MTL on relation classification task and Du et al. (2017) on biomedical semantic indexing. Xing et al. (2018)

exploited domain-invariant knowledge to segment Chinese word in medical text.

The other direction is to apply MTL on different tasks, but the annotations are from a single corpus. Li et al. (2017) proposed a joint model extract biomedical entities as well as their relations simultaneously and carried out experiments on either the adverse drug event corpus (Gurulingappa et al., 2012) or the bacteria biotope corpus (Deléger et al., 2016). Shi et al. (2019) also jointly extract entities and relations but focused on the BioCreative/OHNLP 2018 challenge regarding family history extraction (Liu et al., 2018). Xue et al. (2019) integrated the BERT language model into joint learning through dynamic range attention mechanism and fine-tuned NER and relation extraction tasks jointly on one in-house dataset of coronary arteriography reports.

Different from these works, we studied to jointly learn 8 different corpora from 4 different types of tasks. While MTL has brought significant improvements in medicine tasks, no (or mixed) results have been reported when pre-training MTL models in different tasks on different corpora. To this end, we deem that our model can provide more insights about conditions under which MTL leads to gains in BioNLP and clinical NLP, and sheds light on the specific task relations that can lead to gains from MTL models over single-task setups.

3   Multi-task model

The architecture of the MT-BERT model is shown in Figure 1. The shared layers are based on BERT (Devlin et al., 2018). The input X can be either a sentence or a pair of sentences packed together by a special token [SEP]. If X is longer than the allowed maximum length (e.g., 128 tokens in the BERT's base configuration), we truncate X to the maximum length. When X is packed by a sequence pair, we truncate the longer sequence one token at a time. Similar to (Devlin et al., 2018), two additional tokens are added at the start ([CLS]) and end ([SEP]) of X, respectively. Similar to (Lee et al., 2020; Peng et al., 2019), in the sequence tagging tasks, we split one sentence into several sub-sentences if it is longer than 30 words.

In the shared layers, the BERT model first converts the input sequence to a sequence of embedding vectors. Then, it applies attention mechanisms to gather contextual information. This se-

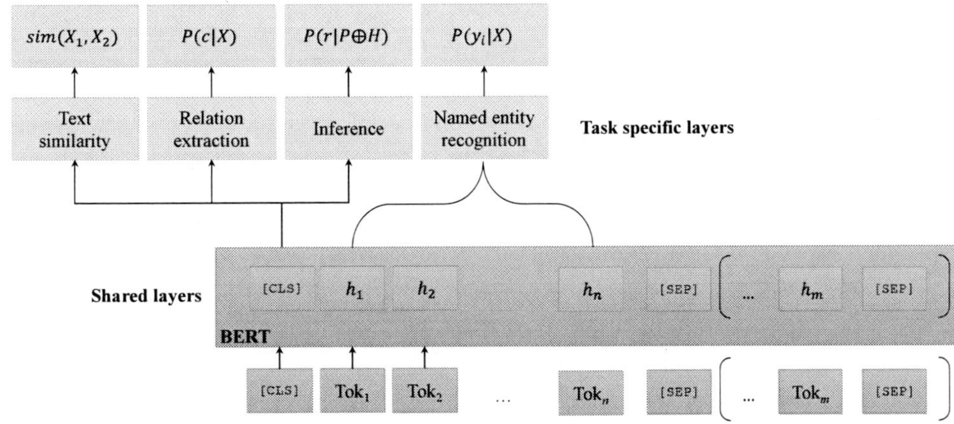

Figure 1: The architecture of the MT-BERT model.

mantic representation is shared across all tasks and is trained by our multi-task objectives. Finally, the BERT model encodes that information in a vector for each token $(h_0, \ldots, h_n)$.

On top of the shared BERT layers, the task-specific layer uses a fully-connected layer for each task. We fine-tune the BERT model and the task-specific layers using multi-task objectives during the training phase. More details of the multi-task objectives in the BLUE benchmark are described below.

3.1 Sentence similarity

Suppose that h_0 is the BERT's output of the token [CLS] in the input sentence pair (X_1, X_2). We use a fully connected layer to compute the similarity score $sim(X_1, X_2) = ah_0 + b$, where $sim(X_1, X_2)$ is a real value. This task is trained using the Mean Squared Error (MSE) loss: $(y - sim(X_1, X_2))^2$, where y is the real-value similarity score of the sentence pair.

3.2 Relation extraction

This task extracts binary relations (two arguments) from sentences. After replacing two arguments of interest in the sentence with pre-defined tags (e.g., GENE, or DRUG), this task can be treated as a classification problem of a single sentence X. Suppose that h_0 is the output embedding of the token [CLS], the probability that a relation is labeled as class c is predicted by a fully connected layer and a logistic regression with softmax: $P(c|X) = softmax(ah_0 + b)$. This approach is

widely used in the transformer-based models (Devlin et al., 2018; Peng et al., 2019; Liu et al., 2019c). This task is trained using the categorical cross-entropy loss: $-\sum_c \delta(y_c = \hat{y}) \log(P(c|X))$, where $\delta(y_c = \hat{y}) = 1$ if the classification $\hat{y}$ of X is the correct ground-truth for the class $c \in C$; otherwise $\delta(y_c = \hat{y}) = 0$.

3.3 Inference

After packing the pair of premise sentences with hypothesis into one sequence, this task can also be treated as a single sentence classification problem. The aim is to find logical relation R between premise P and hypothesis H. Suppose that that h_0 is the output embedding of the token [CLS] in $X = P \oplus H$, $P(R|P \oplus H) = softmax(ah_0 + b)$. This task is trained using the categorical cross-entropy loss as above.

3.4 Named entity recognition

The output of the BERT model produces a feature vector sequence $\{h_i\}_{i=0}^n$ with the same length as the input sequence X. The MTL model predicts the label sequence by using a softmax output layer, which scales the output for a label $l \in \{1, 2, \ldots, L\}$ as follows: $P(\hat{y}_i = j|x) = \frac{\exp(h_i W_j)}{\sum_{l=1}^{L} \exp(h_i W_j)}$, where L is the total number of tags. This task is trained using the categorical cross-entropy loss: $-\sum_i \sum_{y_i} \delta(y_i = \hat{y}_i) \log P(y_i|X)$.

3.5 The training procedure

The training procedure for MT-BERT consists of three stages: (1) pretraining the BERT model,

(2) refining it via multi-task learning (MT-BERT-Refinement), and (3) fine-tuning the model using the task-specific data (MT-BERT-Fine-Tune).

3.5.1 Pretraining

The pretraining stage follows that of the BERT using the masked language modeling technique (Devlin et al., 2018). Here we used the base version. The maximum length of the input sequences is thus 128.

3.5.2 Refining via Multi-task learning

In this step, we refine all layers in the model. Algorithm 1 demonstrates the process of multi-task learning (Liu et al., 2019c). We first initialize the shared layers with the pre-trained BERT model and randomly initialize the task-specific layer parameters. Then we create the dataset by merging mini-batches of all the datasets. In each epoch, we randomly select a mini-batch b_t of task t from all datasets D. Then we update the model according to the task-specific objective of the task t. Same as in (Liu et al., 2019c), we use the mini-batch based stochastic gradient descent to learn the parameters.

Algorithm 1: Multi-task learning.

Initialize *model parameters* θ
   Shared layer parameters by BERT;
   Task-specific layer parameters
     randomly;
end
Create D by merging mini-batches for each
  dataset;
for *epoch in* $1, 2, ..., epoch_{max}$ **do**
   Shuffle D;
   for b_t *in D* **do**
     Compute loss: $L(\theta)$ based on task t;
     Compute gradient: $\nabla(\theta)$
     Update model: $\theta = \theta - \eta \nabla(\theta)$
   end
end

3.5.3 Fine-tuning MT-BERT

We fine-tune existing MT-BERT that are trained in the previous stage by continue training all layers on each specific task. Provided that the dataset is not drastically different in context to other datasets, the MT-BERT model will already have learned general features that are relevant to a specific problem. Specifically, we truncate the last layer (softmax and linear layers) of the MT-BERT and replace it with a new one, then we use a smaller learning rate to train the network.

4 Experiments

We evaluate the proposed MT-BERT on 8 tasks in BLUE benchmarks. We compare three types of models: (1) existing start-of-the-art BERT models fine-tuned directly on each task, respectively; (2) refinement MT-BERT with multi-task training (MT-BERT-Refinement); and (3) MT-BERT with fine-tuning (MT-BERT-Fine-Tune).

4.1 Datasets

We evaluate the performance of the models on 8 datasets in the BLUE benchmark used by (Peng et al., 2019). Table 1 gives a summary of these datasets. Briefly, ClinicalSTS is a corpus of sentence pairs selected from Mayo Clinics's clinical data warehouse (Wang et al., 2018). The i2b2 2010 dataset was collected from three different hospitals and was annotated by medical practitioners for eight types of relations between problems and treatments (Uzuner et al., 2011). MedNLI is a collection of sentence pairs selected from MIMIC-III (Shivade, 2017). For a fair comparison, we use the same training, development and test sets to train and evaluate the models. ShARe/CLEF is a collection of 299 de-identified clinical free-text notes from the MIMIC-II database (Suominen et al., 2013). This corpus is for disease entity recognition.

In the biomedical domain, the ChemProt consists of 1,820 PubMed abstracts with chemical-protein interactions (Krallinger et al., 2017). The DDI corpus is a collection of 792 texts selected from the DrugBank database and other 233 Medline abstracts (Herrero-Zazo et al., 2013). These two datasets were used in the relation extraction task for various types of relations. BC5CDR is a collection of 1,500 PubMed titles and abstracts selected from the CTD-Pfizer corpus and was used in the named entity recognition task for chemical and disease entities (Li et al., 2016).

4.2 Training

Our implementation of MT-BERT is based on the work of (Liu et al., 2019c).[1] We trained the model on one NVIDIA® V100 GPU using the PyTorch framework. We used the Adamax

---

Corpus	Task	Metrics	Domain	Train	Dev	Test
ClinicalSTS	Sentence similarity	Pearson	Clinical	675	75	318
ShARe/CLEFE	NER	F1	Clinical	4,628	1,075	5,195
i2b2 2010	Relation extraction	F1	Clinical	3,110	11	6,293
MedNLI	Inference	Accuracy	Clinical	11,232	1,395	1,422
BC5CDR disease	NER	F1	Biomedical	4,182	4,244	4,424
BC5CDR chemical	NER	F1	Biomedical	5,203	5,347	5,385
DDI	Relation extraction	F1	Biomedical	2,937	1,004	979
ChemProt	Relation extraction	F1	Biomedical	4,154	2,416	3,458

Table 1: Summary of eight tasks in the BLUE benchmark. More details can be found in (Peng et al., 2019).

Model	ClinicalSTS	i2b2 2010 re	MedNLI	ShARe/CLEFE	*Avg*
BlueBERT$_{clinical}$	0.848	0.764	0.840	0.771	*0.806*
MT-BlueBERT-Refinement$_{clinical}$	0.822	0.745	0.835	0.826	*0.807*
MT-BlueBERT-Fine-Tune$_{clinical}$	0.840	0.760	**0.846**	**0.831**	***0.819***

Table 2: Test results on clinical tasks.

Model	ChemProt	DDI	BC5CDR disease	BC5CDR chemical	*Avg*
BlueBERT$_{biomedical}$	0.725	0.739	0.866	0.935	*0.816*
MT-BlueBERT-Refinement$_{biomedical}$	0.714	0.792	0.824	0.930	*0.815*
MT-BlueBERT-Fine-Tune$_{biomedical}$	**0.729**	**0.820**	0.865	0.931	***0.836***

Table 3: Test results on biomedical tasks.

optimizer (Kingma and Ba, 2015) with a learning rate of $5e^{-5}$, a batch size of 32, a linear learning rate decay schedule with warm-up over 0.1, and a weight decay of 0.01 applied to every epoch of training by following (Liu et al., 2019c). We use the BioBERT (Lee et al., 2020), Blue-BERT base model (Peng et al., 2019), and ClinicalBERT (Alsentzer et al., 2019) as the domain-specific language model[2]. As a result, all the tokenized texts using wordpieces were chopped to spans no longer than 128 tokens. We set the maximum number of epochs to 100. We also set the dropout rate of all the task-specific layers as 0.1. To avoid the exploding gradient problem, we clipped the gradient norm within 1. To fine-tune the MT-BERT on specific tasks, we set the maximum number of epochs to 10 and learning rate e^{-5}.

4.3 Results

One of the most important criteria of building practical systems is fast adaptation to new domains.

To evaluate the models on different domains, we multi-task learned various MT-BERT on BLUE biomedical tasks and clinical tasks, respectively. BlueBERT$_{clinical}$ is the base BlueBERT model pretrained on PubMed abstracts and MIMIC-III clinical notes, and fine-tuned for each BLUE task on task-specific data. MT- model are the proposed models described in Section 3. We used the pre-trained BlueBERT$_{clinical}$ to initialize its shared layers, refined the model via MTL on the BLUE tasks (MT-BlueBERT-Refinement$_{clinical}$). We keep fine-tuning the model for each BLUE task using task-specific data, then got MT-BlueBERT-Fine-Tune$_{clinical}$.

Table 2 shows the results on clinical tasks. MT-BlueBERT-Fine-Tune$_{clinical}$ created new state-of-the-art results on 2 tasks and pushing the benchmark to 81.9%, which amounts to 1.3% absolution improvement over BlueBERT$_{clinical}$ and 1.2% absolute improvement over MT-BlueBERT-Refinement$_{clinical}$. On the ShAReCLEFE task, the model gained the largest improvement by 6%. On

[2] https://github.com/ncbi-nlp/bluebert

Figure 2: Pairwise MTL relationships in clinical (left) and biomedical (right) domains.

Model	ClinicalSTS	i2b2 2010 re	MedNLI	ShARe/CLEFE	*Avg*
MT-ClinicalBERT-Fine-Tune	0.816	0.746	0.834	0.817	*0.803*
MT-BioBERT-Fine-Tune	0.837	0.741	0.832	0.818	*0.807*
MT-BlueBERT-Fine-Tune$_{biomedical}$	0.824	0.738	0.824	0.825	*0.803*
MT-BlueBERT-Fine-Tune$_{clinical}$	**0.840**	**0.760**	**0.846**	**0.831**	***0.819***

Table 4: Test results of MT-BERT-Fine-Tune models on clinical tasks.

Model	ChemProt	DDI	BC5CDR disease	BC5CDR chemical	*Avg*
MT-BioBERT-Fine-Tune	**0.729**	0.812	0.851	0.928	*0.830*
MT-BlueBERT-Fine-Tune$_{biomedical}$	**0.729**	**0.820**	**0.865**	**0.931**	*0.836*
MT-BlueBERT-Fine-Tune$_{clinical}$	0.714	0.792	0.824	0.930	*0.815*

Table 5: Test results of MT-BERT-Fine-Tune models on biomedical tasks.

the MedNLI task, the MT model gained improvement by 2.4%. On the remaining tasks, the MT model also performed well by reaching the state-of-the-art performance with less than 1% differences. When compared the models with and without fine-tuning on single datasets, Table 2 shows that the multi-task refinement model is similar to single baselines on average. Consider that MT-BlueBERT-Refinement$_{clinical}$ is one model while BlueBERT$_{clinical}$ are 4 individual models, we believe the MT refinement model would bring the benefit when researchers are in the hassle of choosing the suitable model for new problems or problems with limited training data.

In biomedical tasks, we used BlueBERT$_{biomedical}$ as the baseline because it achieved the best performance on the BLUE benchmark. Table 3 shows the similar results as in the clinical tasks. MT-BlueBERT-Fine-Tune$_{biomedical}$ created new state-of-the-art results

on 2 tasks and pushing the benchmark to 83.6%, which amounts to 2.0% absolute improvement over BlueBERT$_{biomedical}$ and 2.1% absolute improvement over MT-BlueBERT-Refinement$_{biomedical}$. On the DDI task, the model gained the largest improvement by 8.1%.

4.4 Discussion

4.4.1 Pairwise MTL

To investigate which tasks are beneficial or harmful to others, we train on two tasks jointly using MT-BlueBERT-Refinement$_{biomedical}$ and MT-BlueBERT-Refinement$_{clinical}$. Figure 2 gives pairwise relationships. The directed green (or red and grey) edge from s to t means s improves (or decreases and has no effect on) t.

In the clinical tasks, ShARe/CLEFE always gets benefits from multi-task learning the remaining 3 tasks as the incoming edges are green. One factor might be that ShARe/CLEFE is an NER task

Model	BlueBERT _biomedical_	BlueBERT _clinical_	MT-BioBERT Fine-Tune	MT-BlueBERT Fine-Tune_biomedical_	MT-BlueBERT Fine-Tune_clinical_
ClinicalSTS	0.845	**0.848**	0.807	0.820	0.807
i2b2 2010 re	0.744	**0.764**	0.740	0.738	0.748
MedNLI	0.822	0.840	0.831	0.814	**0.842**
ChemProt	0.725	0.692	**0.735**	0.724	0.686
DDI	0.739	0.760	**0.810**	0.808	0.779
BC5CDR disease	**0.866**	0.854	0.849	0.853	0.848
BC5CDR chemical	**0.935**	0.924	0.928	0.928	0.914
ShARe/CLEFE	0.754	0.771	0.812	0.814	**0.830**
Avg	*0.804*	*0.807*	***0.814***	*0.812*	*0.807*

Table 6: Test results on eight BLUE tasks.

that generally requires more training data to fulfill the data-hungry need of the BERT model. ClinicalSTS helps MedNLI because the nature of both are related and their inputs are a pair of sentences. MedNLI can help other tasks except ClinicalSTS partially because the test set of ClinialSTS is too small to reflect the changes. We also note that i2b2 2010 re can be both beneficial and harmful, depending on which other tasks they are trained with. One potential cause is i2b2 2010 re was collected from three different hospitals and have the largest label size of 8.

In the biomedical tasks, both DDI and ChemProt tasks can be improved by MTL on other tasks, potentially because they are harder with largest size of label thus require more training data. In the meanwhile, BC5CDR chemical and disease can barely be improved potentially because they have already got large dataset to fit the model.

4.4.2 MTL on BERT variants

First, we would like to compare multi-task learning on BERT variants: BioBERT, ClinicalBERT, and BlueBERT. In the clinical tasks (Table 4), MT-BlueBERT-Fine-Tune_clinical_ outperforms other models on all tasks. When compared the MTL models using BERT model pretrained on PubMed only (rows 2 and 3) and on the combination of PubMed and clinical notes (row 4), it shows the impact of using clinical notes during the pretraining process. This observation is consistently as shown in (Peng et al., 2019). On the other hand, MT-ClinicalBERT-Fine-Tune, which used ClinicalBERT during the pretraining, drops ~1.6% across the tasks. The differences between ClinicalBERT and BlueBERT are at least in 2-fold. (1) ClinicalBERT used "cased" text while BlueBERT used "uncased" text; and (2) the number of epochs to continuously pretrained the model. Given that there are limited details of pretraining ClinicalBERT, further investigation may be necessary.

In the biomedical tasks, Table 5 shows that MT-BioBERT-Fine-Tune and MT-BlueBERT-Fine-Tune_biomedical_ reached comparable results and pretraining on clinical notes has a negligible impact.

4.4.3 Results on all BLUE tasks

Next, we also compare MT-BERT with its variants on all BLUE tasks. Table 6 shows that MT-BioBERT-Fine-Tune reached the best performance on average and MT-BlueBERT-Fine-Tune_biomedical_ stays closely. While confusing results were obtained when combing variety of tasks in both biomedical and clinical domains, we observed again that MTL models pretrained on biomedical literature perform better in biomedical tasks; and MTL models pretrained on both biomedical literature and clinical notes perform better in clinical tasks. These observations may suggest that it might be helpful to train separate deep neural networks on different types of text genres in BioNLP.

5 Conclusions and future work

In this work, we conduct an empirical study on MTL for biomedical and clinical tasks, which so far has been mostly studied with one or two tasks. Our results provide insights regarding domain adaptation and show benefits of the MTL refinement and fine-tuning. We recommend a combination of the MTL refinement and task-specific fine-tuning approach based on the evaluation results. When learned and fine-tuned on a different domain, MT-

BERT achieved improvements by 2.0% and 1.3% in biomedical and clinical domains, respectively. Specifically, it has brought significant improvements in 4 tasks.

There are two limitations to this work. First, our results on MTL training across all BLUE benchmark show that MTL is not always effective. We are interested in exploring further the characterization of task relationships. For example, it is not clear whether there are data characteristics that help to determine its success (Martínez Alonso and Plank, 2017; Changpinyo et al., 2018). In addition, our results suggest that the model could benefit more from some specific examples of some of the tasks in Table 1. For example, it might be of interest to not using the BC5CDR corpus in the relation extraction task in future. Second, we studied one approach to MTL by sharing the encoder between all tasks while keeping several task-specific decoders. Other approaches, such as fine-tuning only the task specific layers, soft parameter sharing (Ruder, 2017), knowledge distillation (Liu et al., 2019b), need to be investigated in the future.

While our work only scratches the surface of MTL in the medical domain, we hope it will shed light on the development of generalizable NLP models and task relations that can lead to gains from MTL models over single-task setups.

Acknowledgments

This work was supported by the Intramural Research Programs of the NIH National Library of Medicine. This work was also supported by the National Library of Medicine of the National Institutes of Health under award number K99LM013001. We are also grateful to the authors of mt-dnn (`https://github.com/namisan/mt-dnn`) to make the codes publicly available.

References

Emily Alsentzer, John Murphy, William Boag, Wei-Hung Weng, Di Jindi, Tristan Naumann, and Matthew McDermott. 2019. Publicly available clinical BERT embeddings. In *Proceedings of the 2nd Clinical Natural Language Processing Workshop*, pages 72–78.

Joachim Bingel and Anders Søgaard. 2017. Identifying beneficial task relations for multi-task learning in deep neural networks. In *EACL*, pages 164–169.

Rich Caruana. 1997. Multitask learning. *Machine Learning*, 28(1):41–75.

Soravit Changpinyo, Hexiang Hu, and Fei Sha. 2018. Multi-task learning for sequence tagging: an empirical study. In *COLING*, pages 2965–2977.

Qingyu Chen, Yifan Peng, Tiarnan Keenan, Shazia Dharssi, Elvira Agro N, Wai T. Wong, Emily Y. Chew, and Zhiyong Lu. 2019. A multi-task deep learning model for the classification of age-related macular degeneration. *AMIA 2019 Informatics Summit*, 2019:505–514.

Ronan Collobert and Jason Weston. 2008. A unified architecture for natural language processing. In *ICML*, pages 160–167.

Gamal Crichton, Sampo Pyysalo, Billy Chiu, and Anna Korhonen. 2017. A neural network multi-task learning approach to biomedical named entity recognition. *BMC Bioinformatics*, 18:368.

Louise Deléger, Robert Bossy, Estelle Chaix, Mouhamadou Ba, Arnaud Ferré, Philippe Bessières, and Claire Nédellec. 2016. Overview of the bacteria biotope task at BioNLP shared task 2016. In *Proceedings of BioNLP Shared Task Workshop*, pages 12–22.

Jacob Devlin, Ming-Wei Chang, Kenton Lee, and Kristina Toutanova. 2018. BERT: pre-training of deep bidirectional transformers for language understanding. *arXiv preprint: 1810.04805*.

Yongping Du, Yunpeng Pan, and Junzhong Ji. 2017. A novel serial deep multi-task learning model for large scale biomedical semantic indexing. In *IEEE International Conference on Bioinformatics and Biomedicine (BIBM)*, pages 533–537.

Harsha Gurulingappa, Abdul Mateen Rajput, Angus Roberts, Juliane Fluck, Martin Hofmann-Apitius, and Luca Toldo. 2012. Development of a benchmark corpus to support the automatic extraction of drug-related adverse effects from medical case reports. *Journal of Biomedical Informatics*, 45:885–892.

Hrayr Harutyunyan, Hrant Khachatrian, David C. Kale, Greg Ver Steeg, and Aram Galstyan. 2019. Multitask learning and benchmarking with clinical time series data. *Scientific data*, 6:96.

Dan He, David Kuhn, and Laxmi Parida. 2016. Novel applications of multitask learning and multiple output regression to multiple genetic trait prediction. *Bioinformatics (Oxford, England)*, 32:i37–i43.

María Herrero-Zazo, Isabel Segura-Bedmar, Paloma Martínez, and Thierry Declerck. 2013. The DDI corpus: an annotated corpus with pharmacological substances and drug-drug interactions. *Journal of Biomedical Informatics*, 46:914–920.

Muhammad Raza Khan, Morteza Ziyadi, and Mohamed AbdelHady. 2020. MT-BioNER: multi-task learning for biomedical named entity recognition using deep bidirectional transformers. *arXiv preprint: 2001.08904.*

Diederik P. Kingma and Jimmy Ba. 2015. Adam: a method for stochastic optimization. In *International Conference on Learning Representations (ICLR)*, pages 1–15.

Martin Krallinger, Obdulia Rabal, Saber A. Akhondi, Martín Pérez Pérez, Jesús Santamaría, Gael Pérez Rodríguez, Georgios Tsatsaronis, Ander Intxaurrondo, José Antonio López1 Umesh Nandal, Erin Van Buel, Akileshwari Chandrasekhar, Marleen Rodenburg, Astrid Laegreid, Marius Doornenbal, Julen Oyarzabal, Analia Lourenço, and Alfonso Valencia. 2017. Overview of the BioCreative VI chemical-protein interaction track. In *Proceedings of the BioCreative workshop*, pages 141–146.

Jinhyuk Lee, Wonjin Yoon, Sungdong Kim, Donghyeon Kim, Sunkyu Kim, Chan Ho So, and Jaewoo Kang. 2020. BioBERT: a pre-trained biomedical language representation model for biomedical text mining. *Bioinformatics (Oxford, England)*, 36:1234–1240.

Diya Li and Heng Ji. 2019. Syntax-aware multi-task graph convolutional networks for biomedical relation extraction. In *Proceedings of the Tenth International Workshop on Health Text Mining and Information Analysis (LOUHI 2019)*, pages 28–33.

Fei Li, Meishan Zhang, Guohong Fu, and Donghong Ji. 2017. A neural joint model for entity and relation extraction from biomedical text. *BMC Bioinformatics*, 18:198.

Jiao Li, Yueping Sun, Robin J. Johnson, Daniela Sciaky, Chih-Hsuan Wei, Robert Leaman, Allan Peter Davis, Carolyn J. Mattingly, Thomas C. Wiegers, and Zhiyong Lu. 2016. BioCreative V CDR task corpus: a resource for chemical disease relation extraction. *Database (Oxford)*, 2016.

Qingqing Li, Zhihao Yang, Ling Luo, Lei Wang, Yin Zhang, Hongfei Lin, Jian Wang, Liang Yang, Kan Xu, and Yijia Zhang. 2018. A multi-task learning based approach to biomedical entity relation extraction. In *IEEE International Conference on Bioinformatics and Biomedicine (BIBM)*, pages 680–682.

Shikun Liu, Edward Johns, and Andrew J. Davison. 2019a. End-to-end multi-task learning with attention. In *Proceedings of the IEEE Conference on Computer Vision and Pattern Recognition (CVPR)*, pages 1871–1880.

Sijia Liu, Majid Rastegar Mojarad, Yanshan Wang, Liwei Wang, Feichen Shen, Sunyang Fu, and Hongfang Liu. 2018. Overview of the BioCreative/OHNLP 2018 family history extraction task. In *Proceedings of the BioCreative Workshop*, pages 1–5.

Xiaodong Liu, Pengcheng He, Weizhu Chen, and Jianfeng Gao. 2019b. Improving multi-task deep neural networks via knowledge distillation for natural language understanding. *arXiv preprint: 1904.09482.*

Xiaodong Liu, Pengcheng He, Weizhu Chen, and Jianfeng Gao. 2019c. Multi-task deep neural networks for natural language understanding. In *ACL*, pages 4487–4496.

Minh-Thang Luong, Quoc V. Le, Ilya Sutskever, Oriol Vinyals, and Lukasz Kaiser. 2016. Multi-task sequence to sequence learning. In *ICLR*.

Héctor Martínez Alonso and Barbara Plank. 2017. When is multitask learning effective? semantic sequence prediction under varying data conditions. In *EACL*, pages 44–53.

Tahir Mehmood, Alfonso E Gerevini, Alberto Lavelli, and Ivan Serina. 2019. Multi-task learning applied to biomedical named entity recognition task. In *Italian Conference on Computational Linguistics*.

Yifan Peng, Shankai Yan, and Zhiyong Lu. 2019. Transfer learning in biomedical natural language processing: an evaluation of BERT and ELMo on ten benchmarking datasets. In *Proceedings of the Workshop on Biomedical Natural Language Processing (BioNLP)*, pages 58–65.

Sebastian Ruder. 2017. An overview of multi-task learning in deep neural networks. *arXiv preprint: 1706.05098.*

Xue Shi, Dehuan Jiang, Yuanhang Huang, Xiaolong Wang, Qingcai Chen, Jun Yan, and Buzhou Tang. 2019. Family history information extraction via deep joint learning. *BMC medical informatics and decision making*, 19:277.

Chaitanya Shivade. 2017. Mednli – a natural language inference dataset for the clinical domain.

Hanna Suominen, Sanna Salanterä, Sumithra Velupillai, Wendy W. Chapman, Guergana Savova, Noemie Elhadad, Sameer Pradhan, Brett R. South, Danielle L. Mowery, Gareth J. F. Jones, et al. 2013. Overview of the ShARe/CLEF eHealth evaluation lab 2013. In *International Conference of the Cross-Language Evaluation Forum for European Languages*, pages 212–231.

Özlem Uzuner, Brett R. South, Shuying Shen, and Scott L. DuVall. 2011. 2010 i2b2/VA challenge on concepts, assertions, and relations in clinical text. *Journal of the American Medical Informatics Association : JAMIA*, 18:552–556.

Xi Wang, Jiagao Lyu, Li Dong, and Ke Xu. 2019a. Multitask learning for biomedical named entity recognition with cross-sharing structure. *BMC Bioinformatics*, 20:427.

Xiaogang Wang, Cha Zhang, and Zhengyou Zhang. 2009. Boosted multi-task learning for face verification with applications to web image and video search. In *IEEE Conference on Computer Vision and Pattern Recognition (CVPR)*, pages 142–149.

Xuan Wang, Yu Zhang, Xiang Ren, Yuhao Zhang, Marinka Zitnik, Jingbo Shang, Curtis Langlotz, and Jiawei Han. 2019b. Cross-type biomedical named entity recognition with deep multi-task learning. *Bioinformatics (Oxford, England)*, 35:1745–1752.

Yanshan Wang, Naveed Afzal, Sunyang Fu, Liwei Wang, Feichen Shen, Majid Rastegar-Mojarad, and Hongfang Liu. 2018. MedSTS: a resource for clinical semantic textual similarity. *Language Resources and Evaluation*, pages 1–16.

Junjie Xing, Kenny Zhu, and Shaodian Zhang. 2018. Adaptive multi-task transfer learning for Chinese word segmentation in medical text. In *COLING*, pages 3619–3630.

Kui Xue, Yangming Zhou, Zhiyuan Ma, Tong Ruan, Huanhuan Zhang, and Ping He. 2019. Fine-tuning BERT for joint entity and relation extraction in Chinese medical text. In *2019 IEEE International Conference on Bioinformatics and Biomedicine (BIBM)*, pages 892–897.

Jianliang Yang, Yuenan Liu, Minghui Qian, Chenghua Guan, and Xiangfei Yuan. 2019. Information extraction from electronic medical records using multitask recurrent neural network with contextual word embedding. *Applied Sciences*, 9(18):3658.

Qun Zhang, Zhenzhen Li, Dawei Feng, Dongsheng Li, Zhen Huang, and Yuxing Peng. 2018. Multitask learning for Chinese named entity recognition. In *Advances in Multimedia Information Processing – PCM*, pages 653–662.

Yu Zhang and Qiang Yang. 2017. A survey on multitask learning. *arXiv preprint: 1707.08114*.

Sendong Zhao, Ting Liu, Sicheng Zhao, and Fei Wang. 2019. A neural multi-task learning framework to jointly model medical named entity recognition and normalization. In *Proceedings of the AAAI Conference on Artificial Intelligence*, volume 33, pages 817–824.

Jiayu Zhou, Lei Yuan, Jun Liu, and Jieping Ye. 2011. A multi-task learning formulation for predicting disease progression. In *Proceedings of the 17th ACM SIGKDD international conference on Knowledge discovery and data mining*, pages 814–822.

Author Index